U0342365

冶金专业英语

(第3版)

主　编　侯向东
副主编　郝赳赳　孙晓思　王晓鸽

扫码获取
本书数字资源

北　京
冶金工业出版社
2024

内 容 提 要

本书从多渠道、多角度向学生介绍了大量可理解、可接受的冶金专业英语信息，每章提供了大量音频、动画等数字资源，能够让学生快速积累专业英语基础知识，迅速提高专业英语应用能力。

本书可作为冶金专业的英语教材，也可供材料科学与工程等专业选用，并可供其他相关专业及英语爱好者参考。

图书在版编目(CIP)数据

冶金专业英语/侯向东主编.—3版.—北京：冶金工业出版社，2020.12（2024.5重印）

高职高专规划教材

ISBN 978-7-5024-8739-3

Ⅰ.①冶… Ⅱ.①侯… Ⅲ.①冶金工业—英语—高等职业教育—教材 Ⅳ.①TF

中国版本图书馆CIP数据核字(2021)第030954号

冶金专业英语（第3版）

出版发行 冶金工业出版社 **电　话** (010)64027926
地　址 北京市东城区嵩祝院北巷39号 **邮　编** 100009
网　址 www.mip1953.com **电子信箱** service@mip1953.com

责任编辑 杜婷婷 刘林烨 美术编辑 彭子赫 版式设计 禹 蕊
责任校对 郑 娟 责任印制 禹 蕊
三河市双峰印刷装订有限公司印刷
2008年2月第1版，2014年6月第2版，2020年12月第3版，2024年5月第6次印刷
787mm×1092mm 1/16；20印张；480千字；303页
定价49.00元

投稿电话 (010)64027932 投稿信箱 tougao@cnmip.com.cn
营销中心电话 (010)64044283
冶金工业出版社天猫旗舰店 yjgycbs.tmall.com
（本书如有印装质量问题，本社营销中心负责退换）

第3版前言

《冶金专业英语》第 1 版于 2008 年出版，全书内容以钢铁冶金生产、有色冶金生产及材料加工为主线，涵盖钢铁冶金、有色金属冶金、金属材料加工三个方面，从多渠道、多角度向学生介绍大量可理解、可接受的专业英语信息，其以专业性、实用性等特点，得到了社会各界人士，尤其是冶金类院校相关专业师生的一致好评，被评为普通高等教育“十一五”国家级规划教材。2014 年第 2 版对书中的部分内容进行了修订。近年来，面对冶金产业转型升级、数字化发展及职业教育的改革需要，编者开始了第 3 版的修订工作。在第 3 版出版之际，编者对帮助、指导过本书撰写工作的学界前辈、同行及使用和关心本书的读者表示诚挚的感谢。

与前两版相比，第 3 版既沿袭了前两版的特点，又有以下几方面的变化：

（1）每个单元增加课程思政模块，将专业课程中所蕴含的优秀人物品格、大国工匠精神、工程伦理意识、绿色发展等思政元素有效融入教学内容，积极贯彻落实推进党的二十大精神进教材、进课堂、进头脑，落实立德树人根本任务。

（2）每个单元增加延伸阅读栏目，将现代冶金技术发展中的新理论、新技术、新工艺、新标准、新设备、新材料引入教材，传播新知识，同时适应不同层次、不同需求的使用者。

（3）更正了第 2 版中的不足之处，并对部分习题和答案做出更新。同时对第 2 版中的第 1 单元“钢铁冶炼的历史”、第 5 单元“炼铁的新发展”做了较大改动，新增了第 21 单元“钢铁冶金发展趋势”，使全书结构上更具整体性，内容上更贴合时代发展。

（4）本书在第 2 版内容的基础上，每个单元中新增了英文原声音频、动画、互动题库小程序、活页式教材 PPT 课件等多媒体数字化资源，力求在专业知识方面达到严谨、新颖，在语言知识方面更加规范、实用，成为一本集理论性、知识性、新颖性、实用性和时代性为一体的新形态冶金专业英语教材。

本书由侯向东担任主编，郝赳赳、孙晓思、王晓鸽担任副主编。具体编写分工为：第 1~2 单元由李颖编写，第 3、6 单元由侯向东编写，第 4 单元由郝赳赳编写，第 5 单元由史学红编写，第 7 单元由冯捷编写，第 8 单元由薛方编写，第9~10单元由王晓鸽编写，第11~12单元由王艳琴编写，第13~14单元由郭林秀编写，第15单元由胡锐编写，第16单元由陈聪编写，第17单元由苏岱峰编写，第18单元由魏哲编写，第19单元由栗聖凯编写，第20~21单元由孙晓思编写。

由于编者水平所限，书中不妥之处，恳请广大读者批评指正。

编者

2022年11月

第2版前言

《冶金专业英语》自2008年出版以来，以系统性、实用性、可读性、方便性等特点，获得了全国冶金类院校相关专业师生的好评。为了适应我国高等职业教育的发展，更好地满足使用者的需求，编者在保留第1版原有特点的基础上，增加了有色金属冶金方面的课文内容，并对第1版进行了修订和完善。

本书由侯向东担任主编，贾志勤、祁苏燕担任副主编。具体编写分工为：第1~7单元由侯向东编写，第8~10、12、17、18单元由贾志勤编写，第11单元由冯捷编写，第13、14、19、20单元由陈聪编写，第15、16单元由祁苏燕编写。

在本书编写过程中参考了大量的国内外相关资料，得到了许多学界前辈、同行的热心帮助和指导，在此一并表示由衷的谢意。

由于编者水平所限，书中不足之处，恳请读者批评指正。

编者

2014年3月

第1版前言

《冶金专业英语》是冶金技术专业学生学习的专业基础课之一。随着全球一体化进程的加快和中国经济的不断发展，使用外语进行交流、了解行业的最新发展动态已成为冶金专业学生的一项重要技能。

本书为高等教育国家级“十一五”规划教材，是按照教育部高等教育人才的培养目标和规格、应具有的知识能力和素质要求而编写的。全书由18个单元组成，内容涵盖了钢铁冶金、有色金属冶金、材料加工三个方面。安排教学时，可根据培养方向，调用内容合适、难度适中的语篇和相应的练习材料，也可跨专业选用材料，以扩大知识面。

本教材的特点是：(1) 根据职业发展需要，教材内容突出实用性。本书从冶金生产工艺流程入手，编有大量具有针对性的专业阅读材料，同时介绍了科技英语翻译过程中常用的翻译知识。通过学习，学生不仅可以熟悉和掌握本专业常用的单词、短语及其用法，深化本专业的知识，而且可以掌握科技英语的翻译技巧，从而满足其职业发展对英语的基本需求。(2) 注重英语技能训练，提高实际应用能力。教材每单元都设计了大量的练习题，这些项目的内容与课文为同一主题，这有利于学生通过这些循环、交叉、叠加的练习，掌握技巧，形成能力。(3) 改变传统编写思路，便于学生自主学习。为了方便学生自主学习，本教材还配套了参考答案。在参考答案中提供了课文的译文和全部练习答案，这既有利于教师备课和组织教学活动，又扩展了学生的思维空间，便于学生自主学习，进一步锻炼和提高英语自学能力。

本书由侯向东担任主编，贾志勤担任副主编。其中1~7单元由侯向东编写，第8、9、10、12、15、16单元由贾志勤编写，第11单元由冯捷编写，第13、14、17、18单元由陈聪编写。

本书在编写过程中得到了许多同行的帮助，编者在此表示诚挚的谢意。

由于编者水平有限，书中的不足之处，希望读者批评指正。

编者

2007年11月

Contents

Unit 1 From the History of Ironmaking and Steelmaking ········ 1

Part Ⅰ Reading and Comprehension ········ 1
Part Ⅱ Curriculum Ideological and Political ········ 4
Part Ⅲ Further Reading ········ 5
Part Ⅳ Translation Training
翻译标准及过程 ········ 5
Part Ⅴ Exercises ········ 6

Unit 2 Raw Materials for the Production of Iron ········ 8

Part Ⅰ Reading and Comprehension ········ 8
Part Ⅱ Curriculum Ideological and Political ········ 13
Part Ⅲ Further Reading ········ 14
Part Ⅳ Translation Training
选词用字法 ········ 14
Part Ⅴ Exercises ········ 15

Unit 3 The Blast Furnace Plant ········ 17

Part Ⅰ Reading and Comprehension ········ 17
Part Ⅱ Curriculum Ideological and Political ········ 23
Part Ⅲ Further Reading ········ 23
Part Ⅳ Translation Training
词义的引申 ········ 24
Part Ⅴ Exercises ········ 25

Unit 4 Hot Blast Stove ········ 27

Part Ⅰ Reading and Comprehension ········ 27
Part Ⅱ Curriculum Ideological and Political ········ 32
Part Ⅲ Further Reading ········ 32
Part Ⅳ Translation Training
词类转译法（Ⅰ） ········ 32
Part Ⅴ Exercises ········ 33

Unit 5 New Developments in Ironmaking ······ 35
Part Ⅰ Reading and Comprehension ······ 35
Part Ⅱ Curriculum Ideological and Political ······ 41
Part Ⅲ Further Reading ······ 41
Part Ⅳ Translation Training
词类转译法（Ⅱ） ······ 42
Part Ⅴ Exercises ······ 43

Unit 6 Raw Materials of Steelmaking ······ 46
Part Ⅰ Reading and Comprehension ······ 46
Part Ⅱ Curriculum Ideological and Political ······ 51
Part Ⅲ Further Reading ······ 52
Part Ⅳ Translation Training
词类转译法（Ⅲ） ······ 52
Part Ⅴ Exercises ······ 54

Unit 7 Principles of Modern Steelmaking ······ 56
Part Ⅰ Reading and Comprehension ······ 56
Part Ⅱ Curriculum Ideological and Political ······ 60
Part Ⅲ Further Reading ······ 61
Part Ⅳ Translation Training
词类转译法（Ⅳ） ······ 61
Part Ⅴ Exercises ······ 62

Unit 8 The LD Practice ······ 65
Part Ⅰ Reading and Comprehension ······ 65
Part Ⅱ Curriculum Ideological and Political ······ 70
Part Ⅲ Further Reading ······ 71
Part Ⅳ Translation Training
增词（Ⅰ） ······ 71
Part Ⅴ Exercises ······ 72

Unit 9 Electric-arc Furnace Steelmaking Processes ······ 74
Part Ⅰ Reading and Comprehension ······ 74
Part Ⅱ Curriculum Ideological and Political ······ 78
Part Ⅲ Further Reading ······ 78
Part Ⅳ Translation Training
增词（Ⅱ） ······ 79

Part Ⅴ Exercises 80

Unit 10 Secondary Refining 82

Part Ⅰ Reading and Comprehension 82
Part Ⅱ Curriculum Ideological and Political 87
Part Ⅲ Further Reading 87
Part Ⅳ Translation Training
as 的译法 88
Part Ⅴ Exercises 88

Unit 11 Continuous Casting of Steel 91

Part Ⅰ Reading and Comprehension 91
Part Ⅱ Curriculum Ideological and Political 96
Part Ⅲ Further Reading 97
Part Ⅳ Translation Training
省略 97
Part Ⅴ Exercises 98

Unit 12 Types of Steels 101

Part Ⅰ Reading and Comprehension 101
Part Ⅱ Curriculum Ideological and Political 106
Part Ⅲ Further Reading 107
Part Ⅳ Translation Training
正反译与反正译 107
Part Ⅴ Exercises 109

Unit 13 Pyrometallurgical Extraction of Copper from Sulphide Ores 111

Part Ⅰ Reading and Comprehension 111
Part Ⅱ Curriculum Ideological and Political 117
Part Ⅲ Further Reading 117
Part Ⅳ Translation Training
合译法 117
Part Ⅴ Exercises 118

Unit 14 Production of Aluminium 121

Part Ⅰ Reading and Comprehension 121
Part Ⅱ Curriculum Ideological and Political 127
Part Ⅲ Further Reading 127

Part Ⅳ Translation Training
分译法 ······ 127
Part Ⅴ Exercises ······ 129

Unit 15 Zinc Metallurgy ······ 131

Part Ⅰ Reading and Comprehension ······ 131
Part Ⅱ Curriculum Ideological and Political ······ 138
Part Ⅲ Further Reading ······ 138
Part Ⅳ Translation Training
句子成分的转换 ······ 139
Part Ⅴ Exercises ······ 140

Unit 16 The Extraction of Gold ······ 142

Part Ⅰ Reading and Comprehension ······ 142
Part Ⅱ Curriculum Ideological and Political ······ 147
Part Ⅲ Further Reading ······ 147
Part Ⅳ Translation Training
科技英语常用句型 ······ 148
Part Ⅴ Exercises ······ 149

Unit 17 Metal Forming Processes ······ 151

Part Ⅰ Reading and Comprehension ······ 151
Part Ⅱ Curriculum Ideological and Political ······ 157
Part Ⅲ Further Reading ······ 158
Part Ⅳ Translation Training
定语从句的翻译 ······ 158
Part Ⅴ Exercises ······ 160

Unit 18 Rolling ······ 162

Part Ⅰ Reading and Comprehension ······ 162
Part Ⅱ Curriculum Ideological and Political ······ 166
Part Ⅲ Further Reading ······ 167
Part Ⅳ Translation Training
被动句的翻译（Ⅰ） ······ 167
Part Ⅴ Exercises ······ 168

Unit 19 Tube and Wire Rod Making ······ 170

Part Ⅰ Reading and Comprehension ······ 170
Part Ⅱ Curriculum Ideological and Political ······ 174

Part Ⅲ Further Reading ······ 175
Part Ⅳ Translation Training
被动句的翻译（Ⅱ）······ 175
Part Ⅴ Exercises ······ 176

Unit 20 The Heat Treatment of Steel ······ 178

Part Ⅰ Reading and Comprehension ······ 178
Part Ⅱ Curriculum Ideological and Political ······ 182
Part Ⅲ Further Reading ······ 182
Part Ⅳ Translation Training
英语长句的翻译 ······ 183
Part Ⅴ Exercises ······ 184

Unit 21 Development Trend of Iron and Steel Metallurgy ······ 187

Part Ⅰ Reading and Comprehension ······ 187
Part Ⅱ Curriculum Ideological and Political ······ 190
Part Ⅲ Further Reading ······ 191
Part Ⅳ Translation Training
there be 句型的译法 ······ 191
Part Ⅴ Exercises ······ 192

参考答案 ······ 194

第 1 单元 钢铁冶炼的历史 ······ 194
第 2 单元 炼铁原料 ······ 197
第 3 单元 高炉车间 ······ 201
第 4 单元 热风炉 ······ 205
第 5 单元 炼铁的新发展 ······ 208
第 6 单元 炼钢原料 ······ 212
第 7 单元 现代炼钢原理 ······ 216
第 8 单元 转炉生产 ······ 219
第 9 单元 电弧炉炼钢工艺 ······ 223
第 10 单元 炉外精炼 ······ 226
第 11 单元 连续铸钢 ······ 230
第 12 单元 钢的分类 ······ 234
第 13 单元 硫化矿铜的火法精炼 ······ 238
第 14 单元 铝的生产 ······ 242
第 15 单元 锌冶金 ······ 246
第 16 单元 金的提取 ······ 250
第 17 单元 金属的成型工艺 ······ 253

第 18 单元　轧钢简介 …… 258
第 19 单元　管材和线材的生产 …… 262
第 20 单元　钢的热处理 …… 266
第 21 单元　钢铁冶金发展趋势 …… 270

Glossary …… 273

参考文献 …… 303

From the History of Ironmaking and Steelmaking

Part Ⅰ Reading and Comprehension

扫描二维码获取音频

扫描二维码获取 PPT

Worldwide, the iron and steel industry is one of the most significant and, in terms of tradition, one of the oldest sectors of industry. As early as 3000 years ago, iron was serving as a basis of human culture and civilization.

1950s~1960s the steel age begins

By thedawn of the 20th century, steelmaking was a major industry and science was increasingly unlocking the mysteries of steel.

In the 1950s and 1960s, significant developments were made in steel processes, which allowed production to move away from military and shipping to cars and home appliances, which brought a huge growth in the range of steel home appliances that were made available to consumers. Post-war EU trade was also an important factor in the search for resources and the sales of finished goods.

1950s Continuous casting process

The more rapid solidification during the continuous casting process leads to increased homogeneity and better quality.

1951 A Community comes together

The European Coal and Steel Community (ECSC) is formed following the Treaty of Paris (1951) by the inner six: France, Italy, the Benelux countries (Belgium, Netherlands and Luxembourg) and west Germany.

1950s Electric arc furnace (EAF) develops

During the 19th century, a number of men had employed an electric arc to melt iron.

1950s The die is cast

Continuous casting, also called strand casting, is the process whereby molten metal is solidified into a 'semifinished' billet, bloom, or slab for subsequent rolling in the finishing mills.

1959 From blast furnaces to mini mills

Mini mills provided the latest technologies (arc, continuous casting, water-cooling) in smaller plants, which private companies could afford to operate. The rise of mini mills concides with an increase in the availability of scrap.

1967 World of steel

The World Steel Association founded as the International Iron and Steel Institute (IISI) in Brussels, Belgium on 19 October 1967.

1969 Mini mill revolution

When Nucor-which is now one of the largest steel producers in the US-decided to enter the long products market in 1969, they chose to start up a mini mill, which an Electric Arc Furnace as its steelmaking core, a move that was soon followed by other manufacturers during the 1970s.

1970s Innovation in the East-Japan

Pursuing rapid growth in the 1960s and 1970s Japan, followed closely by South Korea, developed massive state-of-the-art integrated facilities.

1990s An industry on the move

The steel industry has been its focus shift towards the emerging economies, as these require a huge amount of steel for urbanisation and industrialisation.

2000s The steel dragon

By the end of 2011, China was by far the world's largest steel producer, with an output of just over 680 millon tonnes.

2000s Arcelor Mittal arrives

ArcelorMittalis first global steel company in 2006 from the take over and merger of Arcelor by Mittal Steel, at the time of its creation, it was the world's largest steel producer.

2010s Large mergers take place

In 2011, Nippon Steel merges with Sumitomo Metal to become NSSMC. In 2016, Baosteel Group merges with Wuhan Group to from China Baowu Group, which is to become the second largest steel company in the world.

Words and Expressions

扫描二维码获取音频

扫描二维码答题

significant [sɪɡ'nɪfɪkənt] *a.* 有重大意义的；显著的

civilization [ˌsɪvəlaɪ'zeɪʃn] *n.* 文明；社会文明

dawn [dɔːn] *n.* 黎明；拂晓；开端；萌芽

increasingly [ɪn'kriːsɪŋli] *ad.* 越来越多地；不断增加地

continuous casting [kən'tɪnjuəs 'kɑːstɪŋ] 连续铸造

solidification 凝固；固化；硬化；凝结过程；凝固作用

homogeneity [ˌhɒmədʒə'niːəti] *n.* 同种；同质
electric arc furnace [ɪ'lektrɪk ɑːk 'fɜːnɪs] 电弧炉；电弧炼钢；电弧熔炉法
die [daɪ] *n.* 模具；冲模；压模
semifinished [ˌsɛmi'fɪnɪʃt] 半加工成形；半成品
slab [slæb] *n.* 厚板；厚片；厚块
blast furnace ['blɑːst fɜːnəs] *n.* （炼铁的）高炉，鼓风炉
mill [mɪl] *n.* 工厂；制造厂；磨粉机；轧机
massive ['mæsɪv] *a.* 巨大的；大而重的；结实的；非常严重的
integrated ['ɪntɪgreɪtɪd] *a.* 各部分密切协调的；综合的；完整统一的
urbanization [ˌɜːbənaɪ'zeɪʃn] *n.* 城市化；都市化
industrialization [ɪnˌdʌstrɪəlaɪ'zeɪʃən] *n.* 工业化
merger ['mɜːdʒə(r)] *n.* （机构或企业的）合并，归并

Proper Names

The European Coal and Steel Community (ECSC) 欧洲煤炭和钢铁共同体（ECSC）
International Iron and Steel Institute (IISI) 国际钢铁协会（IISI）
ArcelorMittal 安赛乐米塔尔公司
Nippon Steel 日本新日铁
Sumitomo Metal 住友金属公司
NSSMC 新日铁住友金属公司
Baosteel Group 宝钢集团有限公司
Wuhan Group 武汉钢铁集团
China Baowu Group 中国宝武钢铁集团

Answer the following questions.

(1) Which countries formed the European Coal and Steel Community (ECSC)?
(2) When was electric arc employed to melt iron?
(3) When and where was the Word Steel Association founded?
(4) Which is the world's largest steel producer?
(5) When did large mergers take place?

(1) Worldwide, the iron and steel industry is one of the most significant and, in terms of tradition, one of the oldest sectors of industry. As early as 3000 years ago, iron was serving as a

basis of human culture and civilization.

钢铁工业是世界上最重要的工业之一，也是传统史上最古老的工业之一。早在 3000 年前，铁就是人类文化与文明的基础。

(2) By the dawn of the 20th century, steelmaking was a major industry and science was increasingly unlocking the mysteries of steel.

在 20 世纪初，钢铁行业成为主要产业，科学正逐步揭开钢铁的神秘面纱。

"increasingly：越来越多地；不断增加地"。例如：It is becoming increasingly clear that such problem will be easily solved（越来越明显的是，这种问题很容易解决）。

(3) The more rapid solidification during the continuous casting process leads to increased homogeneity and better quality.

在连铸过程中凝固速度越快，合金的均匀性越高，质量越好。

"lead to：导致；引起；通向"。例如：Perseverance leads to success（有恒心就能胜利）。

(4) During the 19th century, a number of men had employed an electric arc to melt iron. 在 19 世纪，许多人利用电弧熔化铁。

"a number of：许多的"；"the number of：……的数目"。例如：A number of improvements have been made in this new machine.

这台新机器作了许多改进。

The number of unemployed people in China has decreased a quarter in the last month.

上个月，中国的失业人数降低了 1/4。

Part Ⅱ Curriculum Ideological and Political

【Connecting the world with steel】China steel taking concrete actions to build a better world

In 1949, a steel industry of the New China was born out of the ashes of China's War of Liberation. Starting with an annual output of 158, 000 tons steel.

China's steel enterprises, together with the people's Republic of China started from scratch, yet they blazed a new trail. From ploughshares to bridges. From railroad tracks to "two bombs and one satellite" . The steel industry has committed to advancing the course of New China. In 1996, China's steel production broke through the threshold of 100 million tons for the first time. Since then, it has maintained its position as the world's largest steel producer for more the 28 years.

The steel industry has worked behind the scenes to support various Chinese industries to take off and laid the iron foundation for the Chinese people to move towards prosperity. As the same time, China's state-of-the-art steel products which boast superior quality, advanced technology and wide varieties, also provide premium choices for construction and development all over the world. Today, from Africa to Europe, from East Asia to North America. Chinese steel companies, as a bolster of the industrial civilization have met more than half of the world's iron and steel demand by covering all kinds of high-quality materials. We use steel to invigorate the city. We use steel to connect the World. We use steel to discover the mysteries of the universe. We use steel to shelter our lives. We use steel to light up human civilization.

Part Ⅲ Further Reading

Challenges and future development of the steel industry

The principal contradiction facing Chinese society has become that between unbalanced and inadequate development and the people's ever-growing needs for a better life, and that the contradictions and problems in development are embodied in the quality of development. In this context, the requirements for the steel industry are no longer based on quantity and scale, but on variety quality, green and low-carbon, innovative development. We will focus on the following aspects:

(1) We will continue to push forward supply-side structural reform and maintain the smooth operation of industries;

(2) We will strengthen industrial and supply chains and promote coordinated development of upstream and downstream sectors;

(3) We will adhere to the direction of green and low-carbon development and intensify scientific and technological innovation;

(4) We will accelerate digital construction and promote the development of intelligent manufacturing in the industry.

Part Ⅳ Translation Training

翻译标准及过程

翻译标准是衡量翻译的尺度。一般概括为两点:

(1) 准确。译者必须把原作的内容完整而准确地表达出来，不能篡改歪曲作者的思想内容，要尽可能地保持原作的本来面目。

(2) 流畅。译文必须用词恰当、文理通顺、结构整齐、逻辑清楚，符合汉语语法规范和修辞习惯，使读者明白易懂。

翻译过程是正确理解原文和创造性地用另一种语言再现原文的过程，大体分为三个阶段:

(1) 理解阶段。理解阶段是理解原文词汇含义、句法结构、惯用法，分析理解前后句子及上下段落之间逻辑关系的阶段。对于多义词、短语，要仔细推敲来决定确切译法，然后将前后句子与上下段落联系起来理解，形成对原文的整体印象，真正理解原文的内容。

(2) 表达阶段。表达阶段是译者把自己从原文理解的内容用本族语言重新表达出来的阶段。可以直译，也可以意译。

(3) 校核阶段。校核阶段是理解与表达进一步深化，是对原文内容进一步核实以及对译文语言进一步推敲的阶段。初校，着重内容，对照原文，边看边改，看看有无漏译、错译之处，要特别注意日期、数字。复校，着重润饰文字，脱离原文，避免受原文表达形式的束缚和影响，看看译文是否句简词精、文理通顺、传神达意。定稿，对照原文，对译文再进行一次检查修改，一定要使所有的问题都得以解决，然后定稿。

Part V　Exercises

Ⅰ. Translate the following expressions into English.

(1) 钢铁工业	(2) 文明	(3) 高炉
(4) 模铸	(5) 电弧炉	(6) 城市化
(7) 工业化	(8) 连续铸钢	(9) 凝固

Ⅱ. Fill in the blanks with the words from the text. The first letter of the word is given.

(1) By the d______ of the 20th century, steelmaking was a major industry and science was i______ unlocking the mysteries of steel.

(2) C______ casting, also called strand casting, is the process whereby molten metal is solidified into a "s______" billet, bloom, or slab for subsequent rolling in the finishing mills.

(3) Mini m______ provided the latest technologies.

(4) Pursuing rapid growth in the 1960s and 1970s Japan, followed closely by South Korea, developed m______ state-of-the-art integrated facilities.

(5) The steel industry has been its focus shift towards the emerging economies, as these require a huge amount of steel for u______ and i______.

(6) By the end of 2011, China was by far the world's largest steel p______, with an output of just over 680 millon tonnes.

Ⅲ. Fill in the blanks by choosing the right words form given in brackets.

Early sponge iron was (1) ______ (produce; produced; produces) in bloomery hearths furnaces. The furnace with low pit or shaft furnaces were (2) ______ (made; make) of mud, quarries or erratics. With (3) ______ (improving; improved; improves) utilization of the heat in the lump-producing furnaces, temperatures could be raised to such a degree that the charged material was melted to a liquid state. It is now understood that when the iron is molten in a blast furnace, it (4) ______ (absorb; absorbs; absorbed) three to five per cent of carbon and other impurities as well. This saw the start of the blast furnace. Up to the beginning of the 18th century, blast furnaces were operated with charcoal. In 1709, Abraham · Darby, a yang man succeeded in (5) ______ (smelting; smelted) iron with coke. This innovation resulted in a steep rise in pig iron production.

Ⅳ. Decide whether the following statements are true or false (T/F).

(1) In the 1950s and 60s, significant developments were made in steel processes.　(　)

(2) Mini mills provided the latest technologies which private companies could not afford to operate.　(　)

(3) The more rapid solidification during the continuous casting process leads to increased homogeneity and better quality.　(　)

(4) In the 1960s and 1970s, South Korea pursued rapid growth.　(　)

(5) In 2016, Baosteel Group merges with Wuhan Group to from the largest steel company in the world.　(　)

Ⅴ. Translate the following English into Chinese.

1969 Mini mill revolution——When Nucor-which is now one of the largest steel producers in the US-decided to enter the long products market in 1969, they chose to start up a mini mill, which an Electric Arc Furnace as is steelmaking core, a move that was soon followed by other manufacturers during the 1970s.

Unit 2 Raw Materials for the Production of Iron

Part Ⅰ Reading and Comprehension

扫描二维码获取音频　　扫描二维码获取 PPT

The raw materials for the production of iron in the blast furnace can be grouped as follows: iron-bearing materials, fuels and fluxes.

1. Iron-bearing materials

The major iron-bearing materials are iron ores, sinter and pellets in the blast furnace. Their function is to supply the element iron, which is 93 to 94 per cent of the pig iron produce.

Iron ores are classed by their chemical compositions, such as oxides, sulfides, carbonates, etc (shown in Table 2-1).

Table 2-1 Iron ores classed by chemical composition

Class and Mineralogical Name	Chemical Composition of Pure Mineral	Common Designation
Oxide		
Magnetite	Fe_3O_4	Ferrous-ferric oxide
Hematite	Fe_2O_3	Ferric oxide
Ilmenite	$FeTiO_3$	Iron-titanium oxide
Limonite	$nFe_2O_3 \cdot mH_2O$	Hydrous iron oxide
Carbonate		
Siderite	$FeCO_3$	Iron carbonate
Sulphide		
Pyrite (iron pyrites)	FeS_2	—
Pyrrhotite (magnetic iron pyrites)	FeS	Iron sulphides

Hematite is one of the most widely used ores. If pure, it would give 70 per cent (by mass) iron.The typical reddish color is caused by the iron (Ⅲ). In the case of red iron ore, the compound of iron and oxygen is not so ***tight*** that the hematite is regarded as ***easily reducible*** . Magnetite, a magnetic iron ore, is increasing in use for two reasons. Firstly, it can be separated from the rock by magnetic means; secondly, it has high iron content. Iron and oxygen atoms are very closely combined with each other in magnetite, thus making magnetite ***difficult to reduce***. Limonite is a brown iron ore and contains water, which means that the iron oxides have formed a stable compound with water (water of crystallization). Containing 30 to 40 per cent Fe, siderites are relatively easy to reduce. Most ores contain only 50 to 60 per cent (by mass) iron because they contain 10 to 20 per cent (by mass) gangue (which consists mostly of alumina and silica). If the gangue contains mainly lime, the ore is ***basic***; if silicon acid (SiO_2) predominates, the ore is ***acid***.

The portion of the ore that is too fine to be charged directly into blast furnace is usually agglomerated. The most important processes are: sintering and pelletizing.

The sintering process is in five stages: (1) mixing of the raw materials and fine coal or coke, (2) placing the mixture on a grate, (3) igniting and sintering, air drawn through the mixture burns the fuel at a temperature high enough to frit the small particles together into a cake so that they can be charged into the blast furnace satisfactorily, (4) cooling, (5) crushing and screening before charging to the furnaces. For best results, pulverized flux is added to the sinter mix to combine with the gangue of the ore in the sintering process. Sinter usually contains 50 to 60 per cent (by mass) iron.

During pelletizing, the mixtures made from ultrafine (minus 0.074mm) iron-ore concentrates and binders of grain sizes far less than 1mm are balled to form ***green*** pellets slightly larger than 6mm but smaller than 15mm in diameter. The green pellets are then hardened by firing in a shaft-type furnace or rotary kiln or on a traveling grate. Pellets usually contain from 60 to 67 per cent (by mass) of iron.

Compared with lump ores and sinter, the advantages of pellets are: a narrow size range, constant quality and good permeability during reduction. Furthermore, pellets are well suited for transport and storage. But, any swelling and sticking of pellets during the reduction phase must be avoided.

2. Fuels

The fuels enter the blast furnace as coke, coal, oil or gas. They are used for producing the heat required for smelting, and reducing the iron oxides into metallic iron and carburizing the iron (about 40 to 50 kilograms per ton of iron). In addition, because the coke retains its strength at high temperature, it provides the structural support that keeps the unmelted burden materials from falling into the hearth.

At present, some of the coke in the blast furnace is usually replaced by coal. The blast furnace can inject hard coal, soft coal and mixed coal. BF pulverized coal injection can dramatically reduce coke rate and the dependency on increasing shortage of coke resource, so it is the most effective approach to reducing the ironmaking cost and has become an important part in BF ironmaking technology advances.

3. Fluxes

Fluxes include limestone, dolomite and lime mainly, whose major functions are to combine with the ash in the coke and the acid gangue in the ores to make a fluid slag that can be drained readily from the furnace hearth. The ratio of basic oxides to acid oxides must be controlled carefully to preserve the sulphur-holding power of the slag as the fluidity.

✻ Words and Expressions

扫描二维码获取音频

扫描二维码答题

raw material	原料
iron-bearing material	含铁原料
fuel [ˈfjuːəl] *n.*	燃料
flux [flʌks] *n.*	熔剂，造渣剂
mineralogical [ˌminərəˈlɔdʒikəl] *a.*	矿物学的
sinter [ˈsintə] *v.*	烧结
n.	烧结矿
pellet [ˈpelit] *n.*	球团矿
oxide [ˈɔksaid] *n.*	氧化物
sulphide [ˈsʌlfaid] *n.*	硫化物
sulfur [ˈsʌlfə] *n.*	硫（元素符号 S）
carbonate [ˈkɑːbəneit] *n.*	碳酸盐
magnetite [ˈmægnitait] *n.*	磁铁矿
hematite [ˈhemətait] *n.*	赤铁矿
limonite [ˈlaiməˌnait] *n.*	褐铁矿
ilmenite [ˈilmiˌnait] *n.*	钛铁矿
siderite [ˈsidəˌrait] *n.*	菱铁矿
pyrite [ˈpaiərait] *n.*	黄铁矿
pyrrhotite [ˈpiərəutait] *n.*	磁黄铁矿
ferrous [ˈferəs] *a.*	含铁的，亚铁的
ferric [ˈferik] *a.*	铁的，三价铁的
gangue [gæŋ] *n.*	脉石
titanium [taiˈteinjəm, tiˈteinjəm] *n.*	钛（元素符号 Ti）
hydrous [ˈhaidrəs] *a.*	含水的
reducible [riˈdjuːsəbl] *a.*	可还原的
rock [rɔk] *n.*	岩石

crystallization ['kristəlai'zeiʃən] *n.*	结晶化
water of crystallization	结晶水
alumina [ə'lju:minə] *n.*	氧化铝
charge [tʃɑ:dʒ] *n.*	炉料
v.	装炉，装料
agglomerate [ə'glɔməreit] *v.*	结块，烧结
pelletize ['pelitaiz] *v.*	使……成颗粒状
grate [greit] *n.*	箅条，固定筛
ignite [ig'nait] *v.*	着火，点火
frit [frit] *v.*	熔融（化）
particle ['pɑ:tikl] *n.*	粒子，极小量
pulverize ['pʌlvəraiz] *v.*	使……成粉末
pulverized ['pʌlvəraizd] *a.*	粉状的
ultrafine [ˌʌltrə'fain] *a.*	极其细小的
concentrate ['kɔnsentreit] *n.*	精矿，精煤
binder ['baində] *n.*	黏结剂
green pellet	生球
shaft-type furnace	竖炉
kiln [kiln] *n.*	窑
rotary kiln	回转窑
permeability [ˌpə:miə'biliti] *n.*	渗透，渗透性
storage [stɔridʒ] *n.*	储藏
swell [swel] *n.*	膨胀
stick [stik] *vi.*	粘住，粘贴
metallic [mi'tælik] *a.*	金属的，含金属的
carburize ['kɑ:bjuraiz] *v.*	使……渗碳
strength [streŋθ] *n.*	强度
structural ['strʌktʃərəl] *a.*	结构（上）的
soft coal	烟煤
hard coal	无烟煤
injection [in'dʒekʃən] *n.*	喷吹，喷射
pulverized coal injection	喷煤
dramatically [drə'mætikəli] *ad.*	鲜明地，显著地
coke rate	焦比
approach [ə'prəutʃ] *n.*	方法，途径
advance [əd'vɑ:ns] *n.*	进步，进展
limestone ['laimstəun] *n.*	石灰石
ash [æʃ] *n.*	灰分
drain [drein] *v.*	排出，流掉

sulphur holding power	脱硫能力
fluidity [flu(:)'iditi] *n.*	流动性
portion ['pɔ:ʃən] *n.*	部分，一份
dependency [di'pendənsi] *v.*	依靠，信赖
harden ['hɑ:dn] *v.*	使……变硬，使……坚强
traveling grate	移动床

❈ Answer the following questions.

(1) What are raw materials for the production of iron at the blast furnace?

(2) Why is magnetite increasing in use of iron making?

(3) Can you describe the sintering process?

(4) What are functions of the coke?

(5) What coals can be injected in the blast furnace?

(6) Why must the ratio of basic oxides to acid oxides be controlled carefully?

(7) What are the advantages of pellets compared with lump ores?

(8) What is some of the coke in the blast furnace usually replaced by at present?

(9) What do fluxes mainly include?

(1) The portion of the ore that is too fine to be charged directly into blast furnace is usually agglomerated.

矿石中过细的颗粒不能直接装入高炉，通常需要烧结成块。

“that is too fine to be charged” 是 “The portion of the ore” 的定语从句。“too”（用于形容词或副词之前）意思是“太，过于”。“too…to” 意思是 “太……以致不能”。例如：These pillars are too thin to carry the roof（这些柱子太细，支撑不住屋顶）；He is too careful not to have noticed it（他那么细心，不会不注意这一点的）。

(2) Air drawn through the mixture burns the fuel at a temperature high enough to frit the small particles together into a cake so that they can be charged into the blast furnace satisfactorily.

当空气穿过该混合料时，混合料中的燃料进行燃烧，形成的高温足以使这些细小的原料颗粒烧结成块，以便满足高炉冶炼的要求。

“drawn through the mixture” 是过去分词短语作定语，修饰 “air”。“so that they can be charged into the blast furnace satisfactorily” 是目的状语从句。

(3) During pelletizing, the mixtures made from ultrafine (minus 0.074mm or minus 200mesh) iron-ore concentrates and binders of grain sizes far less than 1mm are balled to form ***green*** pellets slightly larger than 6mm but smaller than 15mm in diameter.

球团生产期间，由极细的铁精矿粉（小于 0.074 毫米或小于 200 目）以及粒度远小于 1 毫米的黏结剂组成的混合物，首先制成直径略大于 6 毫米但小于 15 毫米的生球。

“made from ultrafine (minus 0.074mm or minus 200mesh) iron-ore concentrates and binders of grain sizes far less than 1mm” 是过去分词短语作定语，修饰 “mixtures”。

(4) In addition, because the coke retains its strength at high temperature, it provides the structural support that keeps the unmelted burden materials from falling into the hearth.

此外，由于焦炭在高温环境下仍保持其强度，因此，它还具有防止未熔化物料落入炉缸的结构支撑作用。

“keep” (常与 from 连用) 意思是 “远离，不接触”。例如：Keep away from the scene of the accident “勿靠近事故现场”。

(5) BF pulverized coal injection can dramatically reduce coke rate and the dependency on increasing shortage of coke resource, so it is the most effective approach to reducing the ironmaking cost and has become an important part in BF ironmaking technology advances.

高炉喷煤可以大幅度降低入炉焦比，减少对日益匮乏的焦煤资源的依赖，是炼铁降低成本的最有效手段，已成为高炉炼铁技术进步的一项重要内容。

“BF (Blast Furnace)” 的意思是 “高炉”。

(6) …, whose major functions are to combine with the ash in the coke and the acid gangue in the ores to make a fluid slag that can be drained readily from the furnace hearth.

……它们的主要作用是将焦炭中的灰分和矿石中的脉石结合成可被顺利排出炉缸的液态炉渣。

“whose” 是关系代词引导非限制性定语从句，“that can be drained readily from the furnace hearth” 又是此句中的定语从句，修饰 “fluid slag”。

Part Ⅱ Curriculum Ideological and Political

【Feelings to family and national】Achievements of ancient Chinese steel smelting technology

The characteristics of Chinese ancient steel development were different from those of other countries. Block iron and solid carburizing steel have been used for solid reduction in the world for a long time. While China's cast iron and pig iron steelmaking have been the main methods. Because of the invention and development of cast iron and pig iron steelmaking. China's metallurgical technology has been in the world advanced level before the middle of Ming Dynasty.

In China, the total output of iron and steel has reached an annual output of 1200 tons in the Tang Dynasty. The Song Dynasty was 4700 tons, and the Ming Dynasty reached 40 thousand tons at most. In thirteenth Century, China was the world's largest producer and consumer of iron and remained in the lead until seventeenth Century. From the Han Dynasty to the Ming Dynasty, the Chinese were not only in the lead in quantity, but also the world's most advanced steel smelting technology.

Part Ⅲ　Further Reading

Microwave sintering prereduction technology

Microwave sintering prereduction technology means to realize the sintering of block ore by microwave, rather than the past heating sintering of coking coal powder. After microwave sintering, the sinter still has high temperature waste heat. Through the reduction and heat absorption effect of hydrogen and iron oxide, it not only plays a cooling effect, but also realizes the pre-reduction of sinter, so as to improve the metallization rate of sinter, reduce the consumption of reducing agent of blast furnace after sinter enters the blast furnace, and achieve the purpose of carbon reduction.

Part Ⅳ　Translation Training

选词用字法

选词用字法指的是根据对原文的准确理解来适当地选择用词。由于构词时联想不同、表达方式不同，两种语言表达同一思想时往往用词不同。翻译时，必须在理解的基础上选择恰当的词。英、汉两种语言中都有一词多类、一词多义的现象，在弄清原句的结构后，要善于选择和确定原句中关键词的词义。通常从下面两方面着手。

1. 根据词类选择词义

选择某个词义，首先要弄清这个词在句子中属于哪种词类，然后再进一步确定词义。例如：

（1）英语中的“like”。Like charges repel, unlike charges attract（相同的电荷相斥，不同的电荷相吸），句中“like”为形容词。Like knows like（英雄识英雄），句中的“like”为名词。He likes Metallurgy（他喜欢冶金学），“like”为动词。

（2）英语中的“opposite”。In England you must drive on the opposite side of the road to the rest of Europe（在英国路上开车，要与欧洲其他国家方向相反），句中的“opposite”为形容词。The woman sitting opposite is a detective（坐在对面的那女子是侦探），句中的“opposite”为副词。Light is the opposite of heavy（轻是重的反义词），句中的“opposite”为名词。

2. 根据上下文选择词义

同一个词在不同的场合往往有不同的含义，必须根据上下文的搭配关系或句型来判断确定某个词在特定场合下所具有的词义，例如英语单词“separate”。The magnetite can be separated from the rock by magnetic means（磁铁矿可以通过磁选法从岩石中分离出来）。The branch has separated from the trunk of the tree（这个树枝从树干上脱落了）。The patient should be separated from the others（这个病人应该隔离）。

Part Ⅴ Exercises

Ⅰ. Translate the following expressions into English.

(1) 炼铁原料	(2) 易还原矿石	(3) 酸性脉石
(4) 碱性脉石	(5) 球团矿	(6) 含铁原料
(7) 喷煤	(8) 磁铁矿	(9) 赤铁矿
(10) 焦炭灰分	(11) 烧结矿	(12) 熔剂

Ⅱ. Fill in the blanks with the words from the text. The first letter of the word is given.

(1) The major iron b ______ materials are iron ores, sinter and pellets.

(2) Iron oxides are always mixed with impurities which are called the g ______.

(3) If the gangue contains mainly lime, the ore is b ______.

(4) H ______ is one of the most widely used ores.

(5) L ______ is a brown iron ore.

(6) The portion of the ore that is too fine to be charged directly is usually a ______.

(7) Compared with lump ores, p ______ have a narrow size range, constant quality and good permeability during reduction.

(8) The f ______ enter the blast furnace as coke, coal, oil or gas.

(9) Fluxes include l ______, d ______ and lime mainly.

Ⅲ. Fill in the blanks by choosing the right words form given in the brackets.

Hematite is (1) ______ (one; a; an) of the most widely used ores. If pure, would give 70 per cent iron. The typical reddish color is caused by the iron (Ⅲ). In the case of red iron ore, the compound of iron and oxygen (2) ______ (are; am; is) not so ***tight*** and so the (3) ______ (magnetite; hematite) is regarded as ***easily reducible***.

Iron and oxygen atoms are very closely (4) ______ (combined; combining; combines) with each other in magnetite, thus making magnetite ***difficult to reduce***. Limonite (5) ______ (is; are; was) an brown iron ore. It contains water, which means that the iron oxides have formed a stable compound with water (water of crystallization).

Ⅳ. Decide whether the following statements are true or false (T/F).

(1) Iron ores are classed by their chemical compositions, such as oxides, carbonates, etc. (　　)

(2) If the gangue contains mainly lime, the ore is ***acid***; if silicon acid (SiO_2) predominates, the ore is ***basic***. (　　)

(3) Iron and oxygen atoms are very closely combined with each other in magnetite, thus making magnetite ***easy to reduce***. ( )

(4) The portion of the ore that is too fine to be charged directly is usually agglomerated. ( )

(5) Fluxes include limestone, dolomite, coke and lime mainly. ( )

V. Translate the following English into Chinese.

The most important raw materials for smelting iron and steel are the iron ores. The quality requirements to be satisfied by the ores nowadays are: high iron content, good metallurgical characteristics and reducibility, lower quantities and type of undesired tramp elements, lack of very fine material (less than 5% below 5mm), etc. To achieve these desirable factors, it is usual to treat ores.

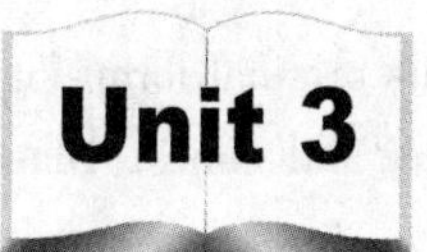

Unit 3 The Blast Furnace Plant

Part Ⅰ Reading and Comprehension

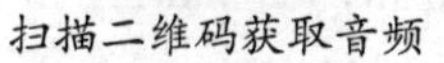

扫描二维码获取音频

扫描二维码获取PPT

A complete blast furnace plant (shown in Fig. 3-1) consequently comprises many components. The most important of these are: the blast furnace, bunkers, charging equipment, the cast house, hot blast stoves, top gas removal and cleaning equipment.

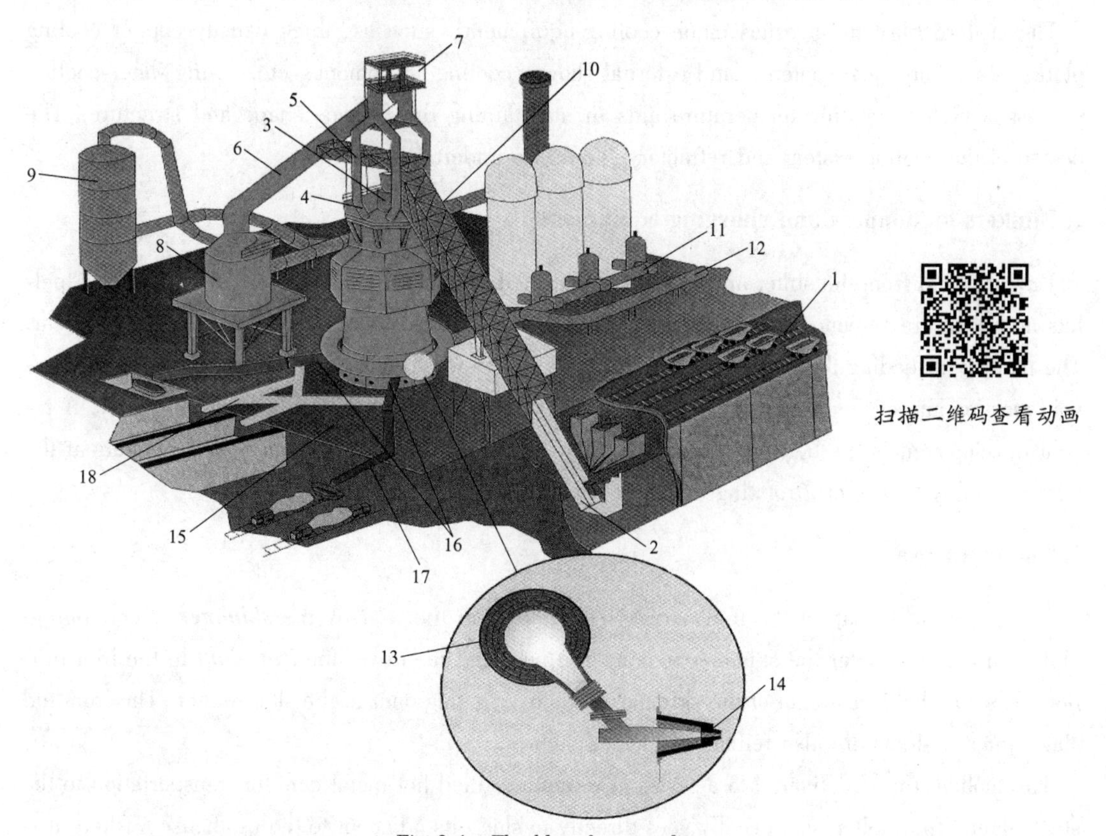

扫描二维码查看动画

Fig. 3-1 The blast furnace plant

1—Stockhouse; 2—Scale car/skip system; 3—System of ***bells*** or a chute bell-less top; 4—Offtakes; 5—Uptakes; 6—Downcomer; 7—Bleeders; 8—Dustcatcher; 9—Gas cleaning system; 10—Clean gas pipe; 11—Hot b alst main; 12—Mixer line; 13—Ring main bustle pipe; 14—Tuyeres; 15—Casthouse; 16—Trough; 17—Iron runner; 18—Slag runner

1. The blast furnace

The blast furnace is a continuously working shaft furnace. In its original form, it was slim and tulip shaped. Today, its silhouette consists of truncated cylinders and cones. Blast furnace is a large refractory-lined steel vessel and the entire furnace requires cooling.

The following is an introduction to the parts and functions of a modern blast furnace. The furnace is built on a foundation of pilings and concrete to support the furnace and burden. The inside profile of the blast furnace is termed (from top to bottom) furnace throat, shaft, belly, bosh and hearth. The lower portion of the furnace is called the hearth. The hearth comprises carbon side wall and a central composite plug. One or more iron notches are located above the hearth, and molten iron and slag are removed through them. Above the iron notch is the slag notch. With modern low slag volume practice, the slag notch is typically for blowing-in or emergency use only. The portion above the hearth is the bosh, and the bosh angle is a critical parameter in the design of the furnace. The belly of the widest point can be cooled by a variety of methods. The bosh and stack usually have ceramic lining. The throat is cylindrical and is referred to as the stockline section. The top of the furnace directs the gas to the offtakes.

The modern blast furnace has much cooling equipment, such as, high density copper cooling plates, cast-iron stave coolers, and external shower cooling equipment, etc. Using water-cooling process at higher smelting temperature aids in maintaining the furnace shape and structure. The design of the cooling system and refractory should be undertaken together.

2. Bunkers or hoppers and charging equipment

Raw material from the sinter and coke plants are fed to a system of bunkers, where sinter, pellets and coke are screened and weighed, before being charged via a belt system or the skip car. The method of feeding the charge into the furnace is somewhat complicated, the object being to avoid wasting any of the blast-furnace gas and to distribute the contents of the furnace evenly. A recent development is the bell-less top in which the charge is distributed from sealed bunkers at the furnace top by means of a rotating chute.

3. The casthouse

Liquid slag will float on the iron and is separated from the iron by the ***skimmer*** after tapping. Molten iron flows under the skimmer into an iron pool and then over the ***iron dam*** to the iron runner. The slag held upstream of the skimmer flows over a slag dam to the slag runner. The iron and slag runner systems are also refractory-lined trenches.

The molten iron is delivered to a series of refractory-lined hot metal cars for transportation to the steel plant. The molten slag usually goes directly to slag pits adjacent to the casthouse, where it is cooled with water sprays. Some facilities lacking the space for local pits use slag ladles to transport the slag to remote pits. There are two critical pieces of equipment in the casthouse: the taphole drill and the mudgun. The taphole drill is used to open the taphole. The mudgun is used to plug the taphole at the end of the cast.

4. Hot blast stoves and blast pipes

Clean cold blast air is supplied from a blower house to the hot blast stoves. Each blast furnace has three or four hot blast stoves. The air is heated as it passes through the stoves to the ***hot blast main***. A ***mixer line*** delivers a portion of the cold blast air to blend with the hot blast air to maintain a constant hot blast temperature to the furnace throughout each stove cycle. The hot blast main connects with a ***ring main bustle pipe*** which encircles the furnace. From the bustle pipe, the hot blast is delivered to the furnace through a series of evenly spaced nozzles called ***tuyeres***.

5. Top gas removal and cleaning equipment

The gas produced by the ironmaking process exits the top of the furnace and is delivered to the gas cleaning system via a series of refractory-lined ducts referred to as ***offtakes***, ***uptakes*** and the ***downcomer***. At the top of the uptakes are a series of pressure relief valves called ***bleeders***. The downcomer delivers the dust-laden gas to the ***dustcatcher*** which will remove about 60% of the particulate matter in the gas. The gas then passes to a ***gas cleaning system*** which will scrub the gas, removing 99.9% of the remaining particulate. At this point, the gas is a suitable fuel for the stoves and other areas.

Words and Expressions

扫描二维码获取音频

扫描二维码答题

comprise [kəm'praiz] *v.* 包括，由……组成
bunker ['bʌŋkə] *n.* 料仓
hopper ['hɔpə] *n.* 漏斗，料斗
cast house 出铁场
hot stove (cowper) 热风炉
slim [slim] *a.* 细长的
tulip ['tju:lip] *n.* 喇叭形
silhouette [ˌsilu(:)'et] *n.* 轮廓
truncated ['trʌŋkeitid] *a.* 截短了的
cylinder ['silində] *n.* 圆筒，圆柱体
cone [kəun] *n.* 锥体
piling ['pailiŋ] *n.* 打桩，打桩工程
throat [θrəut] *n.* （炉）喉
belly ['beli] *n.* 炉腰

bosh [bɔʃ] *n.*	炉腹
composite ['kɔmpəzit] *a.*	复合的，合成的
plug [plʌg] *n.*	炉底砌块，塞子
slag volume	渣量
practice ['præktis] *n.*	实际操作
blow-in	开炉
be referred to as	叫做，称为
notch [nɔtʃ] *n.*	出口，槽口
iron notch	铁口
slag notch	渣口
emergency [i'mə:dʒənsi] *n.*	紧急情况，突然事件
critical ['kritikəl] *a.*	关键性的，紧要的
parameter [pə'ræmitə] *n.*	参数
stack [stæk] *n.*	炉身，堆，一堆
ceramic [si'ræmik] *a.*	陶瓷的
cylindrical [si'lindrikəl] *a.*	圆柱（形，体）的
stockline [stɔklain] *n.*	料线
offtake ['ɔ:fteik] *n.*	煤气导出管，排气管
density ['densiti] *n.*	密度
copper ['kɔpə] *n.*	铜（元素符号 Cu）
stave [steiv] *n.*	板，冷却壁
content [kɔntent] *n.*	里面的东西
evenly ['i:vənlɪ] *ad.*	均匀地
dump [dʌmp] *v.*	倾卸，倾倒
double bell	双料钟
arrangement [ə'reindʒmənt] *n.*	装置，配置
cooling plate	冷却板
external [eks'tə:nl] *a.*	外部的，外面的
undertake [ˌʌndə'teik] *v.*	进行，从事
belt [belt] *n.*	皮带
skip [skip] *n.*	料车
distribute [dis'tribju(:)t] *v.*	分配，分布
bell-less [belles] *a.*	无料钟的
seal [si:l] *v.*	密封
rotating chute	旋转溜槽
liquid ['likwid] *n.*	液体
a.	液体的，液态的
float [fləut] *v.*	飘浮，漂浮
separate ['sepəreit] *v.*	分开，隔离

skimmer ['skimə] *n.* 撇渣器
pool [puːl] *n.* 池
dam [dæm] *n.* 堤，坝
runner ['rʌnə(r)] *n.* 流槽
iron runner 铁沟
trench [trentʃ] *n.* 沟槽，沟道
facility [fə'siliti] *n.* 设施，设备
pit [pit] *n.* 池，深坑
slag pit 渣池
be adjacent to 接近
steel plant 钢厂
water spray 喷水
slag ladle 渣罐
transport [træns'pɔːt] *v.* 传送，运输
remote [ri'məut] *a.* 遥远的，边远的
taphole drill 开（铁）口机
mudgun 泥炮
taphole ['tæphəul] *n.* 出（铁，钢，渣）口
blower house 鼓风机室
blend [blend] *v.* 混合
hot blast main 热风管
mixer line 混风管
ring main bustle 热风围管
encircle [in'səːkl] *v.* 环绕，围绕，包围
deliver [di'livə] *v.* 输送
exit ['eksit] *v.* 排出，离去
a series of 一系列的
tuyere [twiː'jɛə] *n.* 风口，鼓风口，风嘴
uptake ['ʌpteik] *n.* 上升管
downcomer ['daun,kʌmə(r)] *n.* 下降管
bleeder ['bliːdə] *n.* 分压器，放散阀
pressure relief valve 减压阀，溢流阀
scrub [skrʌb] *v.* 使（气体）净化，擦洗
particulate [pə'tikjulit, pə'tikjuleit] *a.* 微粒的

✲ Answer the following questions.

(1) What are the important components of the blast furnace?
(2) What is the lower potion of the blast furnace?
(3) Does the modern blast furnace have much cooling equipment?

(4) What critical pieces of equipment are there in the casthouse?

(5) What will float on the iron and is separated from the iron by the 'skimmer' after tapping?

(6) How many hot blast stoves does each blast furnace have?

(1) Raw material from the sinter and coke plants are fed to a system of bunkers, where sinter, pellets and coke are screened and weighed, before being charged via a belt system or the skip car.

原料从烧结厂和焦化厂被送到料槽系统，在这里，烧结矿、球团矿和焦炭被筛分和称量，然后由皮带或上料车进行上料。

在“where…the skip car”这个定语从句中，“before being…the skip car”是介词短语作时间状语。

(2) The method of feeding the charge into the furnace is somewhat complicated, the object being to avoid wasting any of the blast-furnace gas and to distribute the contents of the furnace evenly.

为了避免浪费高炉煤气，同时使炉内布料均匀，炉料装入炉内的方法有点复杂。

“the object…the furnace evenly”是现在分词作状语。“to avoid…”与“to distribute…”属于并列关系。

(3) A recent development is the bell-less top in which the charge is distributed from sealed bunkers at the furnace top by means of a rotating chute.

高炉最近的发展是无料钟炉顶，它可以借助于旋转溜槽把炉料从顶部密封料仓布入高炉炉内。

“in which”引出的定语从句修饰“the bell-less top”。“by means of”意思是：“通过，用，借助于”。例如：He succeeded by means of perseverance（他依靠坚持获得成功）。

(4) The molten slag usually goes directly to slag pits adjacent to the casthouse, where it is cooled with water sprays.

熔渣通常直接送到出铁场附近的渣池，在那里用水喷射冷却。

“adjacent to the casthouse”是形容词短语，修饰“slag pits. adjacent to sth.”，意思是“与某物邻近的”。例如：Our classroom is adjacent to yours（我们的教室在你们隔壁）；His house is adjacent to the highway（他的住宅靠近马路）。

(5) Some facilities lacking the space for local pits use slag ladles to transport the slag to remote pits.

当一些高炉附近没有渣池时，就用渣罐车把熔渣运送到较远的渣池。

“lacking the space for local pits”是分词短语作定语，修饰“facilities”。“facilities”意思是“设备，装备”，在此指的是高炉。

(6) A ***mixer line*** delivers a portion of the cold blast air to blend with the hot blast air to

maintain a constant hot blast temperature to the furnace throughout each stove cycle.

混风管需要输送部分冷风与热风混合，以便在热风炉循环使用过程中维持（高炉）入炉风温的稳定。

"to maintain…each stove cycle"是不定式短语作状语。

（7）The gas produced by the ironmaking process exits the top of the furnace and is delivered to the gas cleaning system via a series of refractory-lined ducts referred to as ***offtakes***, ***uptakes*** and the ***downcomer***.

炼铁过程中产生的煤气，从高炉顶部排出，经过一系列具有耐火材料内衬的煤气导出管、上升管、下降管被送入煤气净化系统。

"produced by the ironmaking process"是过去分词短语作定语，修饰"the gas"。"referred to as ***offtakes***, ***uptakes*** and the ***downcomer***"又是过去分词短语作定语，修饰"ducts"。"（be）referred to as"意思是"称……为"。例如：The products are referred to as direct reduced iron（这些产品被称为还原铁）；Coal is usually referred to as a fossil fuel（煤通常被称为矿物燃料）；These printouts are sometimes referred to as hard copy（这种打印输出有时也被称为硬拷贝）。

（8）At the top of the uptakes are a series of pressure relief valves called ***bleeders***. The down-comer delivers the dust-laden gas to the ***dustcatcher*** which will remove about 60% of the particulate matter in the gas.

在上升管的顶部有一系列被称为放散阀的煤气减压阀。下降管把含尘煤气送入重力除尘器中，它能除去约60%的灰尘颗粒。

"deliver sth. to sb."意思是"把某物交付给某人"。

Part Ⅱ Curriculum Ideological and Political

【Red figure】The blast furnace guard—Meng tai

In November 1948, Meng Tai returned to Anshan Iron and Steel Plant. At that time, Anshan Iron and Steel Plant was already broken after war. But he did not flinch. He loved the factory as home and worked hard. He picked up the spare parts in the cold and frozen snow. In the face of the stink, he found raw material by picking up the scrap heap. Every day, he picked up iron wires, screws and spare parts with a handful of mud, oil and sweat. Encouraged by his example, the workers of the whole factory had recycled thousands of materials and over ten thousand of spare parts in just a few months. The famous "Mengtai Warehouse" was built and played an important role in restoring production. Then he bravely overcame technical difficulties and solved more than a dozen technical problems.

Part Ⅲ Further Reading

Hydrogen-rich carbon cycle blast furnace

Hydrogen-rich carbon cycle blast furnace refers to replacing carbon with hydrogen in the traditional metallurgical process to greatly reduce greenhouse gas emissions in the iron and steel metallurgical process, until the realization of carbon neutrality in the iron and steel metallurgical production process. The key to the technical characteristics of hydrogen-rich carbon cycle blast furnace is full oxygen. If the blast furnace realizes full oxygen blowing, a large amount of nitrogen in the top gas of the blast furnace will not appear, and the blast furnace gas can easily realize the separation of CO and CO_2. After recycling, the remaining high concentration CO will be transported to the tuyere and furnace stack through pipes, so that CO and H_2 can be enriched into gas with high reduction potential again, and reused to the blast furnace for reducing iron ore. This is called carbon cycle. Through this carbon cycle, carbon chemical energy can be fully utilized. With the gas cycle, the chemical energy of hydrogen will not be wasted when hydrogen-rich substances are used in large quantities, and hydrogen can also be recycled in the blast furnace, thus reducing the consumption of fossil energy in the blast furnace process.

Part Ⅳ　Translation Training

词义的引申

英汉互译时，因为两种语言在词语搭配和句子结构方面有许多差异，总有一些词在英语词典中找不到合适的词义，如果硬套死译便不能表达原文的意思，甚至造成误解。这时就应该根据上下文和逻辑关系，透过原文的表层结构对该词加以引申，选择较适当的词将原文的深层意思表达出来。

引申词义可以从四个方面考虑：

（1）词义转译。例如：The molten slag usually goes directly to slag pits adjacent to the casthouse（熔渣通常直接送到出铁场附近的渣池）。“go to” 词义引申为 “送到”。

（2）词义抽象化。If silica bricks were used, as in the acid process, the lime would attack the furnace lining chemically（如同酸性工艺一样，如果使用硅砖，石灰就会对炉衬产生化学侵蚀）。“attack” 词义引申为 “侵蚀”。例如：Tell the policeman your story（把你知道的情况告诉警察）。“your story” 词义引申为 “情况”。

（3）词义具体化。例如：Some facilities lacking the space for local pits use slag ladles to transport the slag to remote pits（一些高炉附近没有渣池，就用渣罐车把熔渣送到较远的渣池）。“facilities” 词义引申为 “高炉”，Wait a minute, let me put on my things（等等，让我把衣服穿上）。

（4）词的搭配。例如：The oil provides some cooling effects（这种油起到一定的冷却作用）。“provide” 词义引申为 “起到”。

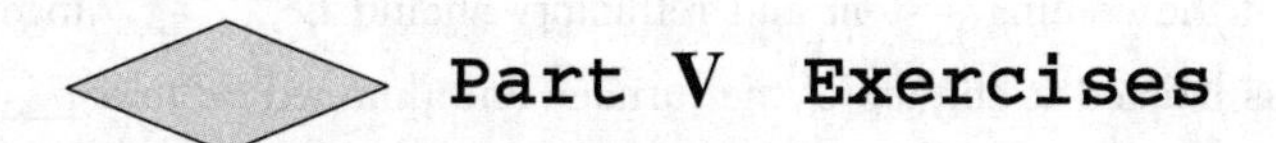

Part Ⅴ Exercises

Ⅰ. Translate the following expressions into English.

(1) 高炉车间	(2) 出铁场	(3) 热风管
(4) 炉身	(5) 炉腹角	(6) 撇渣器
(7) 导出管	(8) 铁沟	(9) 渣池
(10) 泥炮	(11) 除尘器	(12) 无料钟炉顶
(13) 铁口	(14) 旋转溜槽	(15) 皮带系统

Ⅱ. Fill in the blanks with the words from the text. The first letter of the word is given.

(1) A complete blast furnace p ______ consequently comprises many components.

(2) The lower portion of the furnace is called the h ______.

(3) The b ______ is the widest point of the furnace, it can be cooled by a variety of methods.

(4) The small bell opens, admitting the charge onto the large b ______.

(5) Molten iron and slag produced in the blast furnace are removed and separated in the c ______.

(6) There are two critical pieces of equipment in the casthouse: the t ______ drill and the m ______.

(7) The modern blast furnace has much c ______ equipment, such as, high density copper cooling plates, cast-iron staves, external cooling channels, or shower cooling, etc.

(8) The hot blast main connects with a ring main ***b*** ______ ***pipe*** which encircles the furnace.

(9) The gas produced by the ironmaking process exits the top of the furnace and is delivered to the gas c ______ system via a series of refractory-lined ducts referred to as ***offtakes***, ***uptakes*** and the ***downcomer***.

(10) Blast furnace is a large refractory-lined steel v ______ and the entire furnace requires cooling.

Ⅲ. Fill in the blanks with the words listed below.

notch	support	bleeders	undertaken
ceramic	skimmer	double	plug

(1) The furnace is built on a foundation of pilings and concrete to ______ the furnace and burden.

(2) Above the iron notch is the slag ______.

(3) Between the bosh and lower stack usually has a ______ lining.

(4) The design of the cooling system and refractory should be ______ together.

(5) The charge is hoisted to the top of the furnace and dumped on to a ______ bell and hopper arrangement.

(6) Liquid slag will float on the iron and is separated from the iron by the ______ in the cast-house.

(7) The mudgun is used to ______ the taphole at the end of the cast.

(8) At the top of the uptakes are a series of pressure relief valves called ______.

Ⅳ. Decide whether the following statements are true or false (T/F).

(1) The inside profile of the blast furnace is termed (from top to bottom) furnace hearth, shaft, belly, bosh and throat. ( )

(2) Liquid slag will float on the iron and is separated from the iron by the ***skimmer*** in the cast-house. ( )

(3) The molten iron is delivered to a series of refractory-lined hot metal cars for transportation to the steel plant. ( )

(4) Each blast furnace has one hot blast stove. ( )

Ⅴ. Translate the following English into Chinese.

The total amount of water used by the iron and steel industry is great. In the following paragraphs, some specific uses of water in blast furnace are discussed.

The blast furnace requires very large amounts of water for its efficient operation, notably for cooling various parts of the furnace and its auxiliaries. For example, cooling water circulates constantly through the tuyeres, cinder notch, hearth staves, bosh, inwall cooling plates, and stove valves. Gas washing, to remove dust from the gases leaving the top of the furnace, requires considerable water, and additional quantities are used for slag granulation and other purposes.

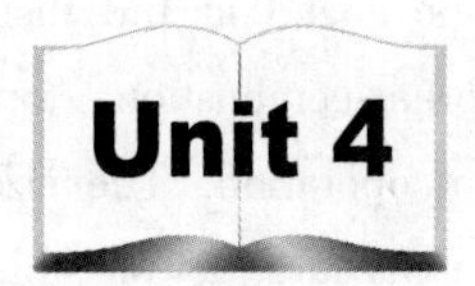

Unit 4 Hot Blast Stove

Part Ⅰ Reading and Comprehension

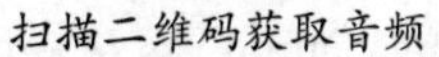

扫描二维码获取音频

扫描二维码获取 PPT

Each hot blast stove is a large exchanger to preheat blast furnace air. It can utilize the heating value of the furnace off-gasses to heat the blast air to 1000 to 1350℃. When well designed and operated, the thermal efficiency of the stoves will be 80% to 85%.

The stove consists of several parts: the shell, the combustion chamber, the checker work, and control valves and lines that regulate and deliver the various gasses (shown in Fig. 4-1). The

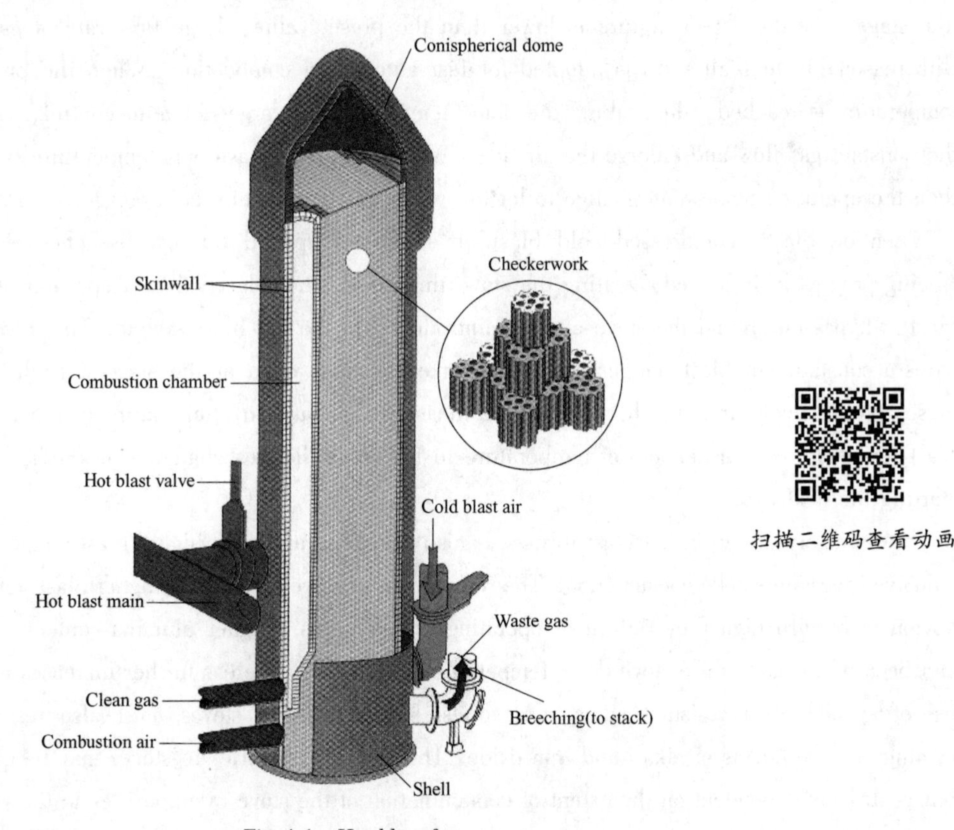

Fig. 4-1 Hot blast furnace

扫描二维码查看动画

shell is a welded steel cylinder 6 to 9m in diameter, and typically 20 to 40m high and its insides are lined with refractory. The combustion chamber is arranged at the inside, top or outside. Therefore, there are three types of hot blast stoves—internal-combustion, top-combustion, external-combustion stoves. They have the same principle of operation. The size of the combustion chamber should be minimized so that the checker mass is as large as possible. The design of the stove burner is critical in assuring good combustion and efficient, stable operation. Internal ceramic burners with high mixing capability of the gas and air streams are recommended to meet this requirement. The checker chamber is packed with checker bricks which provide many small, vertically aligned flues for the high temperature gasses. The efficiency of the stove is improved as the surface area to volume ratio for the checker mass is increased.

Normal operation utilizes three stoves. One is always on blast, while the other two are on gas. When on gas, combustion air and clean blast furnace gas are introduced into the combustion chamber. The blast furnace gas may be enriched by either natural gas or coke oven gas as necessary. The turbulent mixing of the gas and air streams results in a short, intense flame after ignition. Flame temperatures of 1200 to 1400℃ are common. By the time the hot gasses have passed downward through the checker mass, the temperature of the gasses will have been reduced to 300 to 400℃ before being exhausted through the chimney valves.

The stove is PLC controlled with the aim of dome temperature and waste gas temperature. In initial stage, the dome temperature is lower than the preset value, large flow rate of gas and air with reasonable fuel/air ratio is adopted for fast automation combustion. When the preset dome temperature is reached, then taking the dome temperature as target for auto control, i. e. keep the constant gas flow and enlarge the air flow. When the preset waste gas temperature is reached, then the operating mode auto change to let the waste gas temperature as target for auto control.

When on blast, compressed cold blast air is forced upward through the checker chamber (being progressively heated) exiting the stove through the hot blast valve. A portion of the cold blast is bypassed around the stove and is reintroduced to the hot blast system. This blending ensures a constant hot blast temperature. The mixer valve is open at the start of each cycle and closes progressively until the hot air leaving the stove is equal in temperature to the desired hot blast temperature. Further loss of temperature in the stove dictates changing to another stove and starting the next cycle.

The refractory design in hot blast furnace is also changing rapidly as higher blast temperatures are employed to reduce coke consumption. The refractories in stoves producing higher blast temperatures obviously require higher overall stove operating temperatures. Higher alumina-content refractories can be used to safely raise stove dome temperature to 1315℃, whereas further increases require the use of special creep-resistant silica refractories. Refractories in stoves must also resist thermal cycling and the effects of alkali and iron oxide. The life of refractories in stoves may be quite long, but is strongly dependent on the extent of contamination of the stove by impurities in the gas containing significant quantities of iron or alkali backdrafted through the stoves.

Words and Expressions

扫描二维码获取音频

扫描二维码答题

regulate ['regjuleit] *v.*	调节，校准
chamber ['tʃeimbə] *n.*	室，房间
combustion chamber	燃烧室
checkerwork ['tʃekəwəːk] *n.*	格式装置，砌砖格
minimize ['minimaiz] *v.*	将……减到最少
line [lain] *n.*	管线
weld [weld] *v.*	焊接，熔接
cylinder ['silində] *n.*	圆筒，圆柱体
checker mass	蓄热室
burner ['bəːnə] *n.*	燃烧器
air stream	风流，空气流
recommend [rekə'mend] *v.*	使……成为可取，推荐
requirement [ri'kwaiəmənt] *n.*	需求，要求
pack [pæk] *v.*	填塞，塞满
checker brick	格子砖
volume ratio	体积比，容积比
efficiency [i'fiʃənsi] *n.*	效率，效能
assure [ə'ʃuə] *v.*	保证，担保
checker chamber	蓄热室
flue [fluː] *n.*	烟道，通气管
vertically ['vəːtikəli] *ad.*	垂直地
align [ə'lain] *v.*	排列
ratio ['reiʃiəu] *n.*	比，比率，比值
enrich [in'ritʃ] *v.*	使……富足，使……富化
on blast	送风，鼓风
natural gas	天然气
coke oven	焦炉
intense [in'tens] *a.*	强烈的，剧烈的
compress [kəm'pres] *v.*	压缩
progressively [prə'gresivli] *ad.*	逐渐地，渐进地
bypass ['baipɑːs, 'baipæs] *v.*	绕过，使……通过旁道
constant ['kɔnstənt] *a.*	恒定的，不变的
mixer valve	混风阀

equal [ˈiːkwəl] *a.*	相等的，均等的
necessitate [niˈsesiteit] *v.*	成为必要
specification [ˌspesifiˈkeiʃən] *n.*	规格，技术要求
turbulent [ˈtəːbjulənt] *a.*	涡旋的，狂暴的
exhaust [igˈzɔːst] *v.*	排气
n.	废气
chimney [ˈtʃimni] *n.*	烟囱
valve [vælv] *n.*	阀门
aim [eim] *n.*	目标，目的
v.	以……为目标
initial [iˈniʃəl] *a.*	最初的，开始的
preset [ˈpriːˈset] *v.*	事先调整，预调
automation [ɔːtəˈmeiʃən] *n.*	自动控制，自动操作
blend [blend] *v.*	混合
dictate [dikˈteit] *v.*	要求，规定
cycle [ˈsaikl] *v.*	循环
n.	循环
dome [dəum] *n.*	圆顶，圆屋顶
alumina [əˈljuːminə] *n.*	氧化铝，矾土
whereas [(h)weərˈæz] *conj.*	然而，反之
creep [kriːp] *n.*	蠕变，蠕动
creep-resistant [kriːpriˈzistənt] *a.*	抗蠕变的
silica [ˈsilikə] *n.*	硅石，二氧化硅
resist [riˈzist] *v.*	抵抗，承受
thermal [ˈθəːməl] *a.*	热的，热量的
alkali [ˈælkəlai] *n.*	碱
a.	碱性的
dependent on	依靠，由（随）……决定的
contamination [kənˌtæmiˈneiʃən] *n.*	污染，污染物
skinwall [skinwɔːl] *n.*	隔墙
backdraft	倒转，逆通风

✲ Answer the following questions.

(1) How many types of the hot blast stoves are there nowadays?

(2) How many parts do the hot blast stove consists of ?

(3) Where can the combustion chamber be arranged?

(4) When is the efficiency of the stove improved?

(5) What are the common flame temperatures in the hot blast stove?

(6) When is the mixer valve open?

(7) Is the dome temperature lower or higher than the preset value in combustion initial stage?

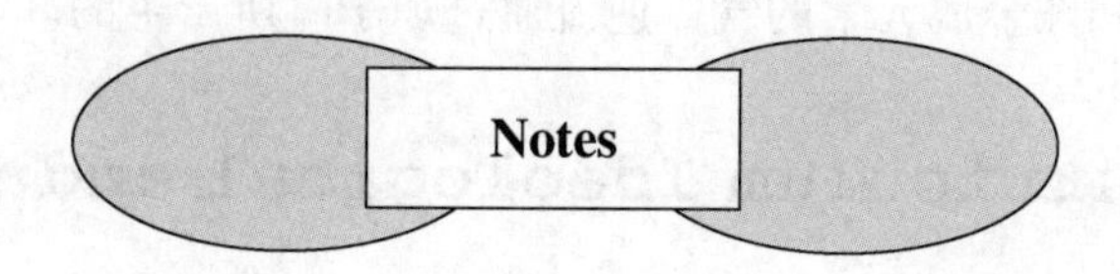

(1) The checker chamber is packed with checker bricks which provide many small, vertically aligned flues for the high temperature gasses.

蓄热室是用格子砖堆砌而成的，它能提供许多直径比较小的高温气体直通管。

“which”是关系代词，它引出的定语从句修饰先行词“checker bricks”。

(2) The efficiency of the stove is improved as the surface area to volume ratio for the checker mass is increased.

当格子砖的表面积与体积比增加时，热风炉的效率就会提高。

“as”作连词，它在句中的意思是“当……时候”。例如：As a child he lived on a farm（他小时候住在一个农场里）。“as a child”在这里等于“when he was a child”。

(3) The mixer valve is open at the start of each cycle and closes progressively until the hot air leaving the stove is equal in temperature to the desired hot blast temperature.

混风阀在每一循环的开始打开，在热风炉送出的风温与所需的风温相等时逐渐关闭。

“equal”常与“to、with”连用，意思是“相等的，相同的”。例如：One li is equal to half a kilometre（一华里等于半千米）；It is equal to me whether he comes or not（他来不来对我都一样）。

(4) The refractory design in hot blast furnace is also changing rapidly as higher blast temperatures are employed to reduce coke consumption.

由于高风温能够降低焦炭的消耗，所以热风炉耐火材料的设计也是变化很快的。

“as”作连词，引出原因状语从句。例如：As he was not feeling well, we all told him to stay at home（由于他感到不舒服，我们都要他待在家里）。

(5) The refractories in stoves producing higher blast temperatures obviously require higher overall stove operating temperatures. Higher alumina-content refractories can be used to safely raise stove dome temperature to 1315℃, whereas further increases require the use of special creep-resistant silica refractories.

高风温热风炉的耐火材料显然承受更高的操作温度。使用高铝质耐火材料可安全承受炉顶1315℃的温度，而温度进一步升高时则需要使用抗蠕变的硅砖。

动词不定式“to safely raise stove dome temperature to 1315℃”是主语补足语。“in stoves producing higher blast temperatures”是介词短语作定语，修饰“the refractories”。“producing higher blast temperatures”修饰“stoves”。

(6) The life of refractories in stoves may be quite long, but is strongly dependent on the extent of contamination of the stove by impurities in the gas containing significant quantities of iron or alkali backdrafted through the stoves.

炉内耐火材料的寿命可以很长，但它主要取决于炉内气体中杂质对炉子污染的程度，这种污染是由含有铁和碱金属元素的气体逆流通过炉子时所产生的。

Part Ⅱ Curriculum Ideological and Political

【Leader in reform】The experience creator of Hangang—Liu Hanzhang

Liu Hanzhang, former chairman and general manager of Handan Iron and Steel Group Co. , Ltd. In the 1990s, he introduced the market mechanism into the internal management of enterprises. In 1991, he established the operation mechanism of "simulating market accounting and implementing cost veto", which enabled the enterprise to quickly reverse the passive situation and embark on the track of sustainable, healthy and rapid development. The "experience of Handan Iron and Steel Group" has set off a revolution in the enterprise management mode nationwide. More than 20000 enterprises and institutions have come to Handan Iron and Steel Group to learn from it, which is known as "a red flag on China's industrial front".

Part Ⅲ Further Reading

The first hot molten iron of WISCO

On September 13, 1958, the first large blast furnace of Wuhan Iron and Steel Corporation, the first large iron and steel enterprise built in New China, discharged the first furnace of molten iron. This hot molten iron not only marks the aspirations and hope of the development of new China's national industry, but also to the whole world, the young People's Republic of China to the world to stand up the backbone of steel. The People's Daily editorialized at the time, "Congratulations on Wuhan Iron and Steel Tapping Iron" saying the incident marked a great victory for the general line of China's socialist construction.

At 10: 09 a. m. on October 14, 2019, the No. 1 blast furnace of WISCO stopped its production permanently. During the past 61 years, it witnessed great changes in China from poverty to steel power, and made great contributions to the construction and development of the national economy.

Part Ⅳ Translation Training

词类转译法（Ⅰ）

英语和汉语属于两个不同的语系，英语属于印欧语系，汉语属于汉藏语系。两种语言表达方式不同，在翻译过程中，有些句子不能逐词对译，需要转译词类，才能使译文通顺自然。

1. 名词的转译（1）

英语中大量由动词派生的名词和具有动作意义的名词以及其他名词，可以转译成汉语的动词；由形容词派生的抽象名词作表语或宾语时，往往可以转译成形容词和副词。例如：

（1）Light from the sun is a mixture of light of many different colours（太阳光是由许多不同颜色的光混合而成的）；Rockets have found application for the exploration of the universe（火箭用来探索宇宙）；Talking with his son，the old man was the forgiver of the young man's past wrong doings（在和儿子谈话时，老人宽恕了年轻人过去所干的坏事）。

（2）Independent thinking is an absolute necessity in study（独立思考对学习是绝对必要的。He had the kindness to show me the way．他好意地给我指路）。

2. 名词的转译（2）

汉语中的名词有时可转换成英语中的动词、形容词、连词等。例如：

（1）His lecture impressed the audience deeply（他的演讲使听众印象很深）；Our bodies are heated by the consumption of sugar in the blood（人体的体温是靠消耗血液中的糖分来维持的）。

（2）Air is perfectly elastic（空气是一种弹性气体）；Radium and uranium are radioactive（镭和铀放射性很强）。

（3）This is why alloys are widely used in industry（这就是工业上广泛使用合金的原因）。

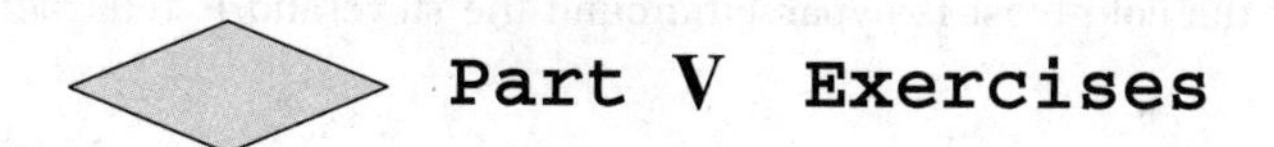

Part Ⅴ Exercises

Ⅰ. Translate the following expressions into English.

（1）内燃式热风炉	（2）燃烧室	（3）格子砖
（4）陶瓷燃烧器	（5）烟囱	（6）冷风阀
（7）蓄热室	（8）高温	（9）天然气

Ⅱ. Fill in the blanks with the words from the text. The first letter of the word is given.

（1）Each hot blast s ______ is a large heat exchanger utilizing the heating value of the furnace off-gasses to heat the blast air to 1000 to 1200℃.

（2）The stove consists of several parts：the shell，the c ______ chamber，the checker work，and control valves and lines that regulate and deliver the various gasses.

（3）The size of the combustion chamber should be minimized so that the checker mass is as l ______ as possible.

（4）Normal operation utilizes three stoves such that one is always on b ______，while the other two are on gas.

(5) The l ______ of refractories in stoves may be quite long, but is strongly dependent on the extent of contamination of the stove by impurities in the gas containing significant quantities of iron or alkali backdrafted through the stoves.

Ⅲ. Fill in the blanks with appropriate prepositions.

Normal operation utilizes three stoves such that one is always on blast, while the other two are (1) ______ gas. When on gas, combustion air and clean blast furnace gas are introduced (2) ______ the combustion chamber. The blast furnace gas may be enriched (3) ______ either natural gas or coke oven gas as necessary. The turbulent mixing (4) ______ the gas and air streams results (5) ______ a short, intense flame (6) ______ ignition. Flame temperatures (7) ______ 1200 to 1400℃ are common. (8) ______ the time the hot gasses have passed downward (9) ______ the checker mass, the temperature of the gasses will have been reduced to 300℃ (10) ______ 400℃ before being exhausted through the chimney valves.

Ⅳ. Decide whether the following statements are true or false (T/F).

(1) Each stove is a large heat exchanger utilizing the heating value of the furnace to off-gasses heat the blast air to 2000℃ to 2300℃. ( )

(2) By the time the hot gasses have passed downward through the checker mass, the temperature of the gasses will have been reduced to 1000 to 1200℃ before being exhausted through the chimney valves. ( )

(3) A portion of the cold blast is bypassed around the stove and is reintroduced to the hot blast system. ( )

(4) The stove consists of several parts: the shell, the combustion chamber, the checker work, and control valves and lines that regulate and deliver the various gasses. ( )

(5) The shell is a welded steel cylinder 20 to 30mm in diameter, and typically 4 to 6mm high and its insides are lined with refractory. ( )

Ⅴ. Translate the following English into Chinese.

Refractories are the primary materials used by the steel industry in the internal linings of furnaces for making steel and iron, in vessels for holding and transporting metal, in furnaces for heating steel before further processing, and in the flues or stacks through which hot gases are conducted. History reveals that refractory developments have occurred largely as the result of the pressure for improvement caused by the persistent search for superior metallurgical processes. The rapidity with which these ever-recurring refractory problems have been solved has been a large factors in the late of advancement of the iron and steel industry.

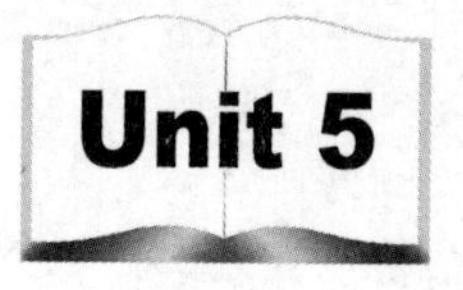

Unit 5 New Developments in Ironmaking

Part Ⅰ Reading and Comprehension

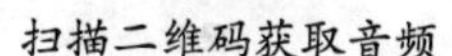

扫描二维码获取音频

扫描二维码获取 PPT

During the past century, many efforts were made to develop processes for producing iron for steelmaking that could serve as alternatives to the conventional blast furnace. This chapter presents a review of the alternative processes that have achieved some measures of commercial success.

1. Direct reduction technology

In the history of iron and steelmaking technology, the first process is the direct reduction process. The blast furnace takes the place of the direct reduction process, which is the great improvement in the metallurgical technology. With the development of iron and steel industry, coke, on which blast furnace have depended for over 200 years, is becoming more expensive every year, and supplies of suitable coking coals are becoming scarcer. The direct reduction process was brought back at the end of the eighteenth century.

Today, technology that produce iron by reduction of iron ore below the melting point of the iron produced is generally classified as direct reduction processes and the products referred to as direct reduced iron (DRI).

According to the reducing agents used, direct reduction technology is commonly classed into two types: gas reduction processes and reduction processes with solid reducing agents. According to the types of reduction reactors used, it is also classified into four main processes—fluidized bed, shaft furnace, retort, rotary kiln processes.

(1) **gas reduction processes.** Reducing gases are carbon monoxide, hydrogen or mixtures of two in gas reduction process. They are normally obtained from natural gas. The carbon monoxide and hydrogen are generated in the methane steam of natural gas reforming processes according to the reaction:

$$CH_4 + H_2O = CO + 3H_2$$

The shaft-furnace process plays an important part in gas reduction processes. More than 50% of all reduction processes are carried out in a shaft furnace. For example, the Midrex process is a typical (shown in Fig. 5-1).

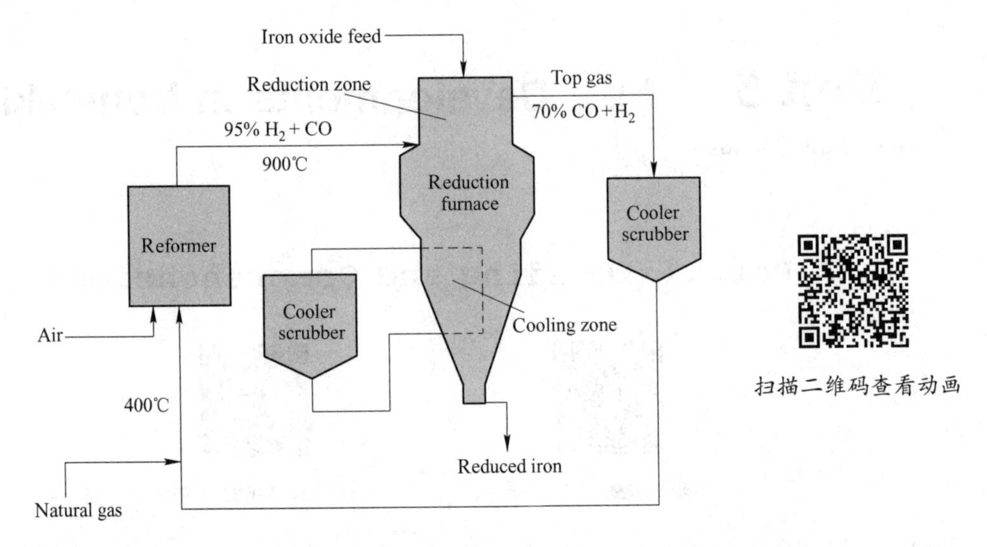

Fig. 5-1　Midrex process

Midrex furnace has developed into a main production form of direct reduction processes, since the first steel plant was built and went into production in 1969. The largest Midrex furnace has an annual capacity of 800000t.

The main components of the process are the shaft furnace, the gas reformer, and the cooling-gas system. Reducing gas enters the reducing furnace through a bustle pipe at the bottom of the reduction zone and flows countercurrent to the descending solids. The charge solids flow continuously into the top of the furnace through seal legs. The reduction furnace is designed for uniform mass movement of the burden by gravity feed, through the preheating, reduction, and cooling zones of the furnace. The cooled DRI is continuously discharged through seal legs at the bottom of the furnace. Cooling gases flow countercurrent to the burden in the cooling zone of the shaft furnace. The reducing top gas flows from an outlet pipe into the top-gas scrubber where it is cooled and its dust particles are removed. The largest portion of the top gas is recompressed, enriched with natural gas, preheated to about 400℃, and piped into the reformer tubes. In the catalyst tubes, the gas mixture is reformed to carbon monoxide and hydrogen. The hot reformed gas is then recycled to the DR furnace.

(2) **reduction processes with solid reducing agents.** The solid reducing agents are coals of any granulation and composition in direct reduction processes. Rotary kiln furnace is important equipment used solid reducing agents. This process operates in a continuously materials flow. The mature processes in coal-based direct reduction are German SL/RN process, British DRC process and French Codir process.

The outcome of all direct reduction techniques is sponge iron. It derives its name from its appearance of high porosity. Sponge iron is exposed to the hazard of reoxidation during handling and storage unless corresponding measures are adopted. Reoxidation can be prevented by hot briquetting, the application of a ***lime shell*** or passivation with water glass.

2. Smelting reduction technology

Smelting reduction technology usually produces hot metal from ore without coke. This reduction process can be subdivided into two main groups: melter gasifiers process and iron bath reactors process. In melter gasifiers processes, the COREX process which is developed together by the Voest Alpine and German Korf has gone into production.

Smelting reduction together with direct reduction techniques represent a significant expansion of the alternatives to classical blast furnace processes.

3. Low carbon ironmaking technology

(1) **Circulation of blast furnace gas**. The recycling of blast furnace top gas is to spray the reduced components (CO and H_2) into the blast furnace through the tuyere or proper position of the furnace stack after dedusting and purifying the top gas and removing CO_2, so as to return to the furnace again to participate in the reduction of iron oxides and strengthen the utilization of C and H. The process is considered as one of the effective measures to improve the performance of blast furnace, reduce energy consumption and reduce CO_2 emissions.

(2) **Blast furnace injection of hydrogen containing substances.** It has become a research hotspot to enhance hydrogen reduction by injecting hydrogen containing substances into blast furnace. The hydrogen containing substances injected into the blast furnace mainly include waste plastics, natural gas, coke oven gas, etc.

(3) **Oxygen enriched blast furnace technology.** Oxygen enriched blast furnace refers to adding industrial oxygen to the blast furnace to make the oxygen content in the blast furnace exceed the oxygen content in the atmosphere. The use of oxygen enriched blast furnace can accelerate the carbon combustion and increase the output under the condition of constant fuel ratio. However, oxygen enriched blast reduces the air volume entering the blast furnace and the heat brought into the blast furnace.

4. Hydrogen metallurgy technology

(1) **Direct reduction of hydrogen.** With the increase of CO_2 emission reduction pressure, hydrogen reduction technology will be more and more valued by the steel industry, ushering in opportunities for vigorous development. Research on hydrogen reduction technology, such as biomass reduction technology and nuclear hydrogen reduction technology, should be the main development direction. Therefore, the source of low-cost hydrogen has become an important issue. It should be an important direction to carry out research on nuclear hydrogen reduction process in cooperation with the nuclear energy industry.

(2) **Direct steelmaking by smelting reduction based on hydrogen metallurgy.** The scheme uses ultrapure iron concentrate for cold impurity removal as raw material to achieve emission reduction at the source. Ultra pure molten steel is obtained by rapid hydrogen reduction at 1200℃

and iron bath smelting reduction with high energy density at 1600℃. After continuous casting and rolling, high-quality and high cleanliness steel materials can be obtained. In this process, iron-making is completely canceled, and the fully continuous and integrated production mode of continuous charging, continuous steelmaking, continuous casting and rolling is realized. The process is simplified and the production efficiency is improved.

❋ Words and Expressions

扫描二维码获取音频　　扫描二维码答题

effort [ˈefət] *n.*	努力，成就
conventional [kənˈvenʃənl] *a.*	常规的，传统的
present [priˈzent] *v.*	介绍，引见
alternative [ɔːlˈtəːnətiv] *n.*	替换物，抉择
a.	选择性的
take the place of	代替
brought back	恢复
be classified as	被分成……类
direct reduced iron (DRI)	直接还原铁
class [klɑːs] *v.*	把……分类
carbon monoxide	一氧化碳
obtain [əbˈtein] *v.*	获得，得到
generate [ˈdʒenəˌreit] *v.*	产生，发生
methane [ˈmeθein] *n.*	甲烷，沼气
play an important part in	在……中起重要作用
descend [diˈsend] *v.*	下来，下降
uniform [ˈjuːnifɔːm] *a.*	均匀的，均质的
scarce [skεəs] *a.*	缺乏的，不足的
go into production	投产，开始生产
gas reformer	气体转化炉
gravity [ˈgræviti] *n.*	重力
discharge [ˈdistʃɑːdʒ] *v.*	排出，流出
seal legs	料封管
hydrogen [ˈhaidrəudʒən] *n.*	氢（元素符号 H）
bustle pipe	环形风管
countercurrent [ˈkauntəˌkʌrənt] *ad.*	相反地
n.	逆流

charge solid	固体炉料
gravity feed	重力给料
burden [ˈbəːdn] *n.*	炉料，配料
outlet [ˈautlet] *n.*	出口，出路
outlet pipe	流出管，排出管
scrubber [ˈskrʌbə] *n.*	气体洗涤器
recompress [rekəmpˈres] *v.*	再压缩
preheat [priːˈhiːt] *v.*	预热
pipe [paip] *v.*	以管输送
n.	管
reformer tube	重整管
catalyst [ˈkætəlist] *n.*	催化剂，触媒
recycle [ˈriːˈsaikl] *v.*	使……再循环，反复应用
rotary kiln furnace	回转窑
mature [məˈtjuə] *a.*	成熟的
outcome [ˈautkʌm] *n.*	结果，成果
sponge iron	海绵铁
derive [diˈraiv] *v.*	来自，源自，出自
appearance [əˈpiərəns] *n.*	外观，外表
porosity [pɔːˈrɔsiti] *n.*	多孔性，有孔性
expose [iksˈpəuz] *v.*	使……面临，暴露
hazard [ˈhæzəd] *n.*	危险，冒险，公害
reoxidation [riɔksiˈdeiʃən] *n.*	重新氧化
storage [ˈstɔridʒ] *n.*	储藏，储存
corresponding [ˌkɔrisˈpɔndiŋ] *a.*	相应的，对应的
briquet [briˈket] *v.*	压块
application [ˌæpliˈkeiʃən] *n.*	应用，运用
lime shell	石灰涂层
passivation [ˌpæsiˈveiʃən] *n.*	钝化
water glass	水玻璃（硅酸钠）
subdivide [ˈsʌbdiˈvaid] *v.*	再分，细分
gasifier [ˈgæsifaiə] *n.*	汽化器，燃气发生炉
melter gasifier	熔融汽化炉
iron bath reactor	铁浴反应炉
represent [reprɪˈzent] *v.*	代表
expansion [iksˈpænʃən] *n.*	发展，扩大，扩展
classical [ˈklæsikəl] *a.*	传统的，古典的

✲ Answer the following questions.

(1) Why was the direct reduction process brought back at the end of the 18th century?

(2) How many types are the direct reduction processes classed into according to the reducing agents used?

(3) What are the reducing gases in gas reduction processes?

(4) Which process plays an important part in gas reduction processes?

(5) What are the main components of the Midrex processes?

(6) What is the outcome of all direct reduction techniques?

(7) How many main groups can smelting reduction technology be subdivided into?

(1) During the past century, many efforts were made to develop processes for producing iron for steelmaking that could serve as alternatives to the conventional blast furnace.

在过去的一个多世纪里，为了找到能够取代传统高炉的炼铁技术，人们做了许多努力。

“an alternative to M” 的意思是 “M 的替换物”。

(2) Today, technology that produce iron by reduction of iron ore below the melting point of the iron produced is generally classified as direct reduction processes.

现在，那些在低于铁的熔点温度下用还原铁矿石的方法来炼铁的技术通常称为直接还原法。

“the iron produced, produced” 为过去分词，修饰 “iron”。“that produce iron by reduction of iron ore below the melting point of the iron produced” 为定语从句，修饰 “technology”。

(3) Reducing gas enters the reducing furnace through a bustle pipe at the bottom of the reduction zone and flows countercurrent to the descending solids.

还原气体通过位于还原带底部的环形风管进入还原炉内，迎着下降炉料向上运动。

“at the bottom of the reduction zone” 为介词短语作定语，修饰 “a bustle pipe”。

(4) The reduction furnace is designed for uniform mass movement of the burden by gravity feed, through the preheating, reduction, and cooling zones of the furnace.

还原炉的设计使得炉料能够依靠自身的重力作用均匀地通过预热带、还原带、冷却带。

(5) The largest portion of the top gas is recompressed, enriched with natural gas, preheated to about 400℃, and piped into the reformer tubes.

大部分炉顶气体被压缩后与天然气混合预热到约 400℃，并导入重整管。

“recompressed, enriched, preheated, and piped” 作并列成分。“pipe” 作动词，意思是 “用管道输送”。例如：Piped into the reformer tubes（导入重整管）。

(6) Sponge iron is exposed to the hazard of reoxidation during handling and storage unless

corresponding measures are adopted.

如果不采取相应的措施，海绵铁在处理和储存过程中就面临重新氧化的危险。

“be exposed to” 意思是 “面临，容易受到，暴露在……，招致……”。例如：Be exposed to accident possibilities（容易发生事故）；Be exposed to the weather（放在露天）。

（7）Smelting reduction together with direct reduction techniques represents a significant expansion of the alternatives to classical blast furnace processes.

熔融还原和直接还原一起代表了传统高炉工艺的重要发展趋势。

“together with” “as well as” “and also” 意思是 “和，连同”。例如：These new facts together with the other evidence prove the prisoner's innocence（这些事实连同其他证据已证明在押者无罪）。句子的主语与 “together with” 短语前的主语一致。

Part Ⅱ Curriculum Ideological and Political

【Green development】The report of the 20th National Congress of the Communist Party of China | Working actively and prudently toward the goals of reaching peal carbon emissions and carbon neutrality

Reaching peak carbon emissions and achieving carbon neutrality will mean a broad and profound systemic socio-economic transformation. Based on China's energy and resource endowment, we will advance initiatives to reach peak carbon emissions in a well-planned and phased way in line with the principle of building the new before discarding the old. We will exercise better control over the amount and intensity of energy consumption particularly of fossil fuels, and transition gradually toward controlling both the amount and intensity of carbon emissions. We will promote clean, low-carbon, and high-efficiency energy use and push forward the clean and low-carbon transition in industry construction transportation, and other sectors.

Part Ⅲ Further Reading

Building cultural confidence and strength—The Report of the 20th National Congress of the Communist Party of China

Enhancing civility throughout society. We will continue the civic morality campaign, carry forward traditional Chinese virtues, foster stronger family ties, values, and traditions, and raise the intellectual and moral standards of minors. We will build public commitment to the greater good, public morality, and personal integrity. These efforts will help raise public moral standards and enhance public civility.

To promote cultural-ethical progress, we will take coordinated steps to raise awareness, apply principles, and develop initiatives and advance efforts in both urban and rural areas. We will foster an ethos of work, enterprise, dedication, creativity, and frugality throughout society and cultivate new trends and new customs for our times.

We will increase people's knowledge of science and encourage everyone to read. The system and working mechanisms for volunteer services will be improved. We will promote integrity and credibility in society and work to perfect relevant long-term mechanisms.

We will see that Party and state awards and honors play a guiding and exemplary role and that a public atmosphere prevails in which people emulate paragons of virtue, look up to heroes, and strive to become pioneers.

Part Ⅳ　Translation Training

词类转译法（Ⅱ）

3. 形容词的转译

（1）由于两种语言在表达方式上不同，有些英语形容词可译为汉语副词，这种转译法在形容词修饰的名词转译为汉语的动词时较常用。例如：The man is difficult to deal with（这个人很难对付）；We place the highest value on our friendly relations with developing countries（我们高度地重视同发展中国家的友好关系）。

（2）表达心理状态的形容词在联系动词后作表语时，转译为汉语的动词。例如：They were not content with their present achievements（他们不满足他们现在的成就）；Van Leewenhoek was interested in improving the lenses that were used in making microscopes（范·拉乌文胡克对改进用来制作显微镜的透镜感兴趣）。

（3）可以转译为汉语名词的形容词通常有以下两种：一种是加上定冠词表示某一类人或事的形容词；另一种是用来表示特征或性质的形容词，可以根据汉语习惯转译为名词。另外，根据情况，有些形容词可以转译为名词。例如：They had deep sympathy for the injured（他们对于受伤的人有很深的同情心）；The more carbon the steel contains, the harder and stronger it is（钢的含碳量越高，强度和硬度就越高）；He was eloquent and elegant, but soft（他有口才、有风度，但很软弱）。

4. 副词的转译

英语中的副词使用频繁而且种类繁多，有的表示时间、地点、方式，用来修饰动词；有的表示程度，用来修饰形容词或其他副词；有的起连接作用引导从句，用来修饰动词或整个句子。英语中的副词在句子中的位置比较灵活，有时在句首，有时在句末，有时在句中。应弄清楚它的修饰关系，正确理解其含义，根据汉语习惯灵活处理。

（1）英语形容词或动词转译为汉语的名词时，原来修饰形容词或动词的副词往往随着转译成为汉语的形容词。例如：Sometimes we have had to pay dearly for mistakes（有时我们不得不为错误付出高昂的代价）；Specialization also enables one country to produce some goods more cheaply than another country（专业化还能使一个国家生产的产品比另一个国家生产的产品更为便宜）；Aluminium is highly resistant to corrosion（铝的耐腐蚀性极强）。

（2）具有动词意思的英语副词作表语或与前面的词搭配有动词意思的英语副词，可转译为汉语的动词。例如：I love having Friday off（我喜欢每星期五休息）；That day he was up before sunrise（那天他在日出前就起来了）。

（3）英语副词可转译为汉语的名词。例如：He is physically weak but mentally sound（他身体虚弱，但思想健康）。

5. 介词的转译

英语中的介词用得相当广泛，常用来表示词语与词语之间的关系。其含义多，使用灵活，搭配关系复杂，翻译时要根据上下文和搭配关系灵活处理。一般情况下，与动词搭配的介词常常省略不译，例如："consist in"意思是"在于"，The casting process consist in pouring molten metals into moulds（铸造过程在于把熔化了的金属浇注到模型中去）；"depend on"意思是"依靠"，All the living things depend on the sun for their growth（万物生长靠太阳）；"amount to"意思是"等于，总计达"，The annual output of steel plant amounts to…tons（这家钢铁厂的年产量达……吨）。具有动词含义的介词或介词短语翻译时，译成汉语的动词，例如：The steel also has to possess certain properties obtained by varying defined amounts of alloying agents（通过加入一定量的不同合金剂，钢材才能具有某些特性）；Measure the current with an ammeter（用安培表测电流）。

Part Ⅴ Exercises

Ⅰ. Translate the following expressions into English.

（1）直接还原法	（2）回转窑	（3）海绵铁
（4）熔融气化炉	（5）反应竖炉	（6）投产

Ⅱ. Fill in the blanks with the words or phrases listed below.

metallurgical	hydrogen	sponge iron	ore
gas	seal legs	equipment	reformer

（1）The blast furnace takes the place of the direct reduction process, which is the great improvement in the ______ technology.

（2）In gas reduction processes, reducing gases are carbon monoxide, ______ and mixtures.

（3）The shaft-furnace process plays an important part in ______ reduction process.

（4）The main components of the Midrex processes are the shaft furnace, the gas ______, and the cooling-gas system.

（5）The cooled DRI is continuously discharged through ______ at the bottom of the furnace.

（6）Rotary kiln furnace is important ______ used solid reducing agents.

(7) The outcome of all direct reduction techniques is ______.

(8) Smelting reduction technology usually produces hot metal from ______ without coke.

Ⅲ. Fill in the blanks by choosing the right words form given in the brackets.

The earth (1) ______ (containing; contains) a large number of metals which are useful to man. One of the most important of these is iron. Modern industry (2) ______ (needed; needs) considerable quantities of this metal, either in the form of iron or in the form of steel. A certain number of non-ferrous metals, (3) ______ (includes; including) aluminium and zinc, are also important but even today the majority of our engineering products (4) ______ (are; is) of iron or steel. Moreover, iron possesses magnetic properties, which (5) ______ (has; have) made the development of electrical power possible. The iron ore (6) ______ (what; which) we find in the earth is not pure. It contains some impurities which we must remove by smelting. The process of smelting consists of heating the ore in a blast furnace with coke and limestone, and (7) ______ (reduces; reducing) it to metal. Blasts of hot air (8) ______ (entering; enter) the furnace from the bottom and provide the oxygen which is necessary for the reduction of ore. The ore becomes (9) ______ (moltened; molten), and its oxide combines with carbon from the coke. The non-metallic constituents of the ore (10) ______ (combined; combine) with the limestone to form a liquid slag. This floats on top of the molten iron, and passes out of the furnace through a tap (渣口). The metal which remains is pig iron.

Ⅳ. Decide whether the following statements are true or false (T/F).

(1) Midrex furnace has developed into a main production form of direct reduction process, since the first steel plant was built and went into production in 1869. ( )

(2) The main components of the processes are the shaft furnace, the gas reformer, and the cooling-gas system. ( )

(3) The reducing gas temperature ranges between 1760 and 1927℃ and flows countercurrent to the descending solids. ( )

(4) In melter gasifiers processes, the COREX process which is developed together by the Voest Alpine and German Korf has gone into production. ( )

(5) Smelting reduction processes usually produce spongy iron from ore without coke. ( )

Ⅴ. Translate the following English into Chinese.

Over the past sixty years the blast furnace process has made tremendous progress and attained remarkable performances. Nevertheless, the blast furnace today is strongly challenged by some new processes which should respond to the objectives of producing metallic iron directly from ore fines and from non-coking coals without polluting the environment. These new processes are mainly direct reduction and smelting reduction.

Direct reduction offers a number of advantages including low investment expenditure and the use of cheap primary energy instead of coke. Consequently, this technique offers many developing countries the possibility of building up a national steel industry. In fact, direct reduction combined with electric arc furnace in the future must be considered among the important steel production techniques alongside the blast furnace and oxygen converter technique.

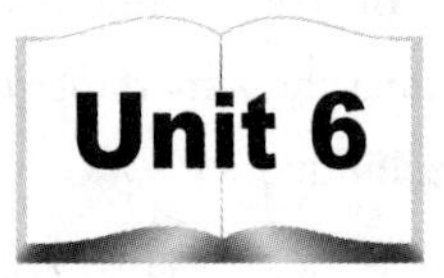

Unit 6 Raw Materials of Steelmaking

Part Ⅰ Reading and Comprehension

扫描二维码获取音频

扫描二维码获取 PPT

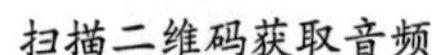

The hot metal tapped from the blast furnace is the principal raw materials used for steelmaking. Besides hot metal, further charge materials are: steel scrap, sponge iron, slag formers, alloying agents and oxidizing agents.

1. Blast furnace iron (Hot metal)

Blast furnace iron consists of the element iron combined with numerous other chemical elements, the most common of which are carbon, manganese, phosphorus, sulphur, and silicon. Pig iron may contain (by mass) 3.0 to 4.5 per cent of carbon, 0.15 to 2.5 per cent or more of manganese, 0.02 to 0.06 per cent of sulphur, 0.3 per cent or more silicon and small amount of phosphorous. Large quantities of tramp elements within the hot metal still have to be removed partly or completely. In fact, this is a most important task of steelmaking.

2. Steel scrap and sponge iron

Scrap and sponge iron can be used as iron supplier and cooling agent.

Worldwide, scrap participates 40 per cent in steel production input and must therefore be considered as an important raw material. Its use in steelmaking varies, depending on the production process applied. In BOF, scrap accounts for about 20 per cent, while in electric-arc furnace, it may be 100 per cent.

Steel scrap may be classified according to its sources:

(1) **Circulating scrap.** This arises during the steelmaking and rolling process in a given works. It consists of sheared ends and rejected materials which is normally returned immediately to the steelmaking vessel.

(2) **Process scrap.** This arises at the customer's works during the manufacture of finished articles. It is usually returned quickly to the steelmaking plants.

(3) **Capital scrap.** This arises from scrapping manufactured goods and equipment. In some cases it may be returned to the works after three or four years'use, or with heavy capital equipment

the life may be fifty years or more, but the average appears to be about twenty years.

Circulating and process scrap usually return to the steelmaker without contamination from undesirable elements, and heavy capital scrap is generally of the same quality. Much ***short life*** capital scrap, however, becomes contaminated with coatings of various kinds and is returned, without the complete separation of non-ferrous components, in the form of pressed bundles. This gives a sharp increase in the residuals in the steel, which detracts from its properties for a number of applications.

In fact, scrap must first be sorted and the impurities elements, e. g. non-ferrous metals, removed in special dressing plants. Nowadays this is frequently done in modern pressing and crushing or shredding equipment. Using the various kinds of scrap (e. g. loose scrap, chips, shredding) helps to promote recycling.

Sponge iron provided the main source of iron and steel for many centuries before the blast furnace was developed around 1300 A. D. In its modern usage, sponge iron is referred to as direct-reduced iron (DRI). Today, the major portion of DRI production is used as a substitute for scrap in the electric arc steelmaking furnace (EAF). DRI derived from virgin iron ore units is a relatively pure material which dilutes contaminants in the scrap and improves the steel quality.

3. Slag formers

Just as in the case of the hot metal, the slag formers are used to produce a reactionable low viscosity slag capable of absorbing undesired elements. Slag formers are used at all stages of iron and steel production, such as refining, pretreatment, post-treatment, and in steel casting.

Slag formers consist of lime, dolomite, fluorspar, etc. Lime (CaO) and dolomite ($CaCO_3$, $MgCO_3$) are the two primary fluxes. Lime is obtained by calcining the carbonate minerals in rotary kilns. Some typical analyses are shown in Table 6-1.

Table 6-1 Typical analyses of lime, fluorspar

Lime	CaO	SiO_2	Al_2O_3	MgO
Per cent	93.0	1.7	1.2	0.7

Fluorspar	CaF_2	SiO_2	S
Per cent	75~85	10.0 (maximum)	1.0 (maximum)

As slag formers, a special limitation is that dusty materials must be avoided, since dust is carried off easily by waste gases.

4. Alloying agents and deoxidizing agents

The steel also has to possess certain properties obtained by varying defined amounts of alloying agents. These properties might be, for example, corrosion resistance, machinability, strength at elevated temperatures.

The important alloying agents used in the production of steel are nickel, ferrochromium, ferrotitanium, ferrotungsten, ferrovanadium, ferrosilicon, and ferromolybdenum, etc.

The deoxidizing agents—the additions for binding the oxygen dissolved in the liquid steel—are often added immediately after the refining process. They are usually classified among the alloying additions.

5. Oxidizing agents

Oxidizing agents consist of oxygen, iron ores and scale, etc. They play the most important role in the production of steel.

✲ Words and Expressions

扫描二维码获取音频　　扫描二维码答题

tap [tæp] *v.*　使……流出
principal ['prinsip(ə)l] *a.*　主要的，首要的
alloying agent　合金剂
slag former　造渣剂
scrap [skræp] *n.*　废钢，废料
v.　扔弃，报废，废弃
process scrap　加工废钢，边角废料
deoxidation [diːˌɔksi'deiʃən] *n.*　脱氧，还原
scale [skeil] *n.*　氧化铁皮，锈皮
manganese ['mæŋgəniːz] *n.*　锰（元素符号 Mn）
shear [ʃiə] *v.*　剪切，修剪
sheared end　切头，切尾
tramp element　杂质元素
participate [pɑː'tisipeit] *v.*　分担，含有，带有
input ['input] *n.*　输入，进料量
BOF (basic oxygen furnace)　碱性氧气转炉
account for　（指数量等）占
rejected material　废品，废料
finished article　成品
detract [di'trækt] *v.*　降低，减损
manufacture [ˌmænju'fæktʃə] *n.*　制造，加工
arise [ə'raiz] *v.*　来源于，出现
capital ['kæpitəl] *a.*　重要的
n.　资金，资产

circulate [ˈsəːkjuleit] *v.*　(使……) 运行，(使……) 循环
contamination [kənˌtæmiˈneiʃən] *n.*　污染，污染物
contaminate [kənˈtæmineit] *v.*　弄脏，污染
separation [sepəˈreiʃən] *n.*　分离，分开
bundle [ˈbʌndl] *n.*　捆，束，包
residual [riˈzidjuəl] *n.*　剩余，残留，残渣
sort [sɔːt] *v.*　分类，拣选
dressing plant　选矿厂
crush [krʌʃ] *v.*　破碎，压碎，碾碎
press [pres] *v.*　压，按
shred [ˈʃred] *v.*　切割，切碎
commercial [kəˈməːʃəl] *a.*　商业的，商务的
loose scrap　粗钢
be referred to as　叫做，称为
substitute [ˈsʌbstitjuːt] *n.*　替代品，代用品
virgin [ˈvəːdʒin] *a.*　原始的，未被玷污的
dilute [daiˈljuːt, diˈljuːt] *v.*　冲淡，稀释
in the case of　在……的情况
stage [steidʒ] *n.*　阶段
fluorspar [ˈflu(ː)əspɑː] *n.*　萤石，氟石
calcine [ˈkælsain] *v.*　焙烧
contaminant [kənˈtæminənt] *n.*　致污物，污染物
viscosity [visˈkɔsiti] *n.*　黏度，黏性
analyse [ˈænəlaiz] *n.*　分析
v.　分析
limitation [ˌlimiˈteiʃən] *n.*　限制，局限性
dust [dʌst] *n.*　粉尘，粉末
possess [pəˈzes] *v.*　拥有，占有
define [diˈfain] *v.*　限定，规定
corrosion [kəˈrəuʒən] *n.*　腐蚀，侵蚀
machinability [məʃiːnəˈbiliti] *n.*　机械加工性，切削性
elevated [ˈeliveitid] *a.*　提高的，严肃的
ferroalloy [ˌferəuˈælɔi] *n.*　铁合金
nickel [ˈnikl] *n.*　镍 (元素符号 Ni)
ferrochromium [ˌferəuˈkrəumiəm] *n.*　铬铁 (合金)
ferrotitanium [ˌferəutaiˈteiniəm] *n.*　钛铁 (合金)
ferrotungsten [ˌferəuˈtʌŋstən] *n.*　钨铁 (合金)
ferrovanadium [ˌferəvəˈneidiəm] *n.*　钒铁 (合金)
ferrosilicon [ˌferəuˈsilikən] *n.*　硅铁 (合金)

ferromolybdenum [ˌferəuməˈlibdinəm] *n.*　钼铁（合金）
deoxidize [diːˈɔksidaiz] *v.*　脱氧，还原
bind [baind] *v.*　联合
alloying addition　合金剂

✲ Answer the following questions.

(1) What are the raw materials used for steelmaking?
(2) What are the chemical elements in the hot metal?
(3) What is the most important task of steelmaking according to the text?
(4) What does DRI stand for?
(5) What do slag formers consist of ?
(6) What are the important alloying agents used in the production of steel?
(7) What do oxidizing agents consist of ?

(1) Blast furnace iron consists of the element iron combined with numerous other chemical elements, the most common of which are carbon, manganese, phosphorus, sulphur, and silicon.

高炉铁水是由铁及许多其他化学元素组成的，最常见的元素有碳、锰、磷、硫和硅。

the most common of which 引出非限制性定语从句。

(2) Large quantities of tramp elements within the hot metal still have to be removed partly or completely.

铁水中大量的杂质元素必须部分或全部除去。

“quantity”意思是“量，数量”，其后可跟可数名词和不可数名词。例如：He ate a small quantity of rice（他吃了少量的米饭）；Quantities of food were on the table（桌上摆了大量食物）；They have a large quantity of water pipes（他们有大批水管）。

(3) In BOF, scrap accounts for about 20 per cent, while in electric-arc furnace, it may be 100 per cent.

在碱性氧气转炉炼钢过程中，废钢量约占20%；而在电炉生产中，它可以占100%。

“account for”意思是“（指数量等）占，解释，说明”。例如：The production of raw materials accounts for a considerable proportion of the national economy（原料生产在国民经济中占相当大的比重）；She could not account for her mistake（她无法解释其错误）；We can not account for his failure in the English examination（我们无法解释他为什么在英语考试中未考好）。

(4) It consists of sheared ends and rejected materials which is normally returned immediately to the steelmaking vessel.

循环废钢以各种切头和废品形式存在，通常会很快回到炼钢炉中。

“consist”与“of”连用，意思是“组成，构成，包括，由……组成”。例如：The United Kingdom consists of Great Britain and Northern Ireland（联合王国包括大不列颠与北爱尔

兰)；His job consists of helping old people who live alone（他的工作包括帮助无人照顾的独居老人）；“consist”与“in”连用，意思是“在于”。例如：The beauty of the plan consists in its simplicity（这个计划妙就妙在简明扼要）；True charity doesn't consist in almsgiving（真正的慈善不在于施舍）。

（5）Much ***short life*** capital scrap，however，becomes contaminated with coatings of various kinds and is returned，without the complete separation of non-ferrous components，in the form of pressed bundles.

而许多寿命短的基建废钢则由于受到各种涂层的污染，在没有完全与非铁成分分离的情况下以打包的形式返回。

“in the form of”意思是“以……的形式，呈……状态”。例如：water in the form of ice（水呈冰的状态）。

（6）DRI derived from virgin iron ore units is a relatively pure material which dilutes contaminants in the scrap and improves the steel quality.

由天然铁矿石冶炼而成的直接还原铁是一种相对纯净的产品，它的杂质比废钢中的少，从而提高了钢的质量。

“derived from virgin iron ore units”为过去分词短语作定语，修饰“DRI”。“which…the steel quality”是定语从句，修饰material。

（7）Just as in the case of the hot metal，the slag formers are used to produce a reactionable low viscosity slag capable of absorbing undesired elements.

就如铁水的（生产）情况一样，使用造渣剂会产生一种反应性好、能吸收杂质元素的低黏度炉渣。

“in the case of”意思是“在……的情况”。“capable of absorbing undesired elements”为形容词短语作定语，修饰“viscosity slag”。

（8）The deoxidizing agents—the additions for binding the oxygen dissolved in the liquid steel— are often added immediately after the refining process.

脱氧剂，即与液态钢中溶解的氧进行结合的添加物，常常是在精炼之后立即加入。

句中“the additions for binding the oxygen dissolved in the liquid steel”是“The deoxidizing agents”的同位语。

Part Ⅱ Curriculum Ideological and Political

【Steel carrier】100-million-ton Baowu strives

In 2020，China Baowu Steel Group Corporation（hereinafter referred to as China Baowu）produced 115 million tons of steel，realizing the historic leap of “100 million tons Baowu” and becoming the top steel enterprise in the world. China Baowu has completed the R&D and manufacture of a series of key materials in many fields，such as aerospace，energy and electric power，transportation，and national major projects，and solved a large number of difficult problems of "bottleneck" materials, effectively supporting the high-quality development of our national economy. In the fields of automobiles, home appliances, and construction, China Baowu keeps climbing, surpassing, innovating and upgrading, launches a series of premium iron and steel products that meet the needs of the development of the new era, constantly contributing Baowu's strength to realizing the people's yearning for a better life, and becoming a well-deserved "pillar of the country".

Part Ⅲ Further Reading

Steelmaking raw materials affected by emission peak

Peak carbon emission is the point at which carbon dioxide emissions stop growing and peak, and then gradually fall back. Peak carbon emission and carbon neutral, referred to as "double carbon". Scrap steel is one of the main raw materials for steelmaking. Under the background of peak carbon emission and carbon neutrality, the green transformation, energy conservation and emission reduction of the steel industry have a long way to go. As a kind of energy which can be recycled, the recycling of scrap steel is a key point in the development of steel industry. In October 2021, the State Council issued "Action Plan for Carbon Dioxide Peaking Before 2030", which pointed out that to promote the steel industry carbon peak, continue to reduce steel production capacity, improve the recycling level of scrap resources, the implementation of all scrap furnace process. In July 2021, the National Development and Reform Commission issued the "14th Five-Year Plan for Circular Economy Development", which pointed out that the amount of steel scrap used in China in 2020 was about 260 million tons, and clarified the target task of steel scrap use to reach 320 million tons by 2025.

Part Ⅳ Translation Training

词类转译法（Ⅲ）

6. 动词的转译（1）

（1）动词不定式作主语时，有时必须转译为汉语的动宾结构或状语。例如：It is necessary to decide on which type of fuel is most suitable（必须决定哪种燃料最合适）；To understand one's environment today and to be able to adapt oneself to it demand some appreciation of scientific attitude（为了了解人们当今所处的环境，并使自己适应这种环境，就要求具有一定的科学素养）。

（2）现在分词短语作定语时，表示的是被修饰人或事的动作，并置于被修饰词的后面；若分词短语较长或与所说明的关系不紧密，可转译为汉语的并列关系分句。例如：Some facilities lacking the space for local pits use slag ladles to transport the slag to remote pits（一些高炉附近没有渣池，就用渣罐车把渣运到较远的渣池）；Only the most important chemical changes taking place in the blast furnace have been outlined（上面提及的仅仅是高炉中发生的一些最重要的化学变化）。

现在分词短语还可以作时间、条件、原因状语，可转译为相应的汉语分句。例如：Being an important aspect of inter cultural communication，translation has an important role to play in our epoch（作为不同文化间进行交流的一个重要方面，翻译在这个时代起着重要的作用）。现在分词短语作方式或伴随状语时，表示主语正在进行的另一动作，可转译为汉语的并列谓语或分句。例如：A drawback of the method is that the oxidation loss of carbon additions given to the ladle may vary，resulting in variations of the carbon content of the steel，depending on the duration of tapping，the quantity and composition of slag passed to the ladle，metal temperature，etc（这种方法的缺点是：由于向钢包中加入增碳剂的氧化损失不同，导致钢中的含碳量不同，这取决于出钢时间、进入钢包的渣量、炉渣成分以及钢水温度等）。

（3）英语中的过去分词短语如果用逗号与前面句子成分分开，往往作状语，位于句中或句末，对谓语表示的动作加以修饰，在很多情况下都说明谓语发生的背景或情况。翻译时可译成汉语的并列分句。例如：The coal tar，combined with other substances，has given us more than 20000 products（煤焦油与其他物质化合，为我们提供了20多万种产品）；The energy in coal and oil come from the sun，stored there by the plants of years ago（煤和石油的能量来自太阳，是数百万年前的植物储藏起来的）；This rather inefficient performance of the open hearth，compared with the enormous output of the blast furnace，justified the emergence of the highly productive oxygen process for steel（与高炉的巨大生产能力相比，平炉效率不高，这就必然促成了大生产能力的氧气转炉炼钢法的出现）。

（4）英语中的过去分词短语如紧跟在过去分词后面，往往作定语修饰它前面的词。翻译时，可译作"……的……"。例如：Reducing process gas，about 95 per cent combined hydrogen plus carbon monoxide，enters the reducing furnace through a bustle pipe and tuyeres located at the bottom of the reduction zone（直接还原气为含有95%氢气和一氧化碳的混合物，它通过位于还原带底部的环形风管和风口进入还原炉内）；The tramp elements contained in the hot metal——carbon，silicon，manganese and sulphur must be removed from the liquid iron by oxidation（铁水中含有的杂质元素——碳、硅、锰、硫必须通过氧化从铁水中除去）。

Part V Exercises

Ⅰ. Translate the following expressions into English.

(1) 废钢	(2) 直接还原铁	(3) 合金剂
(4) 造渣剂	(5) 基建废钢	(6) 选矿厂
(7) 生产工艺	(8) 循环废钢	(9) 铬铁

Ⅱ. Fill in the blanks with the words from the text. The first letter of the word is given.

(1) Scrap and sponge iron can be used as iron supplier and cooling a ______.

(2) In BOF steelmaking, scrap accounts for about 20 p ______.

(3) In fact, scrap must first be s ______ and the impurities and undesired tramp elements, e. g. non-ferrous metals, must be removed in special dressing plants.

(4) In its modern usage, s ______ iron is referred to as direct-reduced iron (DRI).

(5) Sponge iron provided the main source of iron and steel for many centuries before the blast f ______ was developed around 1300 A. D.

(6) Slag formers are used at all s ______ of iron and steel production, such as refining, pre-treatment, post-treatment, and in steel casting.

(7) A special limitation is that dusty materials must be avoided, since dusty is carried off easily by w ______ gases.

(8) Agents consist of o ______, iron ores and scale. They play the most important role in the production of steel.

(9) The hot metal tapped from the blast furnace is the principal raw m ______ used for steelmaking.

(10) Circulating scrap a ______ during the steelmaking process in a given works.

Ⅲ. Fill in the blanks by choosing the right words form given in the brackets.

Pig iron consists (1) ______ (in; of) the element iron combined with numerous other chemical elements, the most common of which are carbon, manganese, phosphorus, sulphur, and silicon. Depending upon the composition of the raw materials (2) ______ (used; using) in the blast furnace (3) ______ (principally; principal) iron ore, coke and limestone…and the manner (4) ______ (at; in) which the furnace is operated, pig iron may contain 3. 0 to 4. 5 per cent of carbon, 0. 15 to 2. 5 per cent or more of manganese, as much as 0. 2 per cent of sulphur, 0. 025 to 2. 5 per cent of phosphorus, and 0. 5 to 4. 0 per cent of silicon. In refining pig iron to convert it into steel, all five of these elements must either be removed almost entirely or at (5) ______ (most; least) reduced drastically in amount.

Ⅳ. Decide whether the following statements are true or false (T/F).

(1) Pig iron may contain 1.0 to 1.5 per cent of carbon, 0.15 to 2.5 per cent or more of manganese, 0.02 to 0.06 per cent of sulphur, 0.5 per cent or more silicon and small amount of phosphorous. ( )

(2) Scrap and sponge iron can be used as iron supplier and reduction agent. ( )

(3) As slag formers, a special limitation is that lump materials must be avoided, since lump is carried off easily by waste gases. ( )

(4) Fluorspar is basically silicon fluoride with varying impurities, and sometimes contains lead. ( )

(5) Important metals and ferroalloys used in the production of steel are nickel, ferrochromium, ferrotitanium, ferrotungsten, ferrovanadium, ferrosilicon, and ferromolybdenum. ( )

Ⅴ. Translate the following English into Chinese.

Major reducing agent in the BF is the carbon monoxide gas generated by the oxidation of the carbon in fuel and carbon in coke. Iron-bearing materials containing iron oxides can be reduced to molten iron (pig iron) in the blast furnace by using the reducing agent. In the process, pig iron absorbs from 3.0 to 4.5 per cent of carbon (by mass). Most of the iron produced in blast furnaces is transported to the steelmaking shop while it is still liquid and is then used directly for the manufacture of steel.

Unit 7 Principles of Modern Steelmaking

Part Ⅰ Reading and Comprehension

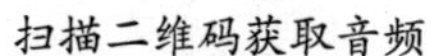

扫描二维码获取音频　　扫描二维码获取 PPT

Currently, there are two major steelmaking processes , the more popular being the oxygen processes (also called converter processes) based on hot metal from a blast furnace and scrap, and the less popular being the electric-arc process which is suitable for making steel from high-quality industrial scrap or from pre-reduced pellets. (The open-hearth process, once responsible for almost 100% of raw steel production, has now dwindled to negligible proportions.) Oxygen steelmaking processes are concerned mainly with the refining of a metallic charge consisting of hot metal (molten pig iron) and scrap through the use of high-purity oxygen to rapidly produce steel of the desired carbon content and temperature. Various steelmaking fluxes are added during the refining process to reduce the sulphur and phosphorus contents of the metal bath to the desired level. The oxygen top-blown process (LD or LD/AC) is currently the more common, but is gradually giving way to combined blowing process in some new plants. The principle of the electric-arc process is simple; scrap is melted by the crude action of heating using an electric-arc struck between the carbon electrodes and the steel charge. Refining is carried out by interaction with a basic slag as in the oxygen vessels described, but the refining times are longer.

The most important chemical reactions during refining are decarburization, slagging of tramp elements (desiliconization, demanganization, dephosphorization, desulphurization) and deoxidation (removal of residual oxygen by ferrosilicon and aluminum).

The equations formulate the principle of chemical reactions during the transformation of hot metal or sponge iron into steel.

$$[C] + [O] = CO$$

$$[Si] + 2[O] + 2[CaO] = (2CaO \cdot SiO_2)$$

$$[Mn] + [O] = (MnO)$$

$$2[P] + 5[O] + 3(CaO) = (3CaO \cdot P_2O_5)$$

$$[S] + (CaO) = (CaS) + [O]$$

$$[Si] + 2[O] = (SiO_2)$$

$$2[Al] + 3[O] = (Al_2O_3)$$

Decarburization is the most important reaction. During this, the added oxygen reacts with the carbon inside the hot metal to form carbon monoxide and escapes as a combustible waste gas. Some of the heat formation remains in the metal or slag. This carbon monoxide produces only about one third of the potential heat of the carbon, the remainder being evolved when it is fully burned to carbon dioxide.

The transfer of the other tramp elements from the hot metal or the scrap is performed in two stages: In the first stage, the tramp elements are oxidized. They are not soluble in the liquid iron. In the second stage, they rise to the metal surface and combine with the added lime to form slag. All the heat generated by their oxidation is available, directly, to the metal and slag.

As silicon is oxidized, it forms silica and produces heat. Silica accelerates dissolution of lime in slag and the process of slag formation as a whole, while the liberated heat is utilized for melting scrap. Silica is very stable, once formed, which is not again reduced in any of the basic processes.

Manganese is unavoidably oxidized to a considerable degree during converter blowing. Manganese oxides can lower the melting point of basic slag and accelerate slag formation. Under certain conditions a little may be reduced from slag to metal, but this is not of much importance to the process.

Sulphur will normally react directly with the burnt lime to form calcium sulphide, although some will escape as a gas. The proportion leaving as gas is strongly influenced by the gaseous atmosphere above the slag surface. The other factor necessary for sulphur removal is slag volume. The higher this is, the less sulphur remains in the steel. This occurs because calcium sulphide has a fixed solubility in a given slag and therefore the greater the slag volume per unit of metal, the greater is the weight of sulphur it can absorb from the metal.

Phosphorus may be oxidized to form phosphoric oxide (P_2O_5). It will remain in the slag provided it contains sufficient lime and iron oxide. If, however, reducing conditions arise, the temperature of the slag increases or the slag becomes less basic, then the P_2O_5 may be reduced and the phosphorus returns to the metallic bath.

When refining is performed with surplus oxygen, some of the oxygen stays dissolved. During deoxidation, silicon or aluminum is often added to the liquid steel forbidding the residual oxygen. The deoxidization products settle out in the slag.

❋ Words and Expressions

扫描二维码获取音频

扫描二维码答题

dwindle ['dwindl] *v.*	减少，缩小
negligible ['neglidʒəbl] *a.*	可忽略的，不重要的
proportion [prə'pɔːʃən] *n.*	比例，部分
crude [kruːd] *a.*	天然的，未加工的
arc [ɑːk] *n.*	弧，弓形
strike [straik] *v.*	打动，穿透
electrode [i'lektrəud] *n.*	电极
carbon electrode	石墨电极
interaction [ˌintər'ækʃən] *n.*	互相作用，互相影响
refining [ri'fainiŋ] *n.*	精炼
slagging ['slægiŋ] *n.*	造渣
desiliconization [diːsilikɔnai'zeiʃən] *n.*	脱硅
dephosphorization [diːfɔsfərai'zeiʃən] *n.*	脱磷
desulphurization [diːsʌlfərai'zeiʃən] *n.*	脱硫
aluminum [ə'ljuːminəm] *n.*	铝（元素符号 Al）
equation [i'kweiʃən] *n.*	等式，方程式
formulate ['fɔːmjuleit] *v.*	用公式表示
transformation [ˌtrænsfə'meiʃən] *n.*	变化，转化
escape [is'keip] *v.*	逸出
combustible [kəm'bʌstəbl] *a.*	易燃的，可燃的
potential [pə'tenʃ(ə)l] *a.*	潜在的
n.	势
remainder [ri'meində] *n.*	残余，剩余物
evolve [i'vɔlv] *v.*	放出，发出
perform [pə'fɔːm] *v.*	完成
dioxide [dai'ɔksaid] *n.*	二氧化物
transfer ['trænsfəː(r)] *n.*	转移，迁移
stage [steidʒ] *n.*	阶段，时期
oxidize ['ɔksiˌdaiz] *v.*	（使……）氧化
soluble ['sɔljubl] *a.*	可溶的，可溶解的
accelerate [æk'seləreit] *v.*	加速，促进
dissolution [disə'ljuːʃən] *n.*	分解，解散
as a whole	总的说来
liberated heat	释放的热量
utilize [juː'tilaiz] *v.*	利用
stable ['steibl] *a.*	稳定的
considerable [kən'sidərəbl] *a.*	相当大的，相当多的
calcium ['kælsiəm] *n.*	钙
sulphide ['sʌlfaid] *n.*	硫化物

calcium sulphide 硫化钙
influence ['influəns] *v.* 影响，改变
gaseous ['gæsiəs] *a.* 气体的，气态的
atmosphere ['ætməsfiə] *n.* 空气，气氛
volume ['vɔlju:m, 'vɔljəm] *n.* 体积，大量
slag volume 渣量
solubility [ˌsɔlju'biliti] *n.* 溶解性
provided [prə:'vaidid] *conj.* 假若，倘若，倘使
sufficient [sə'fiʃənt] *a.* 充足的，足够的
surplus ['sə:pləs] *a.* 过剩的，剩余的
forbid [fə'bid] *v.* 禁止，阻止
settle out 沉积

✻ Answer the following questions.

(1) Are there two major steelmaking processes? What are they?

(2) What are the most important chemical reactions during refining?

(3) How is the excess carbon inside the hot metal removed?

(4) What does it form as silicon is oxidized?

(5) What chemical element is unavoidably oxidized to a considerable degree during converter blowing?

(6) What will sulphur normally react directly with the burnt lime to form?

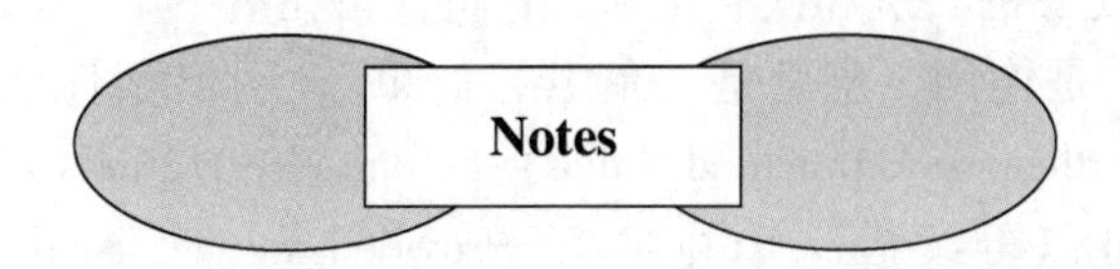

(1) Oxygen steelmaking processes are concerned mainly with the refining of a metallic charge consisting of hot metal (molten pig iron) and scrap through the use of high-purity oxygen to rapidly produce steel of the desired carbon content and temperature.

氧气炼钢工艺主要是通过使用高纯度的氧气，对以铁水（熔融生铁）和废钢组成的金属炉料进行精炼，以便快速生产含碳量和温度合适的钢水。

"be concerned with、be about sth." 意思是"牵涉某事物，与某事有关"。例如：Her latest documentary is concerned with youth unemployment（他最近的一部纪录片是关于青年人失业问题的）。"consisting of hot metal (molten pig iron) and scrap" 是分词短语作定语，修饰 "metallic charge"。"to rapidly produce steel…temperature" 是动词不定式短语作目的状语。

(2) Refining is carried out by interaction with a basic slag as in the oxygen vessels described, but the refining times are longer.

精炼是靠金属与碱性炉渣之间的相互作用实现的，其过程与氧气转炉精炼过程一样，

但是精炼时间比较长。

句中“as”是连词，意思是“像……一样”，引出方式状语从句。

（3）The equations formulate the principle of chemical reactions during the transformation of hot metal or sponge iron into steel.

下面这些方程式示出了把铁水和海绵铁转变成钢的化学反应原理。

（4）This carbon monoxide produces only about one third of the potential heat of the carbon, the remainder being evolved when it is fully burned to carbon dioxide.

形成一氧化碳产生的热量大约占碳燃烧潜热的三分之一，当它完全燃烧成二氧化碳时，剩余的热量将全部释放出来。

“the remainder being evolved”是分词短语作状语，“when it is fully burned to carbon dioxide”又是分词短语作状语中的时间状语。“one third”是分数词，意思是“三分之一”。再如，“two thirds”意思是“三分之二”。

（5）This occurs because calcium sulphide has a fixed solubility in a given slag and therefore the greater the slag volume per unit of metal, the greater is the weight of sulphur it can absorb from the metal.

发生这种情况是因为硫化钙在炉渣中有固定的溶解度，所以单位金属的渣量越大，它从金属中吸收的硫量就越多。

句中“the greater…, the greater…”是形容词比较级重叠形式，放在句首，意思是“越……，越……”。例如：Actually, the busier he is, the happier he feels（事实上，他越忙越高兴）。

（6）It will remain in the slag provided it contains sufficient lime and iron oxide.

如果炉渣中含有足够的石灰和铁氧化物，它将保留在渣中。

句中“provided”是连词，意思是“倘若，如果”，引出条件状语从句。例如：You may keep the book a further week provided (that) no one else requires it（倘若这本书没有其他人想借的话，你可以再续借一个礼拜）；Provided we get good weather, it will be a successful holiday（如果天气良好，我们的假日将过得非常好）。

（7）The deoxidization products settle out in the slag.

脱氧产物进入渣中。

句中“settle out”意思是“沉积”，引申为“进入”。

Part Ⅱ　Curriculum Ideological and Political

【Master footprint】Wei Shoukun, father of Chinese metallurgical physical chemistry

In 1936, Academician Wei Shoukun returned to the motherland after completing his overseas studies and devoted himself to metallurgy. In the 1940s, he proposed the dephosphorization procedure of small Bessemer furnace and the double operation method of Bessemer furnace and Martin furnace to guide dephosphorization of small converter. In order to solve the problem of refractory materials, he carried out “Research on calcium removal and magnesium extraction from dolomite in Sichuan” and “Research on the preparation of magnesia bricks by artificial magnesia firing”, proposed the measure of selective dissolution of calcium removal and magnesium extraction by carbon dioxide, which made the purity of magnesia reach 99.5%, meeting the requirements of manufacturing high-quality magnesia bricks, and clarified the mechanism of purifying magnesia. He devoted his life to rigorous scholarship, imparting knowledge and educating people, and made pioneering contributions to the development of China's higher education and metallurgical science.

Part Ⅲ Further Reading

The mysterious iron-carbon diagram

The iron-carbon phase diagram is the most basic tool for the study of iron-carbon alloys. It is actually the Fe-Fe_3C phase diagram, and the basic components of the iron-carbon alloy should also be pure iron and Fe_3C. Iron has an allotropic transformation, that is, it has a different structure in the solid state. Different structures of iron and carbon can form different solid solutions, and the solid solutions on the Fe-Fe_3C phase diagram are all interstitial solid solutions. Due to the different characteristics of the pores in the α-Fe and γ-Fe lattices, the carbon dissolving capabilities of the two are also different. There are three phases in iron-carbon alloys namely ferrite, austenite and cementite.

Ferrite: The interstitial solid solution formed by C dissolved in α-Fe is called ferrite and is represented by the symbol F or α. The strength and hardness of ferrite are not high, but it has good plasticity and toughness.

Austenite: The interstitial solid solution formed by C dissolved in γ-Fe is called austenite, which is represented by the symbol A or γ. The mechanical properties of austenite are related to its dissolved carbon content and grain size. Generally, it has good plasticity and toughness. Therefore, iron-carbon alloys with austenite as the matrix are easy to be forged and formed.

Cementite: A metal compound formed by Fe and C, an interstitial compound with a complex lattice, represented by Fe_3C. The hardness of cementite is very high, while the plasticity and toughness are almost zero, and the brittleness is very large.

Part Ⅳ Translation Training

词类转译法（Ⅳ）

7. 动词的转译（2）

（1）英语中一些由名词派生或由名词转用的动词，其概念很难用汉语动词来表达。翻译时可将该英语动词转译为汉语中的名词。例如：I weigh less than I used to（我的体重减轻了）；The electronic computer is chiefly characterized by the accurate and rapid computation

（电子计算机的主要特点是计算迅速准确）。

（2）英语中被动句译为汉语“受到……+名词”或“加以……+名词”结构时，该英语动词则转译为汉语名词。例如：Throw a tennis ball on the floor. The ball bounces back. In the same way，when light falls on certain things，it bounces back. When this happens，the light is said to be reflected（将网球扔在地板上，球会弹回来。同样，光照到某个物体上时也会弹回来。这种情况称为光受到反射）；Satellites，however，must be closely watched，for they are constantly being tugged at by the gravitational attraction of the sun，moon and earth（由于经常受到太阳、月球及地球引力的影响，卫星运动必须加以密切的观察）。

（3）英语中有些动词具有副词含义，翻译时可转译为汉语的副词。例如：He is full in vigour，but fails in carefulness（他劲头十足，但不够仔细）；I succeeded in persuading him（我成功地说服了他）。

Part V　Exercises

Ⅰ. Translate the following expressions into English.

（1）脱磷	（2）石墨电极	（3）相互作用
（4）碱性炉渣	（5）方程式	（6）废气
（7）脱硅	（8）转炉工艺	（9）精炼

Ⅱ. Fill in the blanks with the words from the text. The first letter of the word is given.

（1）All of the modern steelmaking processes p ______ molten（liquid）steel.

（2）D ______ is one of the most important reaction.

（3）All the heat generated by their oxidation is available，directly，to the metal and s ______.

（4）Silica is very s ______，once formed，which is not again reduced in any of the basic processes.

（5）Manganese is unavoidably oxidized to a considerable degree during converter b ______.

（6）Phosphorus may be oxidized to form p ______ oxide.

（7）Modern steelmaking processes are divided into two general classes from the c ______ point of view：acid processes and basic processes.

（8）B ______ processes are the most important processes for producing steel.

（9）The tramp element contained in the hot metal——carbon，silicon，manganese and sulphur must be r ______ from the liquid iron by oxidation.

（10）Manganese oxides can lower the m ______ point of basic slag and accelerate slag formation.

Ⅲ. Fill in the blanks by choosing the right words form given in the brackets.

The Siemens-Martin ***open-hearth furnace*** was so called because the molten lies in a comparatively shallow pool on the furnace bottom or hearth. Usually scrap steel is previously charged and (1) ______ (heating; heated) in the furnace and the liquid pig iron is added (2) ______ (to; on) it, thus the impurities in the pig iron are diluted and the refining process (3) ______ (did; does) not take so long as if the entire charge were pig iron. Then iron oxide, in the form of iron ore (4) ______ (or; and) scale, is added; this, together with oxygen in the furnace gases, (5) ______ (oxidizing; oxidizes) the impurities, the carbon being removed as carbon monoxide. The silicon and manganese are also changed (6) ______ (into; from) their oxides, and these react with (7) ______ (adding; added) sand or lime to form a slag which is removed separately. At the end of the process, when the impurities have been brought down to the required level and metal has been tapped, additions of ferro-manganese and ferro-silicon are made (8) ______ (to bring; bring) the steel to the correct composition; later a small addition of aluminium is made to deoxidize the metal further.

Ⅳ. Decide whether the following statements are true or false (T/F).

(1) Modern steelmaking processes are divided into two general classes from the chemical point of view: acid processes and basic processes with the use of acid processes greatly predominant. ( )

(2) In the oxygen steelmaking processes, clean air is blown under pressure through, onto, or over a bath containing hot metal, steel scrap, and fluxes to produce steel. ( )

(3) As silicon is oxidized, it forms silicon oxide and produces heat. ( )

(4) Sulphur will normally react directly with the silicon to form silica, although some will escape as a gas. ( )

(5) The other factor necessary for sulphur removal is slag volume. The higher this is, the less sulphur remains in the steel. ( )

(6) Today, however, when building new steel plants, the technique of combined blowing is frequently preferred to either top or bottom blowing. ( )

(7) Phosphorus may be oxidized to form phosphoric acid. ( )

(8) Principle of the electric arc process is simple; scrap is melted by the crude action of heating using an electric arc struck between the carbon electrodes and the steel charge. Refining is carried out by interaction with a basic slag as in the oxygen vessels, but the refining times are shorter. ( )

(9) If, however, reducing conditions arise, temperature of the slag increases or the slag becomes less basic, then the P_2O_5 may be reduced and the phosphorus returns to the metallic bath. ( )

(10) This carbon monoxide produces only about one third of the potential heat of the carbon, the remainder being evolved when it is fully burned to carbon dioxide. ( )

Ⅴ. Translate the following English into Chinese.

Today, there are two fundamental routes taken in the production of steel: (1) blast furnace—converter; (2) electric-arc furnace. In the first process via, the purpose of reducing the iron ore in blast furnace is to produce pig iron as raw material of steelmaking stage. In the second route, scrap or direct reduced iron (DRI) is melted into liquid steel. The steel produced in these ways is then finish refined by means of secondary metallurgy processes.

Unit 8 The LD Practice

Part Ⅰ Reading and Comprehension

扫描二维码获取音频

扫描二维码获取 PPT

Oxygen top - blowing converter is a pear - shaped converter similar in shape to the Bessemer and Thomas converters. The converter is basic lined usually with magnesite or dolomite. The combination of devices for top blowing of oxygen, including the lance, a standby lance, and lifting and transfer mechanisms, is rather complicated structure (shown in Fig. 8-1). The LD process is shown in Fig. 8-2.

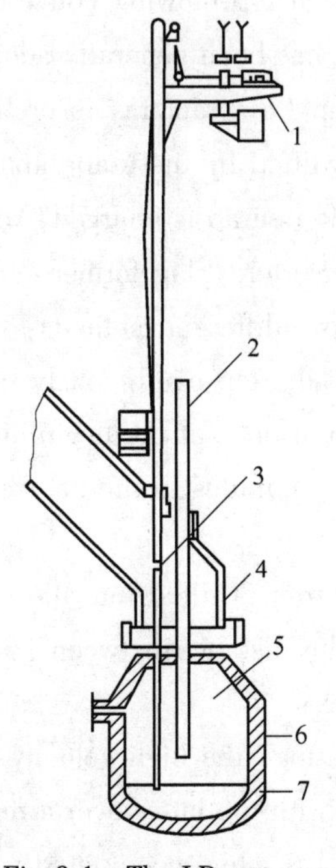

Fig. 8-1 The LD structure

1—Lifting device; 2—Oxygen lance; 3—Standby lance; 4—Top gas catcher; 5—Converter; 6—Furnace shell; 7—Furnace lining

扫描二维码查看动画

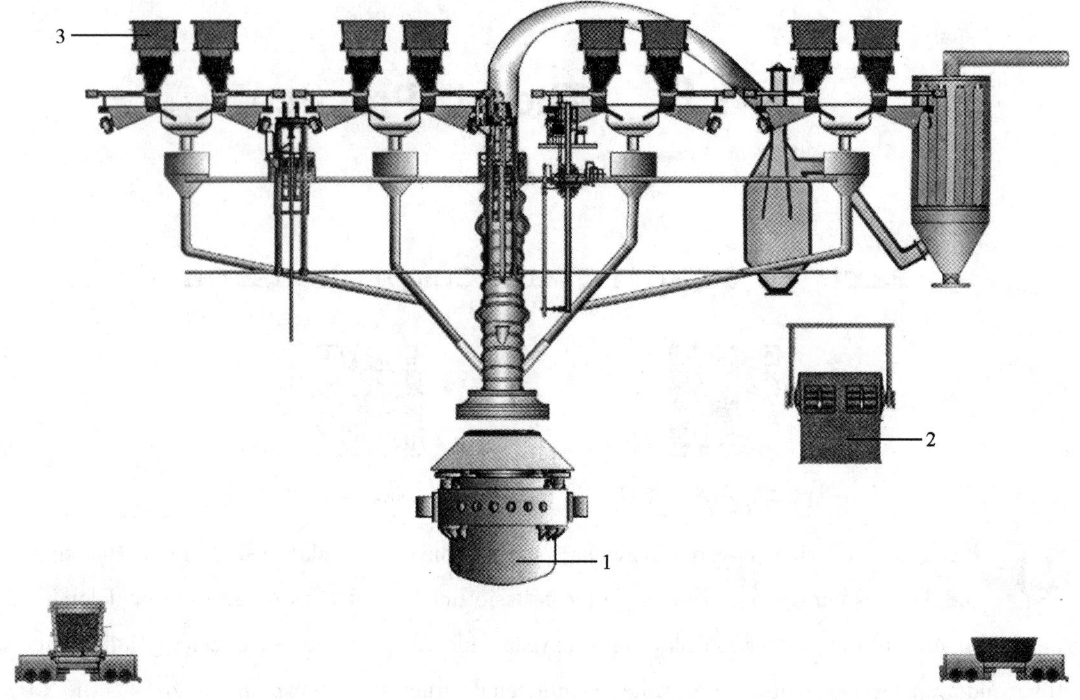

Fig. 8-2　The LD process

1—Converter; 2—Ladle; 3—Bunker

扫描二维码
查看动画

By many years, experience of oxygen top-blowing converter operation, the following sequence of converter charging has been generally adopted. Upon tapping at converter, it is first charged with scrap (a scrap tray is preliminarily weighed and delivered to the converter; the tray is lifted by the crane and inclined to dump the scrap into the vessel). Immediately after scrap is charged, the required quantity of molten pig iron is poured into the converter. Slag formers and other additions (with the particle size of 20 to 25mm) are charged by two different methods:

(1) These materials are given from the top continuously in the course of blow;

(2) Part of these materials is given at once at the beginning of blow and the remaining quantity is introduced continuously during a few minutes in the course of blow. The second method is most popular.

Upon pouring-in pig iron, the converter is tilted into the vertical position and the lance is lowered in the vessel to start blowing. The distance between lance and metal varies during the blow from 7ft to 2ft 6 inch in various plants.

During particular periods of the heating, the high velocity oxygen jet must penetrate as deep as possible into the metal bath and react with the latter over a relatively small area. Circulation of the bath commences from this hot spot and is later accentuated when major carbon monoxide evolution commences, thus an intimate mixture of slag and metal occurs which contributes very markedly to the high speed of the steelmaking operation.

During other periods, on the contrary, the jet must act on the bath surface, rather than on deeper layers, so as to accelerate the dissolution of lime in slag and form free running slag of desired basicity. The effect of oxygen jets on the bath surface should not involve splashing of metal and slag which might be harmful for the lining.

Blowing takes between 10 to 20 minutes. On its completion, a sample is taken to compare the alloying composition with the prescribed specification; at the same time, the temperature of the steel bath is measured. During tapping, the steel bath should have a temperature of 1600 to 1650℃. The sample analysis takes a little more than a minute.

The steel is tapped by tilting the converter. It flows through the tapping hole into the pouring ladle. The slag floating on the surface remains in the converter during and after tapping. The converter is tilted to the other side in order to remove the slag, so that it can flow out over the rim of the converter. In special cases, the converter is equipped with special pouring devices for holding back the slag.

Frequently not all of the slag has been removed on completion of the blowing process and so part of it is still here for the next blowing cycle.

In the oxygen process typically, blowing periods of 10 to 20 minutes, plus filling and emptying, plus temperature measuring and sampling together result in tap-to-tap times of about 30 to 50 minutes. For example, the duration of a heat in a modern 100t converter is around 42 minutes on the average, this time being distributed as follows:

Scrap charging and pig iron pouring	10min
Oxygen blowing	17min
Turning-down and sampling	6min
Tapping of metal and slag and inspection of lining	9min

✻ Words and Expressions

扫描二维码获取音频

扫描二维码答题

pear-shaped 梨形的
magnesite ['mægnəsait] *n.* 菱镁矿
standby ['stændbai] *a.* 备用的
standby lance 副枪
mechanism ['mekənizəm] *n.* 机械装置，机件
sequence ['siːkwəns] *n.* 顺序，序列
tray [trei] *n.* 底板，托（支）架
preliminarily [priˌlimi'nærili] *ad.* 预先地
crane [krein] *n.* 起重机

incline [in'klain] *v.*	使……倾斜，倾向，倾斜
dump [dʌmp] *v.*	倾倒，卸下，丢下
charge [tʃɑːdʒ] *v.*	装料，装满
addition [ə'diʃən] *n.*	增加物，添加剂
tilt [tilt] *v.*	（使……）倾斜
vertical ['vəːtikəl] *a.*	垂直的，直立的
lower ['ləuə] *v.*	降低，跌落
vary ['vɛəri] *v.*	变化，不同
jet [dʒet] *n.*	射流，气流
penetrate ['penitreit] *v.*	穿透，渗透
latter ['lætə] *n.*	（两者中）后者
circulation [ˌsəːkju'leiʃən] *n.*	循环，流通
commence [kə'mens] *v.*	开始，着手
accentuate [æk'sentjueit] *v.*	增强，使……更明显，加重
intimate ['intimit] *a.*	密集的，致密的
evolution [ˌiːvə'luːʃən, ˌevə'luːʃən] *n.*	放出（气体），形成
markedly [mɑːkidli] *ad.*	显著地，明显地
on the contrary	（与此）相反，反之
layer ['leiə] *n.*	层，阶层
free running slag	易流动渣
splash [splæʃ] *v.*	飞溅，斑点
basicity [bə'sisiti] *n.*	碱度，碱性
involve [in'vɔlv] *v.*	包括，涉及
sample ['sæmpl] *n.*	样品，例子
prescribe [pris'kraib] *v.*	指示，规定
tapping hole	出钢口
pouring ladle	钢包
rim [rim] *n.*	边，轮缘
equip [i'kwip] *v.*	装备，配备
completion [kəm'pliːʃ(ə)n] *n.*	完成
typically ['tipikəli] *ad.*	代表性地，作为特色地
plus [plʌs] *v.*	加上
filling ['filiŋ] *n.*	装料
tap-to-tap time	出钢时间
duration [djuə'reiʃən] *n.*	持续时间，为期
heat [hiːt] *n.*	炉次
duration of a heat	冶炼（一炉）时间
inspection [in'spekʃən] *n.*	检查，修补

the LD practice　　　　　氧气顶吹转炉工艺

Answer the following questions.

(1) What is the converter basic lined usually with?

(2) What does the combination of devices for top blowing of oxygen include?

(3) What has the sequence of converter charging been generally adopted?

(4) How long does the sample analysis take?

(5) What is the temperature of the steel bath during tapping?

(6) Has all of the slag been removed on completion of the blowing process?

(7) How long is the duration of a heat in a modern 100t converter on the average?

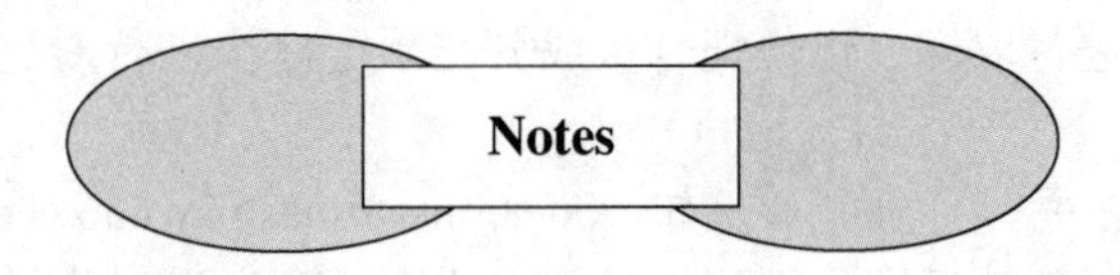

(1) Oxygen top-blowing converter is a pear-shaped converter similar in shape to the Bessemer and Thomas converters.

氧气顶吹转炉是一种梨形转炉，形状与贝塞麦和托马斯转炉相似。

"be similar to" 意思是"与……相似，类似于……"。例如：My new dress is similar to the one you have（我的新衣服和你的那件相似）。"but" 在此是介词，意思是"除……以外"。例如：Yesterday evening, he did nothing but repair his computer（昨天晚上，他除了修理计算机外，没有做其他事）。

(2) Upon tapping at converter, it is first charged with scrap (a scrap tray is preliminarily weighed and delivered to the converter; the tray is lifted by the crane and inclined to dump the scrap into the vessel).

上炉一出钢完毕，首先装入废钢（废钢预先称量，并被运至转炉，由起重机吊起，倾斜倒入炉中）。

"Upon…at or immediately after the time or occasion of" 意思是"就在某时或某场合（之后）"。例如：Upon arriving home, I discovered the burglary（我一回到家就发现家中被盗）；Upon (my) asking for the information I was told I must wait（我一打听，说我得等着）。

(3) During particular periods of the heat, the high velocity oxygen jet must penetrate as deep as possible into the metal bath and react with the latter over a relatively small area.

在供热期间，高速氧射流应尽可能穿入金属熔池中，在相对小的区域与铁水作用。

"the latter" 意思是"后者"，这里指金属熔池中的铁水。例如：Did he walk or swim? The latter seems unlikely（他是散步还是游泳？后者看来不太可能）。

(4) Circulation of the bath commences from this hot spot and is later accentuated when major carbon monoxide evolution commences, thus an intimate mixture of slag and metal occurs, which contributes very markedly to the high speed of the steelmaking operation.

熔池循环从这一热点开始，当大量的 CO 开始产生时，被进一步加强。这样渣-钢混合物就产生了，这极有助于加快炼钢操作的速度。

“which” 引出的定语从句修饰 “an intimate mixture of slag and metal”。“contribute to” 意思是 “有助于，促成，贡献出”。例如：His work has contributed enormously to our understanding of this difficult subject（他的著作极有助于我们对这个困难问题的了解）；Drink contributed to his ruin（酗酒促使他毁灭）；The Song Dynasty contributed three great inventions to world civilization（宋朝为世界文明贡献出三大发明）。

（5）The effect of oxygen jets on the bath surface should not involve splashing of metal and slag, which might be harmful for the lining.

氧气射流对熔池表面的作用应避免金属和渣的喷溅对炉衬造成的危害。

（6）On its completion, a sample is taken to compare the alloying composition with the prescribed specification; at the same time, the temperature of the steel bath is measured.

吹炼一完成就要进行取样，以便把合金成分与预定标准进行对比；与此同时，还要测量钢液的温度。

“on” 意思是 “在……后立即”。例如：Some magazines pay on acceptance, others on publication（一些杂志采用稿件后即付稿酬，另外一些则要到发表后才付）。

（7）In the oxygen process typically, blowing periods of 10 to 20 minutes, plus filling and emptying, plus temperature measuring and sampling together result in tap-to-tap times of about 30 to 50 minutes.

在典型的氧气顶吹工艺中，吹炼时间为 10~20 分钟，加上装料与出钢、出渣、测温以及取样时间，每炉次生产时间为 30~50 分钟。

“result in” 是句中谓语动词，“result”（常与 in 连用）意思是 “造成，导致”。例如：The accident resulted in three people being killed（这次事故造成 3 人死亡）；Acting before thinking always results in failure（做事不先考虑总会导致失败）。

Part Ⅱ　Curriculum Ideological and Political

【Steel spirit】Shuangliang spirit

In the 1950s, Li Shuangliang was known as the “industrial slag blasting expert” in the national metallurgical industry. After his retirement in 1983, Li Shuangliang volunteered to lead the workers in the slag yard to manage the slag mountain without any money. After 10 years' effort, he removed slag mountain (the 23 meters high, 2.3 square kilometers in area and 10 million cubic meters in total), which had slept for more than half a century, and had recovered 1.309 million tons of waste steel. In addition, he created his own equipment to produce a variety of waste residue extension products, led the staff to build a dust-proof slope wall around the Zhashan Mountain, with flower beds, rockeries, fish ponds, and more than 70000 flowers and trees planted in the wall. Since then, the Zhashan Mountain has transformed into a large colorful garden. “Contemporary Yugong” has become a synonym for Li Shuangliang, spreading all over the country. “Shuangliang spirit” has also been identified as the enterprise spirit of TISCO, and has become a valuable spiritual wealth of TISCO people.

Part Ⅲ Further Reading

Application of artificial intelligence in intelligent manufacturing of iron and steel industry

Steel industry is the basic industrial sector of the national economy, which is one of the fundamental elements to measure a country's level of development. China has become the world's largest steel manufacturing country, but its overall innovation capability and product competitiveness are large but not strong. It has become a major strategic task for China's economic and social development in the new period to accelerate the transformation and up grading of steel manufacturing industry to a powerful steel manufacturing country. Intelligent manufacturing is the deep integration of next generation information and communication technology and advanced manufacturing technology, and it runs through a 11 stages of the manufacturing process. Intelligent manufacturing based on the new generation of artificial intelligence technology is the key development direction of the future high end equipment manufacturing industry, which is also a new opportunity for the innovation and development of China's steel industry.

Part Ⅳ Translation Training

增词（Ⅰ）

增词指的是增加原文中虽无其词但有其意的词。英语和汉语在用词、造句和表达等方面有许多不同，在翻译时需要增加一些原文中没有的词，使译文流畅，忠实表达原文的意思。增词翻译有两种：第一，根据意义和修饰的需要。例如：英语中没有量词和助词，英译汉时根据上下文的需要增加量词和助词；汉语名词没有复数概念，动词没有时态的变化，翻译时应增加表示复数和时态的词；英语中一些句子使用不及物动词就能表达完整的意思，而译成汉语时，需要增加不及物动词隐含的宾语意义的词等。第二，根据句法的需要，增加原文省略的词。

根据意义和修饰的需要增词

1. 增加动词

根据意义的需要，可以在名词前增加动词。例如：He favored the efforts to improve relations with all peace-loving countries（他赞成为了同所有爱好和平的国家改善关系而进行努力）；Mutual understanding is the basis for state-to-state relations（相互了解是发展国与国关系的前提）。

2. 增加形容词

根据修辞的需要在某些名词前增加适当的形容词。例如："It'll make a man of him."

said Jack，"College is a place."（"一定把他造就成一个堂堂的男子汉"，杰克说，"就是应该上大学。"）；"This is grasping at straws，I know." said the hopeless man（"我知道，这是在抓救命的稻草。"他无可奈何地说）。

3. 增加名词

（1）根据修辞的需要在某些形容词前增加名词。例如：This computer is indeed cheap and fine. 这台计算机真是价廉物美。She was wrinkled and black，with scant gray hair. 她满脸皱纹，皮肤黝黑，头发灰白稀疏。

（2）在不及物动词后面增加名词，把不及物动词隐含的宾语表达出来。例如：Day after day he came to his work—sweeping，scrubbing，cleaning. 他每天来干活——扫地，擦地板，收拾房间。He began to see things and to understand. 他开阔了眼界，并懂得了一些。Iron and steel products are often coated less they should rust. 钢铁产品常常涂上油漆以免生锈。

Part V　Exercises

Ⅰ. Translate the following expressions into English.

（1）梨形转炉	（2）副枪	（3）氧枪
（4）试样分析	（5）相反	（6）出钢时间
（7）浇注钢包	（8）出钢口	（9）氧射流

Ⅱ. Fill in the blanks with the words from the text. The first letter of the word is given.

（1）The converter is basic lined usually with magnesite or d ______.

（2）By many years' experience of oxygen top-blowing converter operation，the following s ______ of converter charging has been generally adopted.

（3）Upon tapping at converter，it is first charged with s ______.

（4）Upon pouring-in pig iron，the converter is tilted into the v ______ position.

（5）During particular periods of the heat，the high velocity oxygen jet must p ______ as deep as possible into the metal bath and react with the latter over a relatively small area.

（6）The effect of oxygen jets on the bath surface should not involve splashing of metal and slag which might be h ______ for the lining.

（7）During tapping，the s ______ bath should have a temperature of 1600 to 1650℃.

（8）In special cases，the converter is equipped with special pouring devices for h ______ back the slag.

（9）The steel is tapped by t ______ the converter.

（10）The slag floating on the s ______ remains in the converter during and after tapping.

Ⅲ. Fill in the blanks by choosing the right words form given in the brackets.

The refining of steel by Bessemer and open-hearth processes removes impurities（1）______ (among; from) pig iron by the oxygen of the air, most of the impurities (2) ______ (are; being) taken into the slag. Bessemer himself (3) ______ (has; had) envisaged the use of oxygen but of course he (4) ______ (can; could) not obtain sufficient amounts, even for experimental purposes. In the 1960s steel-making (5) ______ (taking; took) a leap forward, thanks (6) ______ (for; to) the production of oxygen on such a scale that it is (7) ______ (measuring; measured) by the tonne (about 700 cubic meters) and (8) ______ (in; at) a fraction of its former cost. Plants have been built adjacent to the large steel-works, each capable of providing several hundred tonnes of high-purity oxygen a day.

Ⅳ. Decide whether the following statements are true or false (T/F).

(1) The distance between lance and metal varies during the blow from 17ft to 12ft 6 inch in a various plants. ( )

(2) In the oxygen process typically, blowing periods of 10 to 20 minutes, plus filling and emptying, plus temperature measuring and sampling together result in tap-to-tap times of about 30 to 50 minutes. ( )

(3) During tapping, the steel bath should have a temperature of 600 to 650℃. The sample analysis takes a little more than a minute. ( )

(4) Oxygen top-blowing converter is a pear-shaped converter similar in shape to the Blast furnace and Thomas converters. ( )

(5) The combination of devices for top blowing of oxygen, including the lance, a standby lance, and lifting and transfer mechanisms, is rather complicated structure. ( )

Ⅴ. Translate the following English into Chinese.

(1) The system of converter for top blowing of oxygen consists of a solid-bottom, brick-lined, oxygen-blowing lance with a water-cooled. The converter is charged with an exact amount of steel scrap, and molten pig iron. The oxygen lance is lowered into the furnace and a high-speed flow of oxygen gas is blown into the molten metal. During the oxygen blow, lime is added as flux to carry off the oxidized impurities. Alloy is added at the end of process.

(2) The special advantages of combined blowing are: homogeneous melt caused by rapid dissolution of the scrap; Acceleration of blowing cycle by about 25%; higher yield of iron and alloying elements; increased accuracy in achieving the chemical composition of the melt; improved converter life. From both process and metallurgical points of view, combined oxygen blowing contains further development potential.

Electric-arc Furnace Steelmaking Processes

Part Ⅰ Reading and Comprehension

扫描二维码获取音频

扫描二维码获取 PPT

The steelmaking processes described in the preceding chapters function satisfactorily for the production of plain carbon and some low-alloy steels, but are not readily adapted for the manufacture of steels containing large amounts of alloy elements which form oxides more stable than iron. For high-grade alloy steel such as cutting tools, die steels, and stainless steel, the metal must be refined and melted under rigidly controlled conditions and in such a way that impurities are reduced to a minimum. When a fuel is burnt in the furnace, some contamination is unavoidable, and this lets steel-makers to realize that electric melting was likely to be technically more desirable than the methods of the open hearth and Bessemer processes.

The essential components of the electric-arc furnace are the furnace shell with tapping device and work opening, the removable roof with the electrodes and tilting device (shown in Fig. 9-1). The hearth can be either acid or basic lined (Acid electric-arc furnace is mainly used in steel foundries.

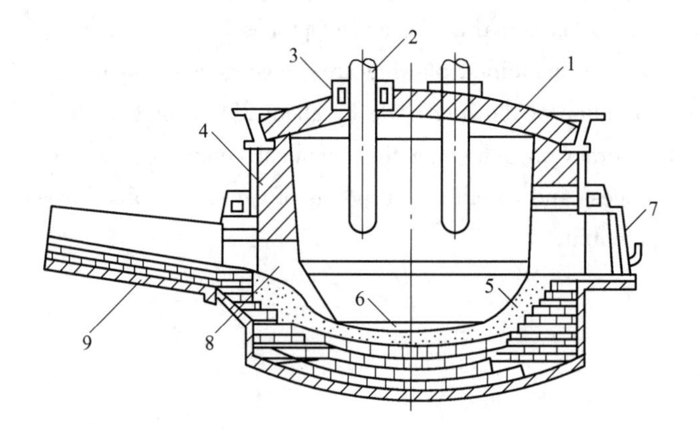

扫描二维码查看动画

Fig. 9-1 Electric-arc furnace

1—Furnace roof; 2—Electrode; 3—Water-cooled conducting collars; 4—Furnace wall; 5—Lining; 6—Bath; 7—Charging door; 8—Tapping hole; 9—Tapping runner

Basic furnaces are now used for making alloy and special steels). The short but very heavy arcs by which the current passes from electrodes to metal are the principal source of heat.

The electric-arc furnace with the basic-lined process generally follows the following pattern:

(1) **Charging.** The bottom of the furnace is covered with lime; Scrap steel of known quality is then put inside. Besides scrap or sponge iron, the charge also includes the ores, fluxes, reducing agents (carbon) and alloying elements in the form of ferroalloys. These can be added through the work opening before or during oxidizing.

When charging has been completed, the bank in front the charging door is built up with refractory material to form a dam to keep the molten metal from slopping out the door. The door is closed and the electrodes are lowered to above the scrap for melting.

(2) **Melting.** The electrodes melt the portion of the charge directly underneath and around them, and continue to bore through the metallic charge, forming a pool of molten metal on the hearth. From the time the electrode bore through the scrap and form the pool of molten metal on the hearth, the charge is melted from the bottom up by heat from the arc. The melting period in the basic electric-arc furnace is the most expensive period in its operation because power and electrode consumption are at the highest rate during this interval.

(3) **Oxidizing or purifying.** During the oxidizing stage, oxygen gas is usually injected into bath. An additional injection can accelerate the melting phase and oxidize directly phosphorus, silicon, manganese, carbon, and iron of metal. During the oxidation period, the reactions that occur in the bath of the basic electric-arc furnace are similar to those in the basic oxygen furnace. When these oxides included in the slag react with the carbon of the bath, this gives rise to the gaseous carbon monoxide which causes the heat to boil and hydrogen, nitrogen and non-metallic compounds escape as gases.

(4) **Deoxidizing or refining (reduction period).** The main tasks of reduction period are deoxidation, desulphurization, the composition and temperature adjustment and alloy added. Slag control is a very important factor in reduction period.

Nowadays, the metallurgical processes of the oxidation and reduction phases are replaced by a downstream phase or secondary metallurgical treatment carried out in the ladle. This enables the high capacity of the electric-arc furnace to be entirely used for melting the scrap.

(5) **The tapping.** In tapping a heat, the electrodes are raised high enough to clear the bath in a tilted position, the taphole is opened, and the furnace is tilted by a control mechanism so that the steel is drained from the furnace into a ladle. The ladle is usually held by a teeming crane during tapping to minimize exposure of the steam to air and minimize erosion of ladle refractory. The slag may be tapped before, with, or after the steel, depending on the particular operation.

✲ Words and Expressions

扫描二维码获取音频

扫描二维码答题

preceding [pri(:)'si:diŋ] *a.* 在前的，前述的

chapter ['tʃæptə] *n.* （书的，文章的）章，回

function ['fʌŋkʃən] *vi.* 发挥作用，运行

satisfactorily [sætis'fæktərili] *ad.* 满意地

plain [plein] *a.* 普通的，平常的

adapted [ə'dæptid] *a.* 适合的

die steel 模具钢

stainless steel 不锈钢

rigidly ['ridʒidli] *ad.* 严格地

Bessemer process 酸性转炉法

foundry ['faundri] *n.* 铸造厂

capacity [kə'pæsiti] *n.* 能力，容量，生产量

heavy ['hevi] *a.* 粗的，迟钝的，沉闷的

current ['kʌrənt] *n.* 电流，水流，气流

ferroalloy [ˌferəu'ælɔi] *n.* 铁合金

bank [bæŋk] *n.* 炉坡，堤

slop [slɔp] *v.* 溢出，溅溢

bore [bɔ:] *v.* 穿井，钻孔

interval ['intəvəl] *n.* 时段，（时间的）间隔

bath [bɑ:θ] *n.* 熔池

melting phase 熔炼期

compound ['kɔmpaund] *n.* 混合物，[化] 化合物

adjustment [ə'dʒʌstmənt] *n.* 调整，调节

adjust [ə'dʒʌst] *v.* 调整，调节

heat [hi:t] *n.* 一炉，熔炼的炉次

taphole ['tæphəul] *n.* 出钢口

teeming crane 铸锭吊车

erosion [i'rəuʒən] *n.* 腐蚀，侵蚀

ladle ['leidl] *n.* 钢包

mechanism ['mekənizəm] *n.* 机械装置，机构

water cooled conducting collar 水冷圈

✻ Answer the following questions.

(1) What are the essential components of the electric furnace?

(2) What are the main tasks of reduction period?

(3) What are the principal source of heat in electric-arc furnace?

(4) What does the charge also include besides scrap or sponge iron?

(5) Can the metallurgical processes of the oxidation and reduction phases in the electric-arc furnace be replaced by a downstream phase or secondary metallurgical treatment carried out in the ladle?

(6) When may the slag be tapped depending on the particular operation?

(1) …, but are not readily adapted for the manufacture of steels containing large amounts of alloy elements which form oxides more stable than iron.

……，但是它们不适合生产含有大量合金元素的钢，因为这些合金元素形成的氧化物比铁氧化物更稳定。

“containing large amounts…than iron” 是现在分词短语作定语，修饰 “steels”，此分词短语中又含一个由 “which” 引出的定语从句。

(2) The essential components of the electric-arc furnace are the furnace shell with tapping device and work opening, the removable roof with the electrodes and tilting device.

电弧炉主要是由带有出钢口和炉门的炉壳、带有电极和移动装置的炉盖组成。

句中有两个由 “with” 引出的介词短语作定语，分别修饰 “the furnace shell” 和 “the removable roof”。

(3) The short but very heavy arcs by which the current passes from electrodes to metal are the principal source of heat.

短而粗的电弧是电炉的主要热源，靠它把电流由电极传给金属。

“by which…to metal” 是定语从句，修饰先行词 “arcs”，“which” 代表 “arcs”。

(4) The melting period in the basic electric-arc furnace is the most expensive period in its operation because power and electrode consumption are at the highest rate during this interval.

在碱性电弧炉操作中，熔化期是其成本消耗最大的阶段，因为此阶段的电能和电极消耗最多。

“at the highest rate” 意思是 “以最快的速度”。

(5) During the oxidation period, the reactions that occur in the bath of the basic electric-arc furnace are similar to those in the basic oxygen furnace.

在氧化期，碱性电弧炉中发生的反应与碱性氧气转炉中的反应相似。

“that occur…electric-arc furnace” 是定语从句，修饰 “the reactions”，句中 “those” 代

替“the reactions”。

（6）When these oxides included in the slag react with the carbon of the bath, this gives rise to the gaseous carbon monoxide which causes the heat to boil and hydrogen, nitrogen and non-metallic compounds escape as gases.

当渣中的氧化物与熔池中的碳反应时就产生了一氧化碳，引起热沸腾，氢气、氮气以及非金属化合物将以气体形式逸出。

句中“included in the slag”是过去分词短语作定语，修饰“oxides”。“which causes the heat to boil”是定语从句，修饰“the gaseous carbon monoxide”。“give rise to/cause (sth.)”意思是“引起，导致（某事物）”，例如：Social practice alone gives rise to human knowledge（只有社会实践能产生人的认识）；Her disappearance gave rise to the wildest rumours（她的失踪引起各种流言蜚语）。

Part Ⅱ　Curriculum Ideological and Political

【Steel ambition】Aim at key and core technologies to solve bottleneck issues

We achieved technical breakthroughs with concentration on bottleneck issues. The preparation process of certain alloy bars with high purity triple smelting has passed the technical review of Aero Engine Corporation of China (AECC), which has preliminarily realized the import substitution and provided a guarantee for the independent control of key materials for China's aero-engine. The development of a super-large solid rocket engine shell has been completed, which has addressed a number of key technical challenges, filled the domestic gap in this field and reached the international frontier technology level.

Part Ⅲ　Further Reading

The new super steel of 2200MPa made in China

The new super steel of 2200MPa was invented under 5 times the pressure of ordinary steel during rolling, and increasing the cooling rate and controlling the temperature strictly. The diameter of super steel grains is only 1 micron, which is 1/10 to 1/20 of ordinary steel. Therefore, the structure of super steel is fine, the strength is high, and the toughness is also high. It can maintain a high strength, even if not adding elements such as nickel and copper. The successful development of super steel is a milestone for our country. With the use of super steel, the comprehensive capabilities of our country's aircraft carriers, warships and nuclear submarines will be improved by leaps and bounds.

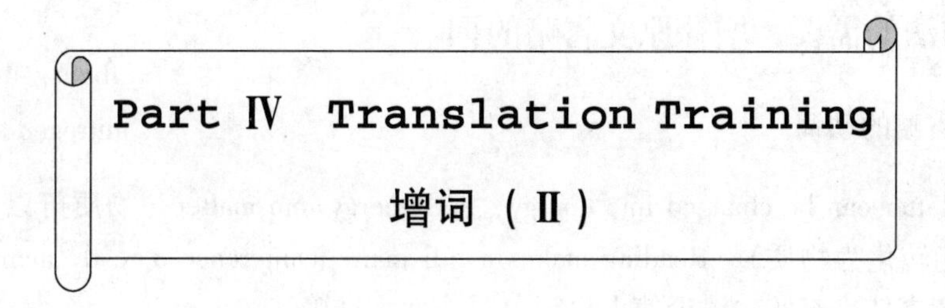

4. 增加数词

根据各方面的知识在某些名词前添加确定的数词，显示原文的明确内涵。例如：You may wonder why the magnet's poles point north and south（你可能会感到奇怪，为什么磁铁的两极会指南北）；The revolution of the earth around the sun causes the changes of the seasons（地球绕太阳旋转，引起四季交替）。

5. 增加相应的词

英语中常用抽象名词表示具体概念，翻译时要根据上下文添加相应的词。例如：The decay of organic matter is a slow oxidation by oxygen in the air（有机物的腐烂是空气中氧引起的一种缓慢氧化过程）；Preparation for the top government officers' visit are almost complete（迎接政府高级官员来访的准备工作差不多已全部完成）。

6. 增加表示名词复数的词

可以添加"们""诸位""各位"等，也可以增加重叠词、数词和其他词来表达。例如：Flowers bloom all over the yard（朵朵鲜花开满庭院）；The lion is the king of animals（狮是百兽之王）。

7. 增加量词

英语中没有量词，可数名词可以和数词与不定冠词连用，翻译成汉语时往往要添加量词。例如：a blast furnace（一座高炉）；a flame-cutter（一台火焰切割机）；an equation（一个方程式）。

8. 增加表示时态的词

英语靠动词的形式变化或加助动词来表示时态，汉语动词没有词形的变化，翻译成汉语时要加上汉语的时态助词或表示时间的词，如"曾""已经""正""在""将""会"等。例如：During the past century, many efforts were made to develop processes for producing iron steelmaking that could serve as alternatives to the conventional blast-furnace（在过去的一个世纪里，为了找到能够取代传统高炉的炼铁技术，人们曾做了许多努力）；Scientists have studied the universe for many centuries（科学家们研究宇宙已经持续了许多世纪）。

根据句法的需要，增补原文省略的词

1. 增补原文省略的动词

例如：Matter can be changed into energy, and energy into matter（物质可以转化成为能量，能量也能转化为物质）；Reading makes a full man; conference a ready man; writing an exact man（读书使人充实，讨论使人机智，写作使人准确）。

2. 增补原文比较句中省略的成分

例如：Better be wise by the defeat of others than by your own（从别人失败中吸取教训比从自己失败中吸取教训更好）。（原句应是：It is better to be wise by the defeat of others than to be wise by the defeat of your own.）

3. 增补原文含蓄条件句中省略的部分

例如：They would not have done such a thing without government approval（如果没有政府的同意，他们不会做这样的事情）。

Part V　Exercises

Ⅰ. Translate the following expressions into English.

（1）电弧炉	（2）工具钢	（3）不锈钢
（4）铸钢厂	（5）氧化阶段	（6）普碳钢
（7）熔化阶段	（8）高级合金钢	（9）控制条件

Ⅱ. Fill in the blanks with the words from the text. The first letter of the word is given.

（1）When a fuel is burnt in the f ______, some contamination is unavoidable.

（2）The short but very heavy a ______ by which the current passes from electrodes to metal are the principal source of heat.

（3）Basic furnaces are now used for making a ______ and special steels.

（4）The bottom of the furnace is covered with l ______; scrap steel of known quality is then put inside.

（5）Besides scrap or sponge iron, the charge also includes the ores, f ______, reducing agents (carbon) and alloying elements in the form of ferroalloys.

（6）During the oxidizing stage, oxygen gas is usually injected into b ______.

（7）Slag control is a very important factor in r ______ period.

（8）In the s ______ slag process, the steel is finished by adjusting the composition and temperature to the desired values.

(9) In tapping a heat, the electrodes are raised high enough to c ______ the bath in a tilted position.

(10) The ladle is usually held by a teeming c ______ during tapping to minimize exposure of the steam to air and minimize erosion of ladle refractory.

Ⅲ. Fill in the blanks by choosing the right words form given in the brackets.

Modern steelmaking processes are (1) ______ (divided; dividing) into two general classes from the chemical point of view: acid processes and basic processes. The acid or basic of the process is in turn decided by the type of refractory lining with which it is in contact. The primary (2) ______ (difference; different) between the two methods of steelmaking is that phosphorus and sulphur can only be (3) ______ (removed; to remove) effectively from the metal under a cover of basic slag. In the acid process there is often slight increase in the concentration of these elements, for the removal of phosphorus and sulphur (4) ______ (requires; require) special conditions that can be met only by the basic processes. Therefore, basic processes are the most important processes for (5) ______ (producing; produced) steel.

Ⅳ. Decide whether the following statements are true or false (T/F).

(1) Where a fuel is burnt in the furnace, some contamination is avoidable. ( )

(2) The bottom of the furnace is covered with scrap. ( )

(3) The electrodes melt the portion of the charge directly underneath and around them, and continue to bore through the metallic charge, forming a pool of molten metal on the hearth. ( )

(4) During the melting stage, oxygen gas is usually injected into bath. ( )

(5) The ladle is usually held by a teeming crane during tapping to minimize exposure of the steam to air and minimize erosion of ladle refractory. ( )

Ⅴ. Translate the following English into Chinese.

Like the open hearth, the electric furnace differs from the Bessemer in that it applies heat by means external to the bath and is, therefore, independent of the reactions occurring in the metal insofar as the control of temperature is concerned. The electric furnace, however, produces its heat by passing electric current through the steel. The metal, therefore, is free from any possibility of contamination by sulphur or other element in the fuel.

Unit 10 Secondary Refining

Part Ⅰ Reading and Comprehension

扫描二维码获取音频

扫描二维码获取PPT

Formerly, steel produced in the refining processes of the converter or electric - arc furnace was considered ***finished*** and ready to be cast and rolled. Nowadays, secondary refining is generally applied after the refining process. The purpose of secondary refining is to produce ***clean*** steel, which satisfies stringent requirements of surface, internal and microcleanliness quality and of mechanical properties. The tasks of secondary refining are: degassing (decreasing the concentration of oxygen, hydrogen and nitrogen in steel), decarburization, removing undesirable nonmetallics (primarily oxides and sulphides), etc.

Secondary refining processes have many methods. They are divided into two broad categories: secondary refining process in vacuum (vacuum treatment process) and secondary refining process in nonvacuum (nonvacuum treatment process). The corresponding techniques are: stirring gas treatment with porous lances, or with the aid of electromagnetic stirring; feeding-refining agent (lance injected solids, wire feeding); heating with cored wire and electric arc; vacuum degassing with the aid of various techniques, etc. Secondary refining processes can be carried out in the ladle, the ladle furnace, in some instances, even in the electric-arc furnace.

Vacuum treatment of steel enjoys high priority because of its versatility and the special advantages in second refining. In this chapter, some vacuum treatment methods will be introduced mainly.

Vacuum treatment is based on the following consideration. Dissolved gases only partly escape while the steel solidifies. For this reason, all steels contain quantities of the gases: hydrogen, nitrogen and oxygen, which cause embrittlement, voids, inclusion, and other undesirable phenomena in the steel after it solidifies. This impoverishes the technological properties of the steel. For example, hydrogen in solution in solid steel has a deleterious effect upon the mechanical properties; ductility is lowered without a corresponding increase in strength and also leads to cracking in highly-stressed components. Nitrogen lowers the ability of steels to undergo deep drawing operations. If the external pressure is lowered, then the gas which is dissolved in the metal can escape.

Besides degassing, additional metallurgical reactions are carried out in the vacuum, such as decarburization. A combination of vacuum treatment and stirring gas accelerates and promotes the metallurgical reactions.

Nowadays, there are many vacuum treatment methods in steelmaking, such as RH process, Finkle process, RH-KTB process, ASEA-SKF process, Finkle-VAD process, VOD process, etc.

RH process is circulation degassing in vacuum chamber. This method is shown in Fig. 10-1. There are two circulation legs under the vacuum chamber and the steel is caused to flow up into evacuated chamber for degassing by the passage of a small but continuous flow of argon gas into one leg of the chamber. Gravity causes it to leave through the other leg and return to the ladle where the outflow creates and adequate degree of circulation. The average circulation rate is usually some 12t per min and 20min are required to fully treat a 100t ladle of steel. The temperature loss will be 40 to 50℃ for about 40 ton.

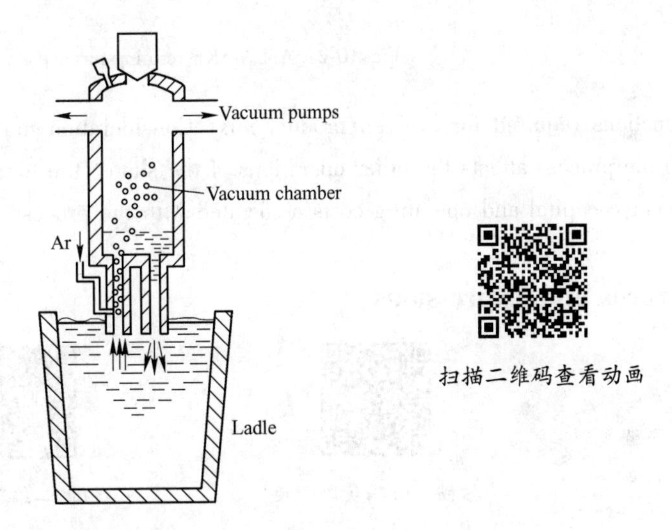

Fig. 10-1 RH circulation degassing

In the case of ASEA-SKF, the equipment comprises a ladle furnace, a mobile induction heater, a steam ejector and a vacuum cover fitted with electrodes for heating. In refining, the ladle is placed in the induction heater and moved to the vacuum degassing station. The atmosphere is reduced to around 200μm Hg. After 15min (30t heat) deoxidizers are added. After vacuum treatment the ladle is moved to the reheating as shown in Fig. 10-2. Here the electrode cover is lowered on to the ladle. Fluxes are added to make a basic slag and alloys added to meet the specification. After heating by arcs the steel is tapped, normally by continuous casting. This method is flexible and well suited for special steels of high quality, and allows for efficient scheduling of casters.

There can be considerable overlap in the metallurgical functions that various secondary steelmaking processes achieve. To determine which process to install, the steelmaker must reflect on the capability and flexibility of the particular process in its ability to perform the various metallurgi-

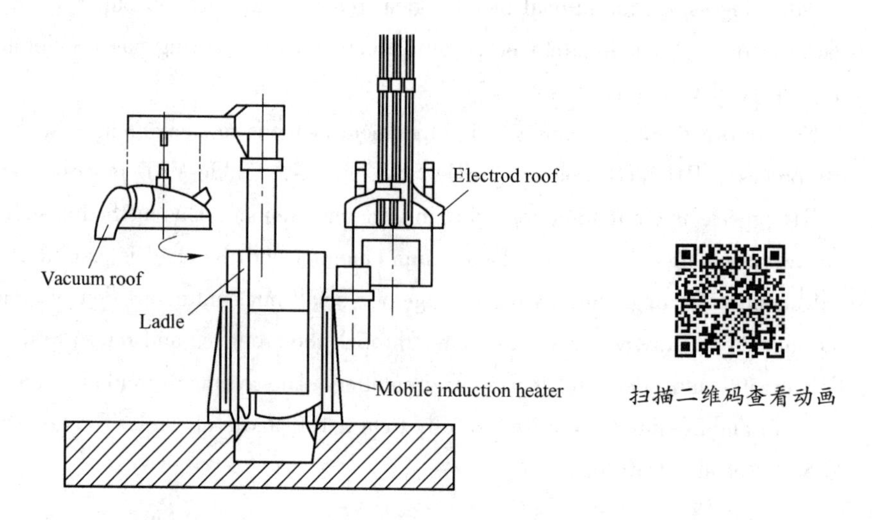

Fig. 10-2　ASEA-SKF refining furnace

cal functions required for a given product mix. Consideration must also be given to the way in which the process affects the other operations of the shop, the tonnage throughput expected, and the relative capital and operating costs associated with the process.

Words and Expressions

扫描二维码获取音频

扫描二维码答题

secondary refining	炉外精炼，二次精炼
roll [rəul] *v.*	轧，轧制
clean steel	纯净钢
stringent ['strindʒənt] *a.*	苛刻的，必须严格遵守的
internal [in'tə:nl] *a.*	内在的，内部的
satisfy ['sætisfai] *v.*	满足，使……满意
nonmetallic ['nɔnmi'tælik] *n.*	非金属物质
a.	非金属的
microcleanliness [maikrəuklen'linis] *n.*	显微清洁（度）
degas [di:'gæs] *v.*	脱气，去氧
nitrogen ['naitrədʒən] *n.*	氮（元素符号 N）
category ['kætigəri] *n.*	种类，范畴
vacuum ['vækjuəm] *n.*	真空，空间
corresponding [ˌkɔris'pɔndiŋ] *a.*	相应的，对应的

stir [stəː] *v.* 搅拌，搅和
porous lance 多孔氧枪
electromagnetic [iˌlektrəumæg'netik] *a.* 电磁的
wire feeding 喂丝，喂线
cored wire 芯钢丝，焊条芯
priority [prai'ɔriti] *n.* 优先，优先权
versatility [ˌvəːsə'tiləti] *n.* 多功能性
solidify [sə'lidifai] *v.* （使……）凝固
embrittlement [em'britlmənt] *n.* 变脆，脆化
void [vɔid] *n.* 空隙，缩孔
inclusion [in'kluːʒən] *n.* 夹杂物
phenomena [fi'nɔminə] *n.* 现象
impoverish [im'pɔvəriʃ] *v.* 使……枯竭，使……贫穷
deleterious [ˌdeli'tiəriəs] *a.* 有害的，有毒的
ductility [dʌk'tiliti] *n.* 延展性，塑性
crack [kræk] *v.* （使……）破裂，裂纹
undergo [ˌʌndə'gəu] *v.* 经受，经历，遭受
external [eks'təːnl] *a.* 外部的，外面的
accelerate [æk'seləreit] *v.* 加速，促进
circulation leg 循环流管
evacuate [i'vækjueit] *v.* 排空，抽空
passage ['pæsidʒ] *n.* 通道，通路
gravity ['græviti] *n.* 重力，地心引力
outflow ['autfləu] *n.* 流出，流出物
deoxidizer [diː'ɔksidaizə] *n.* 脱氧剂，还原剂
flexible ['fleksəbl] *a.* 灵活的
caster ['kɑːstə] *n.* 连铸机
overlap ['əuvə'læp] *n.* 重叠
install [in'stɔːl] *v.* 安装
flexibility [ˌfleksə'biliti] *n.* 灵活性
tonnage ['tʌnidʒ] *n.* 吨，产量
throughput ['θruːput] *n.* 生产量，生产能力
vacuum pump 真空泵
vacuum roof 真空炉顶
mobile induction heater 移动感应加热器

✲ Answer the following questions.

(1) What is the purpose of secondary refining?
(2) What are the tasks of secondary refining?

(3) How many categories are secondary refining processes divided into?

(4) Where can secondary refining processes be carried out?

(5) Why does vacuum treatment of steel enjoy high priority?

(6) What chemical element can lower the ability of steels to undergo deep drawing operations?

(7) What is the average circulation rate and how many minutes are required to fully treat a 100 ton ladle of steel in RH process?

(8) What does the equipment comprise in the case of ASEA-SKF?

(1) Formerly, steel produced in the refining processes of the converter or electric-arc furnace was considered ***finished*** and ready to be cast and rolled.

过去，在转炉或电弧炉精炼过程中生产的钢被称为“成品”，用于铸造和轧制。

“produced in the refining processes of the converter or electric-arc furnace”为过去分词短语作定语，修饰“steel”。

(2) Vacuum treatment of steel enjoys high priority because of its versatility and the special advantages in second refining.

真空处理法由于具有多功能性和特殊优点，在炉外精炼中享有优先权。

“priority”意思是“优先，优先权，需优先考虑的事”。例如：You must give this matter priority（你必须优先处理此事）；The highest priority of governments has been given to the problem of heavy traffic（政府已经优先考虑交通拥挤的问题）。

(3) Nowadays, there are many vacuum treatment methods in steelmaking, such as RH process, Finkle process, RH-KTB process, ASEA-SKF process, Finkle-VAD process, VOD process, etc.

目前炼钢中有多种真空处理法，比如真空循环脱气法、真空吹氩法、川崎顶吹氧真空脱气法、电弧加热真空精炼法、真空电弧加热脱气法、真空吹氧脱碳精炼法等。

“RH”是 Rheinsaht-Heraeus 公司名称的缩写，“RH process”表示“真空循环脱气法”。“Finkle”取自美国 Finkle 公司的名称，“Finkle process”表示“真空钢包吹氩法”。“RH-KTB process”表示“川崎顶吹氧真空脱气法”。“ASEA-SKF”是瑞典 ASEA 和 SKF 公司的名称，“ASEA-SKF process”表示“电弧加热真空精炼法”。“VAD process：Vacuum Arc Degassing process”表示“真空电弧加热脱气法”。“VOD process：Vacuum Oxygen Decarburization process”表示“真空吹氧脱碳精炼法”。

(4) There can be considerable overlap in the metallurgical functions that various secondary steelmaking processes achieve.

许多炉外精炼工艺所达到的冶金功能会相互重叠。

“that”为关系代词引出的定语从句，修饰先行词“functions”。

(5) To determine which process to install, the steelmaker must reflect on the capability and flexibility of the particular process in its ability to perform the various metallurgical functions required for a given product mix.

为了确定采取何种工艺，以实现对给定混合产品所要求的不同的冶金功能，钢厂必须仔细考虑该工艺本身的实际能力和灵活性。

"reflect"常与"on""upon"连用，意思是"仔细考虑"。例如：I need time to reflect on your offer（我需要时间考虑你的建议）。"required for a given product mix"是过去分词作定语，修饰"functions"。

Part Ⅱ Curriculum Ideological and Political

【Will of steel】Paul Kochagin's famous words

Man's dearest possession is life, and it is given to o him to live but once. He must live so as to feel no torturing regrets for years without purpose never know the burning shame of a mean and petty past—so live that, dying he can say: all my life, all my strength were given to the finest cause in all over the world—the fight for the liberation of mankind.

Part Ⅲ Further Reading

The new process for making steel

1. A new high efficiency refining process with powder injection at the bottom of ladle

Development of ladle bottom powder injection technology in secondary refining process. The bottom spraying process has no iron loss, and the mixing dynamic conditions are better than the top spraying process. The supporting technology is mature and easy to realize. The transformation investment is low and the original process is not changed. The production platform of ultra-low sulfur clean steel can be established and good sulfur removal effect can be achieved.

2. Uniform refinement control of microstructure of thick gauge structural steel

The smelting process control implemented by oxide metallurgy technology is combined with the rolling and cooling control in the subsequent rolling process. On the basis of the smelting process control, the implementation of certain final rolling temperature control and cooling rate control can obtain thick steel with full section (uniform) fine grain structure. It has both high strength and toughness as well as large linear energy welding performance, and can be applied to thick plates, heavy H-shaped steel Production of thick wall seamless steel pipes.

Part Ⅳ　Translation Training

as 的译法

英语中，as 是多词性、多词义，可以作连词、介词、关系代词、副词，也可以和其他词搭配构成固定的短语等，其用法非常复杂。英译汉时，首先确定词性，再根据搭配关系或上下文正确选择词义，也要注意其固定结构所具有的特殊意义。

（1）as 作连词时，它的意思是：当……时候，因为，虽然，正如等。例如：Air pressure decreases as altitude increases（大气的压力随着高度的增加而减小）；The cutting speed is in linear terms，as a speed stated in rpm is，by itself，meaningless（切削速度是用线速度来表示的，因为每分钟的转数表示的速度本身没有意义）；Small as they are，atoms are made up of still smaller units（原子虽然小，但却是由一些更小的单位组成的）。as 引导让步状语从句时，位于形容词、副词之后。例如：As rust eats iron，care eats the heart（正如锈能蚀铁，忧愁也能伤人）。

（2）as 作介词时，一般的意思是：作为，以……形式。例如：The energy of the sun comes to the earth as light and heat（太阳的能量主要以光和热的形式传至地球）；Ordinarily we don't consider air as having weight（我们通常认为空气没有重量）；Most bottom-blown processes use methane or propane as the hydrocarbon coolant，while other bottom-blown processes use fuel oil（大多数底吹工艺使用甲烷或丙烷作为碳氢冷却剂，而其他底吹工艺使用燃油）。

（3）as 作为关系代词时，视其具体情况翻译，看它在定语从句中的句子成分以及代替主句什么词。例如：There are no such machines as you mention（没有如你所说的机器），"as" 指代的是 "such machines"；As we know，combined oxygen blowing contains further development potential（人们知道，复吹工艺具有进一步发展的潜力），"as" 指代整个主句的意思。

（4）as 作副词时，意思是：同样地，相同地。例如：He is as strong as a horse（他生得虎背熊腰）；I guessed as much（我料到是这么一回事）。

（5）as 用于固定短语中，一定要弄清它的意思，切忌望文生义。例如，"as regards" 意思是 "至于，关于"；"as yet" 意思是 "到目前为止"；"as a rule" 意思是 "照例，照常"；"as a consequence" 意思是 "因此，从而"。

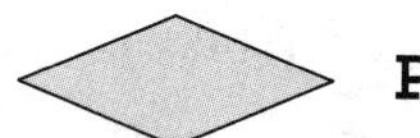

Part Ⅴ　Exercises

Ⅰ. Translate the following expressions into English.

（1）力学性能	（2）纯净钢	（3）真空处理
（4）电磁搅拌	（5）真空吹氩法	（6）感应炉
（7）真空室	（8）外部压力	（9）冶金反应
（10）真空循环脱气法	（11）循环流管	（12）可移动的感应加热器
（13）蒸汽喷射器	（14）真空脱气站	（15）冶金功能

Ⅱ. Fill in the blanks with the words listed below.

overlap	combination	moulds	undergo	applied	outflow
quality	partly	vacuum	dissolved	heater	specification

(1) Nowadays, secondary refining is generally ______ after the refining process.

(2) The metal is also cast into ______ in the vacuum chamber.

(3) Dissolved gases only ______ escape while the steel solidifies.

(4) Nitrogen lowers the ability of steels to ______ deep drawing operations.

(5) If the external pressure is lowered, then the gas which is ______ in the metal can escape.

(6) A ______ of vacuum treatment and stirring gas accelerates and promotes the metallurgical reactions.

(7) RH process is circulation degassing in ______ chamber.

(8) Gravity causes it to leave through the other leg and return to the ladle where the ______ creates and adequate degree of circulation.

(9) In the case of ASEA-SKF, the equipment comprises a ladle furnace, a mobile induction ______, a steam ejector and a vacuum cover fitted with electrodes for heating.

(10) Fluxes are added to make a basic slag and alloys added to meet the ______.

Ⅲ. Fill in the blanks with appropriate prepositions.

Processes (1) ______secondary steelmaking define the various ways (2) ______which the liquid steel is handled in order to achieve the objectives of secondary steel refining. Particular processes are defined partly (3) ______the equipment they use to manipulate the steel and may involve one or more treatments to achieve the objectives of secondary steelmaking, such as producing clean steel. Some processes perform certain functions better than others. There can be considerable overlap (4) ______ the metallurgical functions that various secondary steelmaking processes achieve. To determine which process to install, the steelmaker must reflect (5) ______ the capability and flexibility of the particular process in its ability to perform the various metallurgical functions required (6) ______ a given product mix. Consideration must also be given (7) ______ the way in which the process affects the other operations of the shop, the tonnage throughput expected, and the relative capital and operating costs associated (8) ______ the process.

Ⅳ. Decide whether the following statements are true or false (T/F).

(1) The tasks of secondary refining are: degassing (decreasing the concentration of oxygen, hydrogen and nitrogen in steel), decarburization, removing undesirable non-metallics (primarily oxides and sulphides), etc. (　)

(2) Nitrogen improves the ability of steels to undergo deep drawing operations. (　)

(3) Nowadays, there are many vacuum treatment methods in steelmaking, such as RH process,

Finkle process, AOD process, ASEA-SKF process, CLU process, VOD process, etc.　(　　)

(4) The average circulation rate is usually some 21t per min.　(　　)

(5) After vacuum treatment the ladle is moved to the reheating as shown in Fig. 10-2.　(　　)

V. Translate the following English into Chinese.

Vacuum degassing was first used on a large scale for treating special qualities of killed steel, and later extended to improve the qualities over a wider range of killed steels. More recently, unkilled steels have been subjected to vacuum treatment with very considerable advantage and the major developments will most probably occur in this field. The effects of degassing on both killed and unkilled steels will be considered as follows.

Unit 11 Continuous Casting of Steel

Part Ⅰ Reading and Comprehension

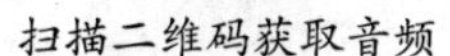

扫描二维码获取音频

扫描二维码获取 PPT

Since steel was first produced in a liquid state over 200 years ago, it has been almost invariably the practice to cast it into rectangular blocks by ingot casting, from which the desired finished shape is obtained by subsequent hot or cold working. These blocks or ingots were originally very small in weight, but as the output of steelmaking increased, so has both the weight to that 15t or 20t ingots are now common practice. These ingots are rolled in primary mills to produce blooms or slabs which subsequently receive further processing.

This method of ingot casting is characterized by the following:

(1) A large amount of capital is invested in moulds, bottom plates, and locomotives.

(2) Considerable capital is also invested in buildings and cranes needed for stripping the ingots and resetting the moulds on the casting cars, while soaking pits and preheating furnaces are necessary to prepare the ingots for rolling, and a primary mill to roll them.

(3) Ingots contain segregation, which is a disadvantage in all steels and needs ***hot tops*** for certain qualities, with a consequent high loss as scrap after rolling in the primary mills. The larger the ingot, the more intense the segregation.

(4) Each ingot needs a discard to be removed at top and bottom after primary rolling.

These requirements have turned the thoughts of steel makers to the possibility of casting molten steel directly slabs, blooms or billets thus elimination much of the capital and working costs outlined above, as well as minimizing the loss of yield and the segregation in the product. Continuous casting of steel was developed in Europe in the mid 1950s and has grown rapidly till the present world capacity is over 1000 million tons per annum, representing about 89.7 per cent of the total tonnage of steel.

The first continuous casting machines for steel were aligned vertically. It involved pouring the metal from a height of over 13 meters above ground level, with the cast section of metal descending vertically and being progressively water cooled (shown in Fig. 11-1). Efforts to reduce building-height first led to continuous-casting systems in which molten metal passed into a vertical mould and solidified completely before being bent and later to the bow-type installation which has a curved mould and is the system most used today.

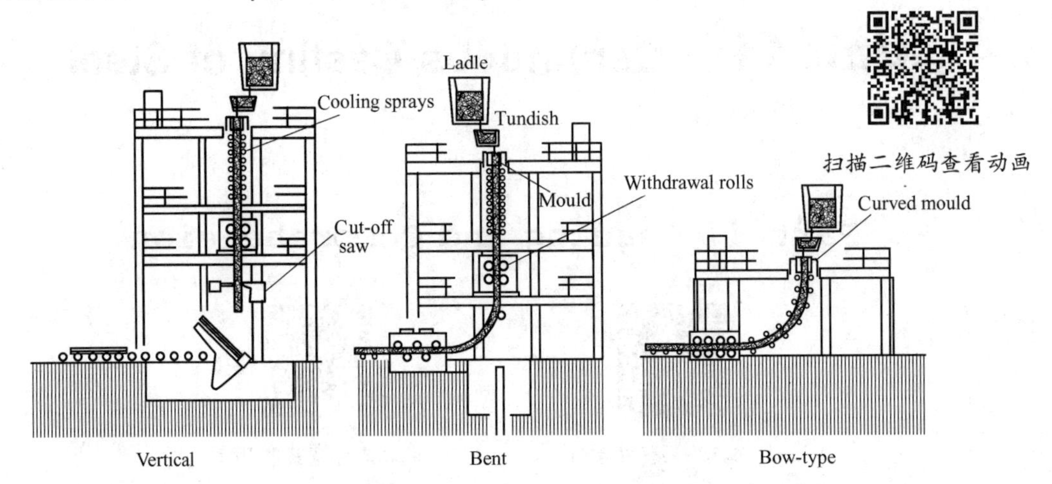

Fig. 11-1 Types of continuous casting machines

In a typical plant, deoxidized steel from an electric furnace (or BOF) is run into a ladle, transported to the casting bay and brought to rest above the Concast machine. Before the start-up, a dummy bar, constructed of chain links, is fitted into the mould bottom and rests between the Ore formed by the machine roller guides. A stopper rod or a slide gate is used to control the flow of steel into a tundish reservoir. Once a predetermined steel level is attained in the tundish, nozzles are opened and the molten steel feeds the open-ended mould set in a frame with internal water-cooling. At the start-up, a dummy bar, constructed of chain links, is fitted into the mould bottom and rests between the arc formed by the machine roller guides. The molten in the mould meets the dummy bar and freezes to it and at a certain level of steel in the mould two operations are immediately initiated; the mould unit begins to reciprocate in a vertical direction and the dummy bar is withdrawn through and down the machine. At the horizontal plane it is supported by a series of rollers which are grouped into segments for operation and maintenance and are cooled down by water spraying.

At the exit of the straightening unit a roll is lowered which separates the dummy bar from the length of cast steel. The dummy bar is either lifted to one side of the machine or hoisted on to the ramp above the machine. The strand of cast metal continues traveling along the discharge rollers to a gas torch cut-off station. At predetermined dimensions this torch will automatically cut the strand to required lengths. Once the start-up complete the concast machine continues to cast sections in accordance with the volume of metal fed from the steel furnaces.

The end products of continuous casting plants are square section billets (up to 250mm × 250mm), and slabs which are about 300mm thick by 2500mm wide. Other shapes are also produced, including round bars and octagonal sections.

Since the 1960s, continuous casting has steadily displaced ingot casting. The types of continuous casting machines, their listing likewise reflecting the historical development of this technique, are: vertical, bending and straightening, circular bow, and oval bow. The evolution from one form to the next has been accompanied by a considerable reduction in overall height. Nowadays, there are further on going process developments in the continuous casting of steel, these include:

Horizontal continuous casting with stationary mould and continuous casting with moving belt or rotating rolls (double rollers), etc.

In brief, the reasons for continuous-casting systems are:

(1) Lower investment outlay compared with that for a blooming train;

(2) More productivity than with conventional ingot-casting;

(3) High degree of consistency of steel composition along the whole length of the strand; better core quality, especially with flat strands; high inherent surface quality, leading to savings on an otherwise expensive surfacing process;

(4) High degree of automation;

(5) Friendlier to the environment.

Words and Expressions

扫描二维码获取音频

扫描二维码答题

continuous casting 连铸
invariably [in'veəriəb(ə)li] *ad.* 不变地，总是
rectangular [rek'tæŋgjulə] *a.* 矩形的，成直角的
block [blɔk] *n.* 毛坯，粗坯
ingot casting 铸锭，模铸锭
subsequent ['sʌbsikwənt] *a.* 随后的，后来的
cold working 冷加工
primary mill 初轧厂
bloom [bluːm] *n.* 大（初轧）方坯
slab [slæb] *n.* （大，初轧）板坯
capital ['kæpit(ə)l] *n.* 资金
bottom plate 底板
locomotive [ˌləukə'məutiv] *n.* 机车，火车头
strip [strip] *v.* 剥，剥去
reset ['riːset] *v.* 重放，重新安排
soaking pit 均热炉
preheating furnace 预热炉
segregation [ˌsegri'geiʃən] *n.* 偏析作用

discard [dis'kɑːd] *v.*	丢弃，抛弃
discharge rollers	卸料辊道辊子
outline ['əutlain] *v.*	略述
yield [jiːld] *n.*	产量，收益
annum ['ænəm] *n.*	年，岁
strand [strænd] *n.*	铸坯
bow-type	弓形的，弧形的
installation [ˌinstə'leiʃən] *n.*	(整套) 装置 (备)，设备 (施)
curve [kəːv] *v.*	弯，使……弯曲
casting bay	连铸跨
stopper rod	棒塞
slide gate	滑动水口
chain links	链环
tundish ['tʌndiʃ] *n.*	中间包，中间罐
mould set	全套锭模，整套锭模
dummy bar	引锭杆
roller ['rəulə] *n.*	辊子
initiate [i'niʃieit] *v.*	开始，发起
reciprocate [ri'siprəkeit] *v.*	(使……) 往复 (运动)，来回
withdraw [wið'drɔː] *v.*	抽出，拉动
horizontal [ˌhɔri'zɔntl] *a.*	水平的
freeze [friːz] *v.*	凝固，凝结
segment ['segmənt] *n.*	弧形，扇形体
straighten ['streitn] *v.*	(使……) 弄直，伸直
straightening unit	矫直装置
hoist [hɔist] *v.*	升起，吊起
ramp [ræmp] *n.*	斜面，斜台
cut-off station	切割站
predetermine ['priːdi'təːmin] *n.*	预定，预先确定
in accordance with	根据，按照
dimension [di'menʃən] *n.*	尺寸，尺度
octagonal [ɔk'tægənl] *a.*	八边形的，八角形的
displace [dis'pleis] *v.*	取代，置换
likewise ['laikˌwaiz] *ad.*	同样地，照样地
oval ['əuvəl] *a.*	椭圆的
evolution [ˌiːvə'luːʃən, ˌevə'luːʃən] *n.*	发展，演变
accompany [ə'kʌmpəni] *v.*	伴随，陪伴
stationary ['steiʃ(ə)nəri] *a.*	固定的
sequence ['siːkwəns] *n.*	顺序，次序

impetus ['impitəs] *n.* 推动力，促进
adoption [ə'dɔpʃən] *n.* 采用
outlay ['autlei] *n.* 花费，支出
conventional [kən'venʃənl] *a.* 传统的，常规的
blooming train 初轧机组
productivity [ˌprɔdʌk'tiviti] *n.* 生产力，生产率
consistency [kən'sistənsi] *n.* 一致性
inherent [in'hiərənt] *a.* 固有的，内在的
otherwise ['ʌðəwaiz] *a.* 其他方面的，另外的
surfacing ['səːfisiŋ] *n.* 表面加工，表面处理

✽ Answer the following questions.

(1) Where and when was continuous casting of steel developed?
(2) What were the first continuous casting machines for steel like?
(3) What are the end products of continuous casting plants?
(4) When has continous casting displaced ingot casting?
(5) How many types of continous casting machines are mentioned in this passage?
(6) Why are continous-casting systems used?

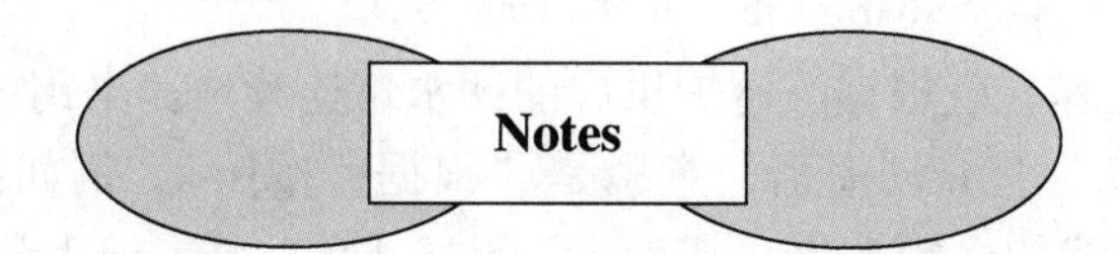

(1) Since steel was first produced in a liquid state over 200 years ago, it has been almost invariably the practice to cast it into rectangular blocks by ingot casting, from which the desired finished shape is obtained by subsequent hot or cold working.

自从200多年前首次生产液态钢以来，几乎一直不变的操作工艺是：通过模铸把钢水浇注成矩形的毛坯，再将毛坯经过随后的冷、热加工获得理想的形状。

"from which" 引出非限制性定语从句，修饰 "ingot casting"。

(2) These requirements have turned the thoughts of steel makers to the possibility of casting molten steel directly slabs, blooms or billets thus elimination much of the capital and working costs outlined above, as well as minimizing the loss of yield and the segregation in the product.

这些要求使炼钢工作者产生了把钢水直接铸造成板坯或方坯的设想，这样就减少了上面提到的投资和加工费用，也能把产量的损失和产品中的偏析减小到最低程度。

"as well as" 意思是"（除了……外）也，还，而且"。例如：He grows flowers as well as vegetables（他既种菜也种花）；She is a talented musician as well as being a photographer（她不但是个摄影师，还是个天才的音乐家）。

(3) Efforts to reduce building-height first led to continuous-casting systems in which molten metal passed into a vertical mould and solidified completely before being bent and later to the bow-type installation which has a curved mould and is the system most used today.

为了降低厂房的高度，首先研制出的连铸系统是将钢水倒入立式结晶器中，并且在弯曲之前让钢水完全凝固。这种系统随后发展为弧形连铸机，这是目前常用的设备。

这个句子比较长，“to reduce building-height”是不定式作定语，修饰“efforts”。“in which molten metal…in the liquid phase”为定语从句，修饰“continuous-casting systems”。“later”后面省去谓语“led”。“which has a…most used today”是定语从句，修饰“the bow-type installation”。

（4）Once a predetermined steel level is attained in the tundish，nozzles are opened and the molten steel feeds the open-ended mould set in a frame with internal water-cooling.

一旦中间包内的钢水达到规定的高度，水口就被打开，钢水注入放在托架上、有内部水冷装置的结晶器内。

“once”是一个多词性、多词义的词，在此它是连词，意思是“一旦，一……就……”。例如：Once the principal contradiction is grasped，all problems will be readily solved（一旦抓住了主要矛盾，一切矛盾就迎刃而解）。

（5）At the horizontal plane it is supported by a series of rollers which are grouped into segments for operation and maintenance and are cooled down by water spraying.

当凝固的铸坯从垂直状态过渡到水平状态时，用一系列辊子托住。为了操作和维修方便，辊子被分组，呈扇形布置。在辊子间还要进一步喷水冷却。

（6）Once the start-up complete，the concast machine continues to cast sections in accordance with the volume of metal fed from the steel furnaces.

一旦连铸机开始工作，就根据炼钢炉供应的钢水量连续地铸出钢坯。

“in accordance with”是介词短语，意思是“根据，按照”。例如：We must act in accordance with the law（我们必须按照法律做事）；“fed from the steel furnaces”为过去分词短语作定语，修饰“the volume of metal”。

（7）The types of continuous casting machines，their listing likewise reflecting the historical development of this technique，are：vertical，bending and straightening，circular bow，oval bow.

下面所列出的连铸机类型同样反映出这一技术的历史发展。连铸机的类型有立式、立弯式、弧形、椭圆形连铸机。

“their listing likewise reflecting…this technique”是现在分词短语作状语，做进一步说明。

Part Ⅱ Curriculum Ideological and Political

【Steel backbone】The C919

The C919 is China's first self-developed truck jetliner in accordance with international airworthiness standards，and owns independent intellectual property rights. The project was approved in 2007 and the C919 airliner accomplished its maiden flight in 2017. After completing all airworthiness certification，the C919 received the type certificate issued by the Civil Aviation Administration of China in September 2022. The first aircraft will be delivered by the end of 2022. The fact that the C919 was successfully developed and received its type certificate marks that China is capable of researching and developing world-class truck jetliners independently，and it represents a major milestone in the development of the country's large aircraft sector. Over the past 15 years, China has successfully blazed a development path that features indigenous design and systematic integration, attracts global public bidding and helps increase the proportion of domestic technologies. The country has also cultivated a team of talents in the large aircraft sector who uphold firm convictions, are willing to devote themselves, have the courage to make breakthroughs and compete in tough battles and have an international vision, achieving fruitful results and gaining valuable experience.

Part Ⅲ Further Reading

Near net shape continuous casting

Near net shape continuous casting is a major high-tech in the development of modern steel industry, including thin slab casting continuous and direct rolling, continuous casting of steel strip, beam blank continuous casting, tube billet continuous casting, wire rod casting and rolling, and spray deposition. The purpose of near net shape continuous casting is to cast strands as close as possible to the dimensions of the final product, in order to reduce intermediate processing steps, save energy, reduce storage and shorten production time. Compared with traditional processes, these technologies have many advantages, such as short process, simple equipment, low construction investment, low energy consumption, high success rate and low production cost.

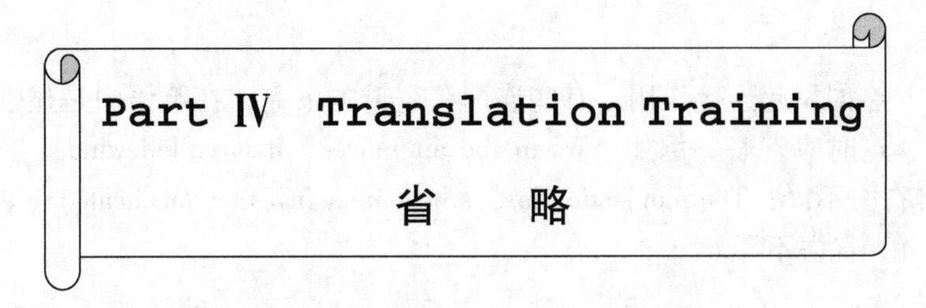

为了符合汉语习惯，将原文中有些词省略不译，因为译文中虽无其词但有其义。省略译主要省去的是英语中因语法上、逻辑上或修辞上的需要而存在，但根据汉语习惯并不需要的词。例如某些动词、冠词、介词、连词、代词等，当其在句子中不具备实际意义时，可以省略。

1. 省略动词

英语中有许多词既是名词又是动词，它们常与另一动词连用，表达的还是动词的含义，翻译时应译出其动词的含义，省去连用的动词。例如：This gives a sharp increase in the residuals in the steel, which detracts from its properties for a number of applications（这使得钢中的杂质急剧增加，有损于钢的一些应用性能）。“give an increase”实际意义是名词的动词意义“增加”。例如：This chapter presents a review of alternative processes in ironmaking（本章将回顾钢铁生产交替使用的方法）。“present a review”实际意义是名词的动词意义“回顾”。

2. 省略代词

英文句子中，用了人称代词的主格或表示人与事物的名词之后又用它的所有格，译成汉语时，这个所有格代词省略不译。例如：Nobel cut his finger on broken glass jar and had the answer to his problem of how to pack his explosive（诺贝尔被一口玻璃缸割破了手指，因而想出了包装炸药的方法），句中两个"his"省去不译；The melting period in the basic electric-arc furnace is the most expensive period in its operation（碱性电弧炉的操作中，熔化期是其成本消耗最大的阶段），句中"its"省去不译。

3. 省略冠词

英语中有冠词，汉语中没有冠词，英译汉时常常可以将其省略；但当不定冠词表示数字"一"时，不省略。例如：A square has four equal sides（正方形四条边相等），"A"省去不译；As early as 300 years ago，iron was serving as a basis of human culture and civilization（早在公元300年前，铁就是人类文化与文明的基础），"a"省去不译。

4. 省略介词

例如：Hydrogen is the lightest element with an atomic weight of 1.008（氢是最轻的元素，原子量为1.008），"with"省去不译；American Society for Metal（美国金属学会），"for"省去不译。

5. 省略连词

在英语中，连词只起连接作用。有些连词在逻辑关系上，在英语中是显性的，而在汉语中是隐性的，这时应省略。例如：When the air moves，it is called wind（空气流动产生风），"when"省去不译；Thermoplastic plastics become soft if they are heated（热塑性塑料受热会变软），"if"省去不译。

6. 省略某些词

原文中逻辑上或修辞上需要的词，如照译出来可能多余或重复，这时可以省略。例如：The treatment did not produce any harmful side effect（这种治疗没有产生副作用），副作用自然是有害的，如果译出"harmful"，则显得多余；Bacteria capable of causing disease are known as pathogenic，or disease-producing（引起疾病的细菌称为致病菌），"disease-producing"是对"pathogenic"的解释，不必译出，如果译出就会重复。

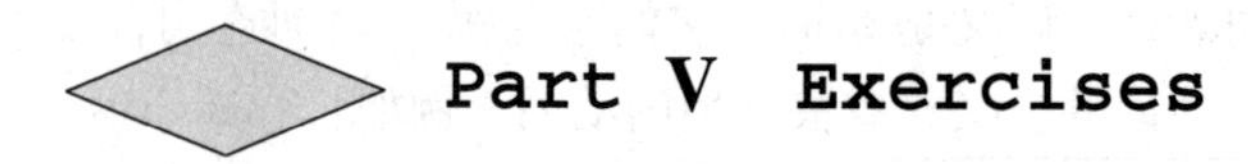

Part Ⅴ　Exercises

Ⅰ. Translate the following expressions into English.

（1）模铸	（2）初轧厂	（3）均热炉
（4）保温帽	（5）弧形连铸机	（6）棒塞

(7) 引锭杆	(8) 出料辊道	(9) 正方形断面方坯
(10) 固定铸模	(11) 传动带	(12) 连铸系统
(13) 初轧机组	(14) 较高的内表面质量	(15) 较好的工作条件

Ⅱ. Fill in the blanks with the words from the text. The first letter of the word is given.

(1) These ingots are rolled in primary mills to produce b ______ or slabs which subsequently receive further processing.

(2) A large amount of capital is i ______ in moulds, bottom plates, and locomotives.

(3) The l ______ the ingot, the more intense the segregation.

(4) Each ingot needs a discard to be removed at top and b ______ after primary rolling.

(5) The first continuous casting machines for steel were aligned v ______.

(6) In a typical plant, deoxidized steel from an e ______ furnace is run into a ladle, transported to the casting bay and brought to rest above the concast machine.

(7) At the exit of the straightening unit a roll is lowered which separates the d ______ bar from the length of cast steel.

(8) The strand of cast metal continues t ______ along the discharge rollers to a gas torch cut-off station.

(9) The evolution from one form to the next has been accompanied by a considerable reduction in o ______ height.

(10) Since the 1960s, continuous casting has steadily displaced i ______ casting.

Ⅲ. Fill in the blanks by choosing the right words form given in the brackets.

Research in recent years has helped to perfect the continuous casting of (1) ______ (melting; molten) steel directly into the form of slabs and billets. While the continuous casting of nonferrous metals has been (2) ______ (succeeded; successful) for many years, the continuous casting of steel has been difficult to (3) ______ (perform; be performed). The important traits of steel, such as a high melting point, high specific heat, and low heat conductivity, are contributing factors (4) ______ (for; to) the difficulty in continuous steel casting. The continuous steel casting has been developed for commercial use. It has been used to cast squares, called billets, and slabs. The casting of rounds, however, has presented more difficult technical problems. Continuous casing of rounds is quite (5) ______ (limiting; limited) at present. Stainless steel, tool steel, alloys, and plain carbon steel have all been cast in continuous billets.

Ⅳ. Decide whether the following statements are true or false (T/F).

(1) Since steel was first produced in a liquid state over 2000 years ago, it has been almost invariably the practice to cast it into rectangular blocks by ingot casting, from which the desired finished shape is obtained by subsequent hot or cold working. (   )

(2) This method of continuous-casting is characterized by the following:

1) A large amount of capital is invested in moulds, bottom plates, and locomotives;
2) … ( )

(3) Since the 1860s, continuous casting has steadily displaced ingot casting. ( )

(4) The reasons for ingot casting systems are:
1) Lower investment outlay compared with that for a blooming train;
2) More productivity than with conventional ingot-casting. ( )

(5) The dummy bar is either lifted to one side of the machine or hoisted on to the ramp above the machine. ( )

V. Translate the following English into Chinese.

The steel is either poured in continuous casting machines, or into ingot moulds whose capacity is governed by ladle size. Liquid steel contains dissolved oxygen which can be removed by adding strong defoliants such as aluminum and silicon. Three basic ingot styles are made depending on the made of solidification which is governed by melt composition and the extent of the use of defoliants. They are known as killed, rimmed and semi-killed. In the production of killed steel, virtually all the dissolved oxygen is removed and the liquid solidifies without any evolution of gas. In contrast to killed steels, the manufacture of rimmed steels takes advantage of the chemical reactions, which take place when a non-deoxidized steel is allowed to solidify. The dissolved oxygen combines with carbon to give carbon monoxide gas, which is released at the solidification front. The third class of product is called semi-killed steel and, as the term indicates, it is only partially deoxidized prior solidification.

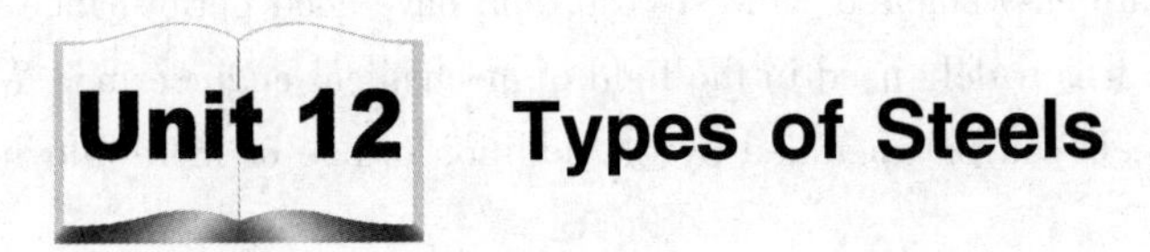

Unit 12 Types of Steels

Part Ⅰ Reading and Comprehension

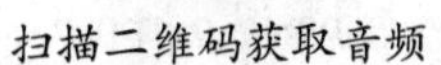

扫描二维码获取 PPT

Steel can be subdivided in non-alloy steel (plain carbon steels) and alloy steels according to chemical composition.

1. Non-alloy Steel

Steel theoretically is an alloy of iron and carbon. When produced commercially, however, certain other elements-notably manganese, phosphorus, sulfur, and silicon are present in small quantities. While the carbon content (by mass) is between 0.021% and 2.11%, the product is referred to as non-alloy steel. Its strength is primarily a function of its carbon content. When the carbon content (by mass) is less than 0.09%, the higher the carbon content, the higher the tensile strength and the greater the hardness to which the steel may be heat-treated. Unfortunately, the plasticity and toughness of plain-carbon steel decreases as the carbon content is increased, and its hardenability is quite low. In addition, the properties of ordinary carbon steels are impaired by both high and low temperature, and they are subject to corrosion in most environments.

Non-alloy steel are generally classed into three subgroups, based on carbon content. Low-carbon steels have less than 0.3 per cent carbon, possess good formability and weldability, but poor hardenability. Their structures usually are ferrite and pearlite, and the material generally is used as it comes from the hot-forming or cold-forming process. Low-carbon steels are usually used for low-strength parts requiring a great deal of forming. Medium-carbon steels have between 0.25 and 0.60 per cent carbon (by mass), and they are used for forgings and other applications where increased strength and a certain amount of ductility are necessary. The structure of low carbon steel and medium carbon steel is composed of ferrite group and pearlite group. However, with the increase of carbon content, the relative ferrite content in the structure decreases while the relative pearlite content increases. High-carbon steels have more than 0.60 per cent carbon (by mass), toughness and formability are quite low, but hardness and wear resistance are high. Severe quenches can form martensite, but hardenability is still poor. High-carbon steels are used for high-strength parts such as springs, tools, and dies.

Non-alloy steels are easy smelted, lowest-cost, and have good performance to meet the needs of many occasions, so it is widely used in the field of mechanical engineering. When improved property is required, steels can be upgraded by the addition of one or more alloying elements.

2. Alloy Steel

Alloy steels are formed by the addition of one or more of the following elements: chromium, nickel, molybdenum, vanadium, tungsten, titanium, and copper, as well as manganese [w(Mn) > 1.0%], silicon [w(Si) >0.5%], and small amounts of other alloying elements. Alloy steels are divided into two types: low alloy steels with under 10 per cent (by mass) of added elements and high alloy steels with over 10 per cent (usually between 15 and 30 per cent) of added elements.

Alloy steels have special properties determined by the mixture and the amount of other metals added. Alloy elements are added to steel for the following basic purposes:

(1) **Chromium.** Addition of chromium imparts hardness, strength, wear resistance, heat resistance, and corrosion resistance to steels. In 1913 Harry Brearley of Sheffield was experimenting with alloy steels for gun barrels, and among the samples which he threw aside as being unsuitable was one containing about 14 per cent (by mass) of chromium. Some months later he noticed that most of the steels had rusted, but the chromium steel remained bright. This led to the development of stainless steels. Chromium is the main alloying element of stainless steel, and the combination with nickel can significantly improve the oxidation resistance, heat resistance and corrosion resistance of steel.

(2) **Nickel.** Nickel is one of the oldest and most common of alloy elements. The addition of nickel increases strength, yield point, hardness, and ductility. It also increases the depth of hardening. Nickel increases corrosion resistance and is one of the major constituents of the "stainless" or corrosion-resisting steels. Nickel steels have been used for components in automobiles and other vehicles, including aircraft, and for a multitude of components such as gears, shafting which require great strength, and other heavy duty products.

(3) **Molybdenum.** Molybdenum is usually used in amounts of less than 0.25 per cent (by mass). Molybdenum, even in extremely small amounts, has considerable effect as an alloying element on the physical properties of steels. Molybdenum improves strength, hardenability, and wear resistance. Molybdenum steels are readily heat-treated, forged and machined.

(4) **Vanadium.** Vanadium is usually used in amounts of less than 0.25 per cent (by mass). As an alloying agent, vanadium improves fatigue strength, ultimate strength, yield point, toughness, and resistance to impact and vibration. Chromium- vanadium steels have good ductility and high strength. They are widely used in heavy forgings, axles, springs, and tools, depending upon their carbon content.

(5) **Tungsten.** Tungsten is used largely with chromium as a high-speed tool steel which contains (by mass) 14.00 to 18.00 percent tungsten and 2.00 to 4.00 percent chromium. This steel processes the characteristic of being able to retain a sharp cutting-edge even though heated to red-

ness in cutting.

Mo and W can be added to roll steel, so that the matrix will harden when the steel is tempered, and the heat resistance and wear resistance can be improved, such as $9Cr_2Mo$.

(6) **Titanium.** Titanium helps to improves strength, hardenability, and wear resistance.

(7) **Copper.** The addition of copper increases corrosion resistance, and improves machinability.

(8) **Manganese.** Manganese helps to reduce certain undesirable effects of sulphur by combining with the sulphur. It also combines with carbon to increase hardness and toughness. Manganese processes the property of aiding in increasing the depth of hardness penetration. It also improves the forging qualities by reducing brittleness at rolling and forging temperatures. Manganese alloys are used mostly for making axles, forgings, gears, shafts, and gun barrels.

(9) **Silicon.** A small amounts of silicon improves ductility. It is used largely to increase impact resistance when combined with other alloys.

In the GB/T 1299—2014 Mold Steel, new types of plastic mold steel were added. On the basis of $3Cr_2Mo$ and $3Cr_2MnNiMo$, 19 new materials, such as $4Cr_2MnlMoS$ and $8Cr_2MnWMoVS$, were added, the plastic mold steel system in China was preliminarily formed, such as $8Cr_2MnWMoVS$ for precision cold stamping die, 2CrNiMoMnV for large and medium-sized mirror plastic mold, etc.

Words and Expressions

扫描二维码获取音频

扫描二维码答题

subdivide [ˈsʌbdiˈvaid] *v.*	再分，细分
plain carbon steel	普碳钢
theoretically [θiəˈretikəli] *ad.*	理论上，理论地
commercially [kəˈməːʃəli] *ad.*	商业上，通商上
notably [ˈnəutəbəli] *ad.*	显著地，特别地
a function of	随……而变
tensile strength	抗拉强度，拉伸强度
percentage [pəˈsentidʒ] *n.*	百分率，百分比
hardenability [ˌhɑːdənəˈbiliti] *n.*	淬透性，可淬性
impair [imˈpɛə] *v.*	削弱，损害
be subject to	受……影响的，易于……
subgroup [ˈsʌbgruːp] *n.*	小群，子群
formability [fɔːməˈbiliti] *n.*	可模锻性，可成型性
weldability [weldəˈbiliti] *n.*	可焊性，焊接性
significant [sigˈnifikənt] *a.*	重大的，重要的

ferrite ['ferait] *n.*	铁素体
pearlite ['pə:lait] *n.*	珠光体
forging ['fɔ:dʒiŋ] *n.*	锻件
toughness ['tʌfnis] *n.*	韧性，坚韧
quench [kwentʃ] *n.*	淬火
v.	淬火
severe [si'viə] *a.*	剧烈（猛烈）的
martensite ['mɑ:tənzait] *n.*	马氏体
limitation [ˌlimi'teiʃən] *n.*	限制，局限性
restrictive [ris'triktiv] *a.*	限制性的
upgrade ['ʌpgreid] *v.*	使……升级，提升
spring [spriŋ] *n.*	弹簧，弹性
die [dai] *n.*	硬模，冲模
chromium ['krəumjəm] *n.*	铬（元素符号 Cr）
molybdenum [mə'libdinəm] *n.*	钼（元素符号 Mo）
vanadium [və'neidiəm] *n.*	钒（元素符号 V）
tungsten ['tʌŋstən] *n.*	钨（元素符号 W）
springiness ['spriŋinis] *n.*	富于弹性，弹性
resistance [ri'zistəns] *n.*	抵抗力，阻力
barrel ['bærəl] *n.*	管状物，圆桶状物
gun barrel	枪管
axle ['æksl] *n.*	轮轴，车轴
bolt [bəult] *n.*	螺栓
stud [stʌd] *n.*	双头螺栓，柱头螺栓
automobile ['ɔ:təməbi:l] *n.*	［美］汽车，车辆
chisel ['tʃizl] *n.*	凿子
drill [dril] *n.*	钻头，锥子
file [fail] *n.*	锉刀
shear [ʃiə] *n.*	大剪刀，剪床
yield [ji:ld] *n.*	屈服（点）
yield point	屈服点
constituent [kən'stitjuənt] *n.*	要素，组成部分，成分
vehicle ['vi:ikl] *n.*	交通工具，车辆
multitude ['mʌltitju:d] *n.*	大量，众多
shafting ['ʃɑ:ftiŋ] *n.*	轴系，制轴材料
heavy duty product	重型产品，大型产品
physical properties	物理性能
fatigue [fə'ti:g] *n.*	疲劳
ultimate ['ʌltimit] *a.*	最后的，根本的

ultimate strength	极限强度
vibration [vaiˈbreiʃən] *n.*	振动，颤动
retain [riˈtein] *v.*	保持，保留
cutting-edge	刀口，刃
penetration [peniˈtreiʃən] *n.*	穿过，渗透
brittleness [ˈbritlnis] *n.*	脆性，脆度
gear [giə] *n.*	齿轮，传动装置
shaft [ʃɑːft] *n.*	传动轴，杆状物
impact resistance	冲击阻力

✲ Answer the following questions.

(1) How many subgroups are plain-carbon steels generally classed into?

(2) What are the structures of plain-carbon steels?

(3) What are low-carbon steels usually used for?

(4) What are high-carbon steels used for?

(5) How are alloy steels formed?

(6) What element is one of the oldest and most common of alloy elements?

(7) How much carbon do medium-carbon steels contain?

(8) How many types are alloy steels divided into?

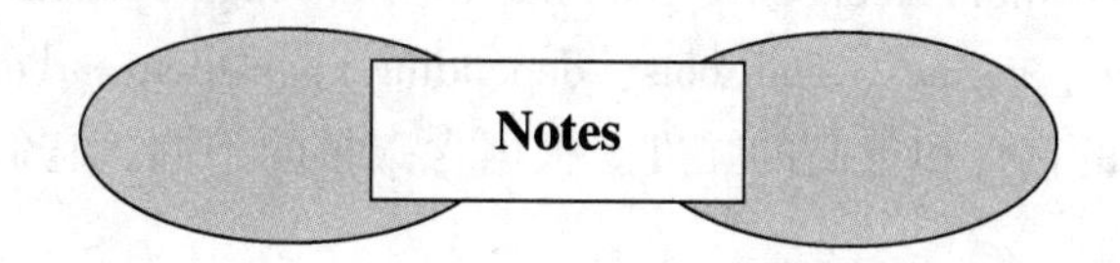

(1) When these four foreign elements are present in their normal percentages, the product is referred to as plain-carbon steel.

当这四种杂质元素以通常的比例存在时，这一产品就称为普碳钢。

“(be) referred to as” 意思是“叫作，称为”。例如：This near-stoichiometric operation is sometimes referred to as one-step reforming（这种接近化学计算比的工艺过程有时被称为一步转化）。

(2) The higher the carbon content, the higher the tensile strength and the greater the hardness to which the steel may be heat-treated.

含碳量越高，抗拉强度就越高，并且热处理后所达到的硬度也越高。

“to which the steel may be heat-treated” 是定语从句，修饰“the hardness”。

(3) In addition, the properties of ordinary carbon steels are impaired by both high and low temperature, and they are subject to corrosion in most environments.

此外，在过高和过低的温度之下，普碳钢的性能也会降低。多数环境下，它们还容易受到腐蚀。

“be subject to” 意思是“易遭受，易发生”。例如：They were subject to great suffering

(他们遭受了巨大痛苦); The prices are subject to change (价格可能有变动)。

(4) Alloy steels have special properties determined by the mixture and the amount of other metals added.

合金钢之所以具有各种特殊的性能，取决于所加入的金属的种类及其数量。

(5) In 1913, Harry Brearley of Sheffield was experimenting with alloy steels for gun barrels, and among the samples which he threw aside as being unsuitable was one containing about 14 per cent of chromium.

1913 年，谢非尔德的哈里 · 布雷尔利试用合金钢制作枪筒，在他认为不适用而丢在一边的试样当中，有一种含铬 14%的钢。

"which he…unsuitable" 是定语从句，修饰 "samples"。"containing about 14 per cent of chromium" 是分词短语，修饰 "one"。

(6) Nickel steels have been used for components in automobiles and other vehicles, including aircraft, and for a multitude of components such as gears, shaftings which require great strength, and other heavy duty products.

镍钢可以被用来制造汽车、其他运载工具，包括飞机，以及大量高强度齿轮、轴系部件和其他重型产品。

"be used for" 意思是 "用于"，例如：A hammer is used for driving in nails (锤子是用来钉钉子的)。"a multitude of" 意思是 "大量的"，例如：A multitude of thoughts filled her mind (她头脑里充满大量的各种想法)。

(7) Chromium-vanadium steels have good ductility and high strength. They are widely used in heavy forgings, axles, springs, and tools, depending upon their carbon content.

铬钒钢具有很好的韧性和很高的强度。根据含碳量的不同，它们广泛用于大型铸件、轮轴、弹簧和工具中。

Part Ⅱ　Curriculum Ideological and Political

【National pillar】Five-hundred-meter aperture spherical radio telescope (FAST)

The scientific name of "China Sky Eye" is a five-hundred-meter aperture spherical radio telescope. It is the world's largest single-aperture radio telescope and owns our country's independent intellectual property rights. It is mainly used to realize scientific objectives such as patrolling neutral hydrogen in the universe, observing pulsars and other application objectives such as space vehicle measurement and communication. In 2016, "China Sky Eye" was completed. It was officially launched through national acceptance in January 2020, and was open to the world at the end of March 2021. By July 2022, the five-hundred-meter aperture spherical radio telescope has discovered more than 660 new pulsars, entering a period of achievement explosion, which has greatly expanded the limit of human observation of the universe.

Part Ⅲ Further Reading

New steelmaking technology

1. Pretreatment technology of high efficiency desulfurization hot metal

The high efficient hot metal pretreatment method of strong stirring and blowing is developed to reduce the sulfur content to a very low level in a short time.

2. Manufacture of high heat input welding steel by oxide metallurgy technology

Carbon manganese steel, HSLA, high strength steel, etc. for large heat input welding can be developed by using oxide metallurgy technology. Contrary to the traditional ideas of "pure purification" and "clean purification", this technology improves the structure of steel in the subsequent solidification, rolling, cooling and use processes by effectively controlling the properties (distribution, composition, size, etc.) of inclusions in the steelmaking process, so as to obtain the required structure and new properties, such as steel for large heat input welding.

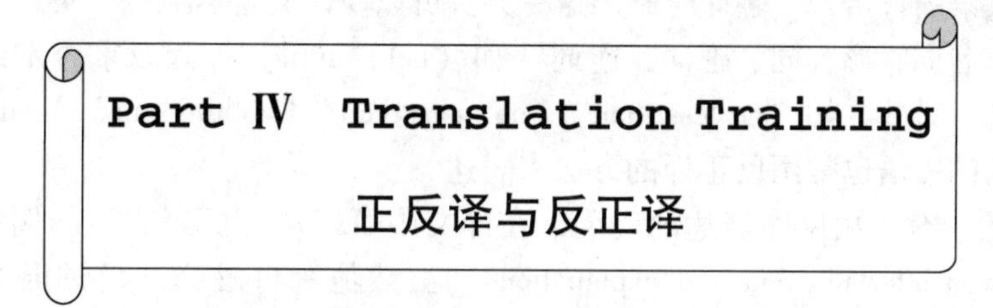

1. 正反译

正反译是指英语从正面表达，译文从反面表达。每一种语言都有自己在表达否定含义时的独到之处，以英语为母语的人与以汉语为母语的人对否定的思维方法和表达方法很不相同。所谓正反译，就是用变换的表达方法把原文的肯定式译为汉语的否定式。例如：招贴用语"Wet paint"可译为"油漆未干"，如果译成"湿油漆"就不符合汉语的习惯。"Keep off the grass"译为"不要践踏草坪"。翻译时运用这一方法可以使译文合乎汉语的规范，更恰当地表达原文的意思。

（1）英语中含否定意义的动词，如"fail""lack""refrain""refuse""neglect""ignore"等，译成汉语时用否定表达，例如：Galileo, among others, recognized the problem, but failed to solve it（伽利略和其他人认识到这一问题，但没能加以解决）。肯定式"fail (to do)"译为否定表达"没能"，例如：They refused us admittance（他们不让我们进去）。肯定式"to refuse"译为否定表达"不让"。

（2）英语中含否定意义的名词，如"absence""failure""ignorance""lack"等。例如：This lack of uniformity of the force of gravity results from a number of factors（引力这种非均匀性是由许多因素引起的），肯定式"lack of uniformity"译为否定表达"非均匀性"；I had to put on the calm act to prevent my worry from spreading（我故作镇静，以免我的不安心情影响他们），肯定式"worry"译为否定表达"不安心情"。

（3）英语中含否定意义的形容词，如“few”“ignorant”“little”“free from”“far from”“short from”等。例如：There is little carbon in this alloy steel（这种合金钢中没多少碳）；How can we obtain water free from these materials?（我们怎样才能得到没有这些物质的水呢?）；The medium onto which the jet strikes becomes inelastic（气流所冲击的介质变为非弹性）。

（4）英语中的一些连词，如“unless”“before”“untill”“（rather）than”“lest”“or”等。例如：A change is not a chemical change unless it results in forming a new substance（一种变化要是不导致生成新的物质，就不是化学变化）；She would rather die than surrender（她宁可死也不投降）。

（5）英语中的一些介词及其短语，如“without”“above”“except”“beyond”“instead of”“in place of”等。例如：It was beyond his power to sign such a contract（他无权签这种合同）；The problem is above me（这个问题我解决不了）。

2. 反正译

反正译是指英语从反面表达，译文从正面表达。反正译是用变换的表达方法把原文的否定式译成汉语的肯定式，翻译时利用这一方法可以使译文自然流畅。英语中一些含否定意义的动词、名词、形容词、副词、连词［如（not）until］等和双重否定结构，常用这种方法来翻译。另外，如“no less than”“no more than”“nothing but”“can not…too”等含有否定结构的短语也常用反正译的方法来表达。

（1）英语中有些从反面表达的动词，在译文中可以从正面来表达。例如：The doubt was still unsolved after his repeated explanations（虽然他一再解释，疑团仍然存在）；The ships seem to have disappeared off the face of the earth（那些轮船似乎已经从地球表面消失了）。

（2）名词。例如：He manifested a strong dislike for his father's business（他对他父亲的行业表现出强烈的厌恶情绪）；Try to control your impatience when any unexpected problem arises（当出现没有预料到的问题时，要尽量控制住你的急躁情绪）。

（3）形容词。例如：All the articles are untouchable（博物馆内一切展品禁止触摸）；The effects of some medicine are not immediate（某些药疗效缓慢）。

（4）副词。例如：He carelessly glanced through the note and got away（他马马虎虎地看了那张纸条就走了）；“I don't know if I ought to have come .” She said breathlessly（“我不知道我是否该来。”她气喘吁吁地说）。

（5）连词。“（not）until”意思是“直到……才，到……之后才”。例如：Metals don't melt until it is heated to a definite temperature（金属加热到一定温度才会熔化）；The machines must not used until they are properly tested（这些机器进行适当的测试之后才能使用）。

（6）含否定结构的短语。例如：She is no less active than she used to be（她和以前一样活跃）；Nothing but a miracle can save her now（现在只有出现奇迹她才能得救）；You cannot be too careful（你越仔细越好）。

（7）英语中的双重否定有时可译为肯定。英语中常常两个否定词并用，构成一个肯定的意思，译成汉语时若保留原来的“否定之否定”但不能得到流畅的译文，则可采用肯定形式。例如：There can be no sunshine without shadow（有阳光就有阴影）；He is not seldom ill（他常常生病）。

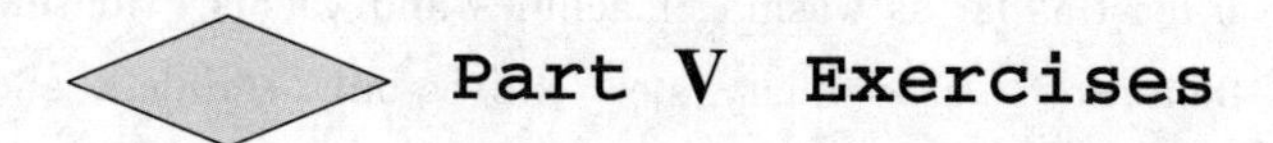

Part Ⅴ Exercises

Ⅰ. Translate the following expressions into English.

（1）抗拉强度	（2）中碳钢	（3）轮轴
（4）耐磨性	（5）力学性能	（6）铬合金
（7）屈服点	（8）耐蚀钢	（9）钼钢
（10）疲劳强度	（11）高速工具	（12）锋利刀刃
（13）锻造质量	（14）枪管	（15）普碳钢

Ⅱ. Fill in the blanks with the words from the text. The first letter of the word is given.

（1）Steel grades can be subdivided according to chemical c ______ and service properties.

（2）Steel theoretically is an alloy of iron and c ______.

（3）Unfortunately, the ductility of plain-carbon steel decreases as the carbon content is i ______, and its hardenability is quite low.

（4）Low-carbon steels have less than 0.3 per cent carbon, possess good formability and w ______, but not enough hardenability to be hardened to any significant depth.

（5）Low-carbon steels are usually used for low-strength parts r ______ a great deal of forming.

（6）High-carbon steels have more than 0.6 per cent carbon, toughness and f ______ are quite low, but hardness and wear resistance are high.

（7）When improved material is required, steels can be u ______ by the addition of one or more alloying elements.

（8）Alloy steels have special p ______ determined by the mixture and the amount of other metals added.

（9）Chromium alloys are commonly used in axles, bolts, s ______, and studs for automobiles.

（10）Nickel is one of the oldest and most common of a ______ elements.

（11）Molybdenum improves s ______, hardenability, and wear resistance.

（12）Vanadium is usually used in a ______ of less than 0.25 per cent.

Ⅲ. Fill in the blanks by choosing the right words form given in the brackets.

Alloy steels are divided into two types: low alloy steels with under 10 per cent of added elements and high alloy steels（1）______（with; within）over 10 per cent（usually between 15 and 30 per cent）of added elements. During the past hundred years, alloy steels have entered more and more into the manufacture of highly stressed components or parts（2）______（what;

which) have to work in corrosive (3) ______ (conditioning; conditions), from the military uses of heavy guns and armour plate, and in the construction of bridges and bicycles, right through the alphabet at least as far as washing machines and yachts. Housewives (4) ______ (turned; turn) up their noses at ordinary steel knives-only stainless alloy steel cutlery is (5) ______ (acceptable; accepted). In order to get the full value of the improvements by the alloying elements, these steels are generally (6) ______ (using; used) in the heat-treated condition. Carbon is an essential constituent of the steels which (7) ______ (are; is) heat-treated, for (8) ______ (with; without) carbon they could not attain their useful mechanical properties; the carbon makes hardening and (9) ______ (tempered; tempering) possible, the alloying elements modify the effect of the (10) ______ (treat; treatment). Various combinations of strength, hardness, springiness and toughness may be achieved by the selection of times and temperatures of heat-treatment for a particular alloy composition.

Ⅳ. Decide whether the following statements are true or false (T/F).

(1) According to chemical composition there are stainless carbon steels and alloy steels. (　　)

(2) Plain-carbon steels are generally classed into three subgroups (low-carbon steels, medium-carbon steels, and high-carbon steels), based on carbon content. (　　)

(3) The addition of nickel increases strength, yield point, hardness, and ductility. (　　)

(4) Titanium helps to low strength, hardenability, and wear resistance. (　　)

(5) Manganese alloys are used mostly for making axles, forgings, gears, shafts, and gun barrels. (　　)

Ⅴ. Translate the following English into Chinese.

The prime function of an alloy element in steel is to improve the physical and mechanical properties of the steel; that is, to increase its corrosion resistance, hardness, wearability, or strength. Since metallurgy became a science, more than one thousand alloys have been created. Each of these alloys has unique properties.

Unit 13 Pyrometallurgical Extraction of Copper from Sulphide Ores

Part Ⅰ Reading and Comprehension

扫描二维码获取音频

扫描二维码获取 PPT

Copper is present in the earth's curst mainly in the form of sulphide minerals, often in conjunction with iron sulphide FeS, e. g.

Chalcocite	Cu_2S	(black)
Chalcopyrite	$CuS \cdot FeS$	(yellowish)
Bornite	$CuS \cdot FeS$	(bluish)

The concentration of the above minerals in an ore-body is low, typical copper contents (by mass) being about 0. 5%~2%.

About 90% of the world's primary copper comes from sulphide ores. The main impurity is iron, which may be accompanied by small amounts of Ni, Au and Ag and traces of Zn, Sn, Pb, Co, As, Sb, Se, Te and Bi (Low-grade ores containing oxidised minerals also occur, but these are nearly always treated by hydrometallurgical methods). Some of the important sources of copper are, Zambia, Katanga (Congo), Chile, Canada, U. S. A (Montana) and Australia.

1. Ore Preparation

As mentioned above, the sulphide ores of copper are lean, with a copper content (by mass) as low as 0. 5%, and froth flotation is used to produce a concentrate containing 20% ~ 30% Cu, after crushing and grinding of the ore.

2. Extraction

At one time copper sulphide concentrate was completely roasted to copper oxide which was then reduced by carbon to the metal. This method was discontinued because the copper produced was very impure, and loss of metal to the slag was very high. The method now used depends on the following relationships at high temperature:

(1) Copper has a greater affinity than iron for sulphur. Iron has a greater affinity than copper for oxygen.

(2) Iron can thus be separated from copper by oxidation to form iron oxide, followed by combination with silica to form iron silicate slag.

There are two main stages in the pyrometallurgical extraction of copper:

(1) Matte smelting to form a molten sulphide melt, which contains all the copper of the charge, and a molten slag free from copper.

(2) Conversion of the matte into blister copper. As a preliminary to matte smelting, partial roasting of the concentrate may be carried out in order to decrease the sulphur content. This operation may be conducted in a multi-hearth or fluo-solids roaster. Which ever roasting unit is used, it is important to ensure that all the copper and part of the iron remain as sulphides in order to generate heat during the converting of the matte.

3. Matte Smelting

As stated above, the reverberatory furnace is being superseded by flash-furnace smelters for matte smelting. The dried, fine concentrate (or calcine) is blown into the furnace, where it is combusted with oxygen or oxygen-enriched air. During the smelting some of the iron sulphide is oxidised to iron oxide and fluxed with silica to form iron silicate slag:

$$2FeS + 3O_2 = 2FeO + 2SO_2$$

$$2FeO + SiO_2 = 2FeO \cdot SiO_2$$

The rest of the iron remains as FeS, which mixes perfectly with copper sulphide to form a mixture of molten sulphide (matte). The slag is lighter than matte, and immiscible with the matte and is tapped off separately. The matte containing (by mass) 25% ~ 50% Cu together with any precious metals (e. g. Au, Ag) present in the ore, is run off into a converter. The combustion reaction provides most of the heat required for heating and melting the matte and slag. The concentration of SO_2 in the effluent gases is high (usually more than 10%) and the SO_2 can be used to produce sulphuric acid, elemental sulphur or liquid SO_2.

4. Conversion of the Matte into Blister Copper

Converting consists of oxidising the molten matte with air or oxygen, resulting in the removal of iron and sulphur and the production of a crude blister/copper containing about 98% Cu (by mass).

Molten matte is poured into a horizontal converter and air is introduced through the tuyeres. The heat generated by the oxidation of the FeS is sufficient to make the process autogenous. Converting is carried out in two stages:

(1) The FeS elimination or slag-forming stage. Since FeO is more stable than Cu_2O, the FeS is oxidized in preference to Cu_2S and the FeO formed is fluxed with SiO_2 to form iron silicate slag.

Any Cu_2O present will react with FeS as follow:

$$Cu_2O + FeS = Cu_2S + FeO$$

The slag is poured off by tilting the converter and more matte is added: the process is repeated until the converter is nearly full of molten copper sulphide.

(2) The blister copper-forming stage. When all the iron sulphide has been oxidised, copper production begins. Firstly, Cu_2S is oxidised to Cu_2O:

$$2Cu_2S + 3O_2 = 2Cu_2O + 2SO_2$$

then the reaction takes place:

$$Cu_2S + 2Cu_2O = 6Cu + SO_2$$

The crude copper is about 98% pure and contains 0.02% ~ 0.1% sulphur (by mass). If allowed to solidify, the metal is porous and blisters on the surface caused by the evolution of SO_2.

Any precious metals present in the matte are not oxidised during conversion and enter the blister copper. The blister copper may then be purified by fire-refining.

5. Fire-refining of Blister Copper

This involves controlled oxidation of impurities, followed by deoxidation of the copper.

Air is blown into the molten blister copper in a reverberatory furnace, and the more reactive impurities (e. g. Fe, Pb, Zn) are oxidised and removed to the slag. The addition of sodium carbonate, sodium nitrate and lime to the slag helps to remove other impurities such as As, Sb and Sn. The copper also starts to oxidise and air blowing is continued until the oxygen content of the copper reaches about 0.9% (by volume) in order to ensure satisfactory removal of sulphur. However, copper having such a high oxygen content would be brittle so that deoxidation is required.

The slag is removed and the copper is stirred with tree trunks——an operation which is called ***poling***—— the hydrocarbons evolved from the green wood reducing the copper oxide. Poling is a critical operation in which the aim is to bring the oxygen content (by volume) down to 0.03% ~ 0.06%. If the oxygen content is not lowered enough, the copper is mechanically weak, while if taken too low the solidified copper is porous due to the formation of steam pockets caused by the interaction of Cu_2O and reducing gases during solidification:

$$Cu_2O + H_2 = 2Cu + H_2O$$

After poling, the copper is said to be in the ***tough pitch*** condition. Tough pitch copper contains 0.03% ~ 0.06% oxygen (by mass) and is very suitable for working. If the copper is over-poled it must be re-oxidised and re-poled.

Final deoxidation of fire-refined copper may be achieved by adding phosphorous in the form of Cu-14%P alloy. However, the small amount of phosphorus [w(P) = 0.05%] remaining in the Cu adversely affects the electrical conductivity and, to avoid this, lithium may be used as the deoxidant.

Fire-refining produces copper which is about 99.5% pure, and electro-refining is necessary to obtain higher purity.

✲ Words and Expressions

扫描二维码获取音频

扫描二维码答题

pyrometallurgical [pairɔˌmetəˈləːdʒikəl] *a.*　火法冶金的
crust [krʌst] *n.*　[地质] 地壳，表层
in conjunction with　与……一道；结合
chalcocite [ˈkælkəsait] *n.*　[矿] 辉铜矿
chalcopyrite [ˌkælkəˈpaiərait] *n.*　[矿] 黄铜矿
bornite [ˈbɔːnait] *n.*　[矿] 斑铜矿
yellowish [ˈjeləuiʃ] *a.*　微黄色的
bluish [ˈbluːiʃ] *a.*　浅蓝色的，带蓝色的
concentration [ˌkɔnsenˈtreiʃən] *n.*　浓缩，浓度，含量
accompany [əˈkʌmpəni] *v.*　伴随，陪伴
trace [treis] *n.*　少许，有点
lean [liːn] *a.*　贫乏的，歉收的
froth [frɔθ, frɔːθ] *n.*　渣滓，废物
flotation [fləuˈteiʃən] *n.*　浮选
grinding [ˈgraindiŋ] *n.*　研磨，制粉
roast [rəust] *v.*　煅烧，焙烧
impure [imˈpjuə] *a.*　不纯的，杂质
affinity [əˈfiniti] *n.*　亲和力，吸引力
matte [mæt] *n.*　锍，冰铜
conversion [kənˈvəːʃən] *n.*　变换，转化
blister [ˈblistə] *n.*　粗铜，气泡
v.　产生气泡
preliminary [priˈliminəri] *a.*　预备的，初步的
n.　准备工作
partial [ˈpɑːʃəl] *a.*　部分的，局部的
multi-hearth　多膛焙烧炉
reverberatory [riˈvəːbərətəri] *a.*　反射的
n.　反射炉
supersede [ˌsjuːpəˈsiːd] *n.*　代替，取代
flash-furnace　闪速熔炼炉，闪速炉

smelter ['smeltə] *n.* 熔炉，熔炼工
immiscible [i'misəbl] *a.* 不能混合的
precious ['preʃəs] *a.* 贵重的，宝贵的
tap off 分出，抽出
effluent ['efluənt] *a.* 发出的，流出的
horizontal converter 卧式转炉
autogenous [ɔː'tɔdʒinəs] *a.* 自生的，自体的
flux [flʌks] *v.* 熔化，使……熔解
pour off 倒出
porous ['pɔːrəs] *a.* 多孔的，有气孔的
sodium ['səudiəm] *a.* 可浸渍的
nitrate ['naitreit] *n.* [化] 钠（元素符号 Na）
[化] 硝酸盐，硝酸根
stir [stəː] *v.* 搅和，搅拌
green wood 生（湿）木材
trunk [trʌŋk] *n.* 树干，躯干
poling ['pəuliŋ] *n.* 插树还原，还原，除气
mechanically [mi'kænikəli] *ad.* 机械地
pocket ['pɔkit] *n.* 凹处，小块地区
interaction [ˌintər'ækʃən] *n.* 互相作用，互相影响
pitch [pitʃ] *n.* 树脂，沥青
tough pitch copper 火法精炼铜，工业纯铜
pole [pəul] *v.* （青木）还原（除气）
over-pole ['əuvəpəul] *v.* 过度还原
phosphorous ['fɔsfərəs] *a.* 磷的，含三价磷的
adversely ['ædvəːsli] *ad.* 逆地，反对地
lithium ['liθiəm] *n.* [化] 锂（元素符号 Li）
electro-refining 电解精炼
fire-refining 火法精炼

✲ Proper Names

Zambia ['zæmbiə] *n.* 赞比亚
Katanga [kə'tæŋgə] *n.* 加丹加（扎伊尔沙巴地区 Shaba 的旧名）
Congo ['kɔŋgəu] *n.* 刚果
Montana [mɔn'tænə] *n.* 蒙大拿州（美国州名）

✻ Answer the following questions.

(1) What ores does about 90% of the world's primary copper come from?

(2) How many main stages are there in the pyrometallurgical extraction of copper?

(3) What does converting consist of ?

(4) When does copper production begin?

(5) Where are some of the important sources of copper?

(1) Copper is present in the earth's crust mainly in the form of sulphide minerals, often in conjunction with iron sulphide FeS.

铜主要是以硫化矿物的形式存在于地壳中，并常常与硫化亚铁并存。

"in conjunction with sb. /sth. (together with sb. /sth.)" 意思是"与某人（某事物）一道，结合"。例如：This section should be studied in conjunction with the preceding three（这一节应连同前三节一起研读）；We are working in conjunction with the police（我们与警方配合进行工作）。

(2) The main impurity is iron, which may be accompanied by small amounts of Ni, Au and Ag and traces of Zn, Sn, Pb, Co, As, Sb, Se, Te and Bi (Low-grade ores containing oxidised minerals also occur, but these are nearly always treated by hydrometallurgical methods).

其主要杂质是铁，可能还伴有少量的镍、金、银和微量的锌、锡、铅、钴、砷、锑、硒、碲、铋（低品位矿石中也含有氧化物，但可以用湿法冶金来去除这些杂质）。

"which may be accompanied…" 是非限制性定语从句，对主句做进一步说明。

(3) Copper has a greater affinity than iron for sulphur. Iron has a greater affinity than copper for oxygen.

铜和硫的亲和力比铁大，铁和氧的亲和力比铜大。

"have an affinity for" 意思是"对……有吸引力，对……有亲和力，喜爱"。例如：She feels a strong affinity for him（她对他很有吸引力）；He has a strong affinity for Beethoven（他酷爱贝多芬的乐曲）；Salt has an affinity for water（盐对水有亲和力）。

(4) Since FeO is more stable than Cu_2O, the FeS is oxidized in preference to Cu_2S and the FeO formed is fluxed with SiO_2 to form iron silicate slag.

由于 FeO 比 Cu_2O 更稳定，因此 FeS 优先于 Cu_2S 氧化，形成的氧化铁与二氧化硅熔融形成硅酸铁渣。

"in preference to sb. /sth." 意思是"优先于某人（某事物），而不要某人（某事物）"。例如：She chose to learn the violin in preference to the piano（她愿学小提琴而不学钢琴）。

(5) …, while if taken too low the solidified copper is porous due to the formation of steam pockets caused by the interaction of Cu_2O and reducing gases during solidification.

但是如果含氧量过低，那么铜凝固后将会是多孔的，因为在凝固过程中还原性气体和氧化亚铜会相互作用而形成气泡。

“due to（because of）”意思是“因为，由……引起，由于”。例如：The accidents happened due to driving at high speed（发生这些交通事故是由于高速开车）；Her illness was due to bad food（她的病是坏了的食物造成的）。

Part Ⅱ Curriculum Ideological and Political

【Looking back on metallurgical history】The Bronze age

The Bronze Age in China lasted more than 1500 years, from the Xia through the Shang and Western Zhou dynasties to the Spring and Autumn Period. Large numbers of unearthed artifacts indicate a high of level of ancient bronze civilization in the country. They feature rich political and religious themes and are of high artistic value. A representative example is the Great Ding for Yu which is now preserved in the Museum of Chinese History. It was cast 3000 years ago during the reign of King Kang of Western Zhou Dynasty.

Part Ⅲ Further Reading

Working guidance for carbon dioxide peaking and carbon neutrality in full and faithful implementation of the new development philosophy—Optimizing and upgrading industrial structures

We will move faster to promote green agricultural development and improve carbon sequestration and efficiency in agriculture. We will create implementation plans for industries and fields including energy, steel, non-ferrous metals, petrochemicals, building materials, transportation and construction. Based on the goals of energy conservation and carbon reduction, we will revise the Catalog for Guiding Industry Restructuring. Authorities will be “looking back” to inspect steel and coal facilities that have cut overcapacity in order to consolidate achievements in this area. We will accelerate innovation in low-carbon industrial processes and the digital transformation of the industrial sector. We will launch the construction of demonstration zones for peaking carbon dioxide emissions. In fields such as goods distribution and information services, green transformations will be accelerated, and low-carbon development will be enhanced in the service sector.

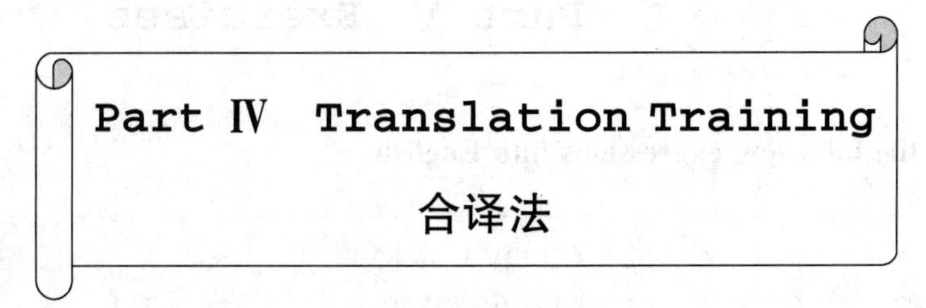

英语和汉语两种语言结构不完全相同，有时套用原文句式往往是行不通的。当原文表

层结构与汉语表达方式不一致时，需要进行句子结构的调整，使译文和原文内容一致。合译指根据译语表达需要，将原作的表达方式进行整合，包括融合、重铸和合并。

1. 合译单句

（1）单句内部融合。单句内部融合表现为某些成分的融合。例如：Electron emission is obtained at temperatures near 1000 degrees centigrade（温度接近1000℃时发射电子），"Emission"与"obtained"在意义上融合起来，在语句上是主谓的融合，显得译文言简意赅；By comparison there occurred an 89 per cent reduction in print-circuit-board area（比较起来，印刷电路板的面积减少了89%），"occurred"与"reduction"融合起来，相当于后者的动词含义。

（2）整个单句重铸。整个单句重铸指完全撇开原句的形式，将其意义用汉语表达出来。例如：The air offers resistance to any body moving through it（在空气中运动的任何物体都要受到阻力）；During the cold working there is a gain in strength（冷加工会提高强度）。

（3）单句合并。当英语中两个或两个以上的简单句关系密切、意义贯通时，可不限于原文的表层结构，将它们合译成一个汉语单句。合译时可使译文的句子意义完整，并避免不必要的重复。例如：There are men here from all over the country. Many of them are from the South（从全国各地来到这里的人中许多是南方人）。He would miss many things and many people. He would miss Celia（他会想念许多往事、许多朋友，也会想念西莉亚）。

2. 合译复合句

将原文中的复合句译成汉语的一个简单句，可以使译文紧凑、简练、语气连贯，合乎汉语的习惯。

（1）合译主从复合句。例如：The temperature at which certain solid melts is known as the melting point of that solid（固体熔化的温度称为熔点），将原文的限制性定语从句译成汉语的"的"字结构；When the metal is heated, it expands（金属受热膨胀），原句主句和从句的主语相同，翻译时合并成一句；The unit should be run at least 10min each week while it is idle（该装置至少每周要空转10分钟），"run"与"while it is idle"融合，成为机器的专有名词"空转"。

（2）合译并列复合句。例如：The time was 10：30, and traffic on the street was light（10点30分时，街上来往车辆稀少）；In 1844 Engels met Marx, and they became friends（1844年，恩格斯与马克思相遇并成为朋友）。

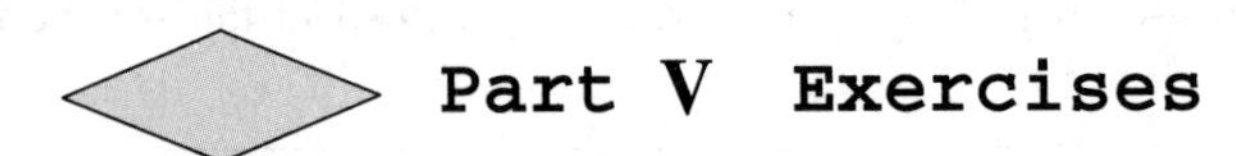

Part V Exercises

Ⅰ. Translate the following expressions into English.

（1）粗铜	（2）生（湿）木材	（3）硫酸
（4）黄铜矿	（5）流态化焙烧炉	（6）氧化铜
（7）卧式转炉	（8）反射炉	（9）电解精炼

Ⅱ. Fill in the blanks with the words from the text. The first letter of the word is given.

(1) I ______ has a greater affinity than copper for oxygen.

(2) As a p ______ to matte smelting, partial roasting of the concentrate may be carried out in order to decrease the sulphur content.

(3) If allowed to solidify, the metal is p ______ and blisters on the surface caused by the evolution of SO_2.

(4) The slag is removed and the copper is stirred with tree t ______—an operation which is called poling.

(5) After p ______, the copper is said to be in the "tough pitch" condition.

(6) Fire-refining produces copper which is about 99.5% pure, and electro-refining is necessary to obtain h ______ purity.

(7) If the oxygen content is not lowered enough, the copper is mechanically w ______.

(8) C ______ is present in the earth's curst mainly in the form of sulphide minerals, often in conjunction with iron sulphide FeS, e. g. Chalcocite, Chalcopyrite, Bornite.

Ⅲ. Fill in the blanks by choosing the right words form given in the brackets.

The slag is (1) ______ (remove; removing; removed) and the copper is stirred with tree trunks—an operation which is called ***poling***—the hydrocarbons evolved from the green wood reducing the copper oxide. Poling is a critical operation in which the aim is (2) ______ (to bring; bring; brought) the oxygen content down to 0.03%~0.06%. If the oxygen content is not lowered enough, the copper is mechanically weak, while if taken too low the solidified copper is porous due to the formation of steam pockets caused by the interaction of Cu_2O and (3) ______ (reducing; reduce; reduced) gases during solidification:

$$Cu_2O + H_2 = 2Cu + H_2O$$

After poling, the copper is said to be in the ***tough pitch*** condition. Tough pitch copper contains 0.03%~0.06% oxygen and is very (4) ______ (suit; suitable; suited) for working. If the copper is over-poled it must be re- (5) ______ (oxidising; oxidise; oxidised) and re-poled.

Ⅳ. Decide whether the following statements are true or false (T/F).

(1) About 90% of the world's primary copper comes from oxide ores. ( )

(2) During the smelting some of the iron sulphide is oxidised to iron oxide and fluxed with silica to form iron silicate slag:

$$2FeS + 3O_2 = 2FeO + 2SO_2$$

$$2FeO + SiO_2 = 2FeO \cdot SiO_2$$

( )

(3) Molten matte is poured into a horizontal converter and air is introduced through the tuyeres. The heat generated by the oxidation of the FeS is sufficient to make the process autogenous. ( )

(4) Copper having such a low oxygen content would be brittle so that deoxidation is required. ( )

(5) The small amount of phosphorus remaining in the Cu (0.05%) adversely affects the electrical conductivity and, to avoid this, lithium may be used as the deoxidant. ( )

V. Translate the following English into Chinese.

For the lower boiling point metals as Pb and Zn, a blast furnace of rectangular cross section is used with tuyeres placed only along the two long sides. This produced a less severe built-up in temperature at the tuyeres and reduces the danger of metal loss through volatilisation of the low boiling point metals. When the blast furnace is used for matter smelting of metals sulphides, only sufficient coke is charged to provide the necessary furnace atmosphere and heat since a reduction reation is not required. The success of the matte smelting operation is judged by the fall and grade of the matte.

Unit 14 Production of Aluminium

Part Ⅰ Reading and Comprehension

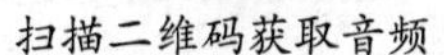

扫描二维码获取音频

扫描二维码获取 PPT

In most countries aluminium is used in five major areas: building and construction; containers and packaging ; transportation ; electrical conductors ; machinery and equipment. The consumption pattern varies widely from country to country depending upon the level of industrialization and economic growth.

Aluminium is obtained from bauxite which is the name given to ores usually containing 40%~60% hydrated alumina together with impurities such as iron oxides, silica and titania. The largest known bauxite reserves are found in Northern Australia, Guyana and Brazil. Production of aluminium from bauxite involves two distinct processes which are often operated at quite different locations. First, pure aluminium (Al_2O_3) is extracted from bauxite almost exclusively by the Bayer process which, essentially, involves digesting crushed bauxite with strong caustic soda solution at temperatures up to 2400℃. Most of the alumina is extracted leaving an insoluble residue known as red mud, consisting largely of iron oxide and silica which is removed by filtration. After cooling, the liquor is seeded with crystals of alumina trihydrate to reverse the chemical reaction, the trihydrate being precipitated and the caustic soda recycled. The whole process can be represented by the chemical reaction:

$$Al(OH)_3 + NaOH = NaAlO_2 + 2H_2O$$

The alumina trihydrate is then calcined in a rotary kiln at 1,200℃ to remove water of crystallization and alumina is produced as a fine powder.

Alumina has a high melting point (2400℃) and is poor conductor of electricity. The key to the successful production of aluminium lies in dissolving the alumina in molten cryolite (Na_3AlF_6) and a typical electrolyte contains (by mass) 80%~90% of this compound and 2%~8% alumina together with additives such as aluminium and calcium fluorides.

The exact mechanism for the electrolytic reaction in the cell is uncertain but it is probable that the current carrying ions are Na^+、AlF_4^-、AlF_6^{3-} and one or more complex ions such as $AlOF_3^{2-}$. At the cathode it is probable that the aluminium fluoride ions are discharged to produce aluminium metal and F^- ions while, at the anode, the complex ions dissociate to liberate oxygen which forms CO_2. The overall reaction can be written as follows:

$$2Al_2O_3 + 3C = 4Al + 3CO_2$$

The first commercial preparation of aluminium occurred in France in 1855 when H. Stainte-Claire Deville reduced aluminium chloride with sodium. The Bayer-Hall-Heroult process is the process used worldwide today, it consists of four steps:

(1) the extraction of pure alumina from bauxite;

(2) the manufacture of carbon anodes from petroleum coke and coal tar pitch;

(3) the reduction of alumina in electrolytic cells to produce molten aluminium;

(4) the melting of recycled scrap with smelted metal followed by purification, alloying and casting.

Alumina for cell feed is produced by caustic or acid digestion of crushed aluminous ores at high temperature and pressure followed by separation to produce pure anhydrous alumina. Digestion with sodium hydroxide preferentially dissolves alumina. Since alumina precipitated during sedimentation is lost with the residue, the inherent stability of supersaturated sodium aluminate solution under pressure is exploited to hold the maximum amount of alumina in solution at the lowest possible temperature. Some excess water is removed by reduction to atmospheric pressure. This ***blow-off*** slurry is clarified in three steps to discard the undissolved residue: (1) sand traps to remove the coarse fraction, (2) large-diameter, shallow tanks to remove most of the remainder by settling sedimentation and (3) pressure filters to remove the fine bauxite residue particles remaining after sedimentation. The result is a clear brown liquor. After cooling, aluminium trihydroxide (gibbsite) of the desired particle size distribution is precipitated from the clarified, supersaturated sodium aluminate solution, fines are recycled and the −0.15mm to 0.05mm fraction (with small amounts of oversize and 8~10 per cent—0.045mm) is calcined in rotary or fluid flash calciners to produce alumina feed for the smelting cells.

In the preferred cell design carbon anodes are made by mixing and forming sized petroleum coke and coal tar pitch into green blocks, baking them and making poured cast iron electrical connections to a performed socket in the anode.

The third step is carried out in smelting cells where alumina is added to molten cryolite (Na_3AlF_6) modified by the addition of aluminium fluoride (AlF_3), calcium fluoride (CaF_2) and other additives. These additives permit operation at the desired temperature and adjust the hardness of the insulating crust that forms on the bath of the smelting cell, as well as the interfacial tension and activities of the ionic species present in the melt to reduce backreaction of the electrode products. Electrolysis at 950~980℃ yields molten aluminium and carbon oxides (CO, CO_2). Molten aluminium is vacuum siphoned from the cells periodically and transferred to open-hearth holding furnaces to be melted with recycled aluminium scrap and alloyed. Oxides, dissolved hydrogen and unwanted trace elements are removed in a flowing molten stream by purification systems in which the metal is filtered and contacted with reactive gases as it is transported to the continuous casting units for ingot production. The carbon oxides containing traces of fluoride compounds, discharged at the anode both as particulate and gas, are diluted with 100 parts of air and collected in a hooded exhaust system to meet environmental requirements. Fluoride values are recovered in wet or dry scrubbing systems and returned to the cell, the clean carbon oxides being discharged to the atmosphere.

Smelting cells are arranged in long rows called potlines. The replaced side by side as close as possible commensurate with anode changing, with the anodes of one cell being electrically connected to the cathodes of the next in series. A plant will consist of between 400 ~ 700 cells. In prebaked potlines anodes are changed once a day per cell. The most primitive form of crane assist for this operation requires the operator to hand-break the crust around an anode with a jackhammer, remove the clamp and attach the crane cable for both insertion and removal of the anode. The most sophisticated crane allows the operator to sit in an air-conditioned cab and go through these operations by manipulating robot arms.

Words and Expressions

扫描二维码获取音频

扫描二维码答题

container [kən'teinə] *n.* 容器（箱，盆，罐，壶，桶，坛子）
packaging ['pækidʒiŋ] *n.* 包装，包装物
electrical conductor 导电体
bauxite ['bɔːksait] *n.* 铝矾土，铝土矿，铁铝氧石
hydrate ['haidreit] *n.* 氢氧化物，水合物
titania [tai'teiniə] *n.* 二氧化钛，氧化钛
reserve [ri'zəːv] *n.* 储备（物），储藏量
distinct [dis'tiŋkt] *a.* 清楚的，明显的
exclusively [ik'skluːsivli] *ad.* 排外地，专有地
digest [di'dʒest, dai'dʒest] *v.* 浸煮，蒸煮，煮解
caustic ['kɔːstik] *a.* 腐蚀性的，苛性的
soda ['səudə] *n.* 碳酸水，纯碱，氢氧化钠
residue ['rezidjuː] *n.* 剩余物，残余，残渣
filtration [fil'treiʃən] *n.* 过滤，筛选
liquor ['likə] *n.* 液，液体，母液

be seeded with	做孕育处理，使孕育
crystal ['kristl] *n.*	晶体，结晶，水晶
trihydrate [trai'haidreit] *n.*	[化] 三水合物
alumina trihydrate	氢氧化铝
reverse [ri'vəːs] *v.*	颠倒，倒转，倒退
precipitate [pri'sipiteit] *v.*	使……沉淀，沉淀
conductor [kən'dʌktə] *n.*	导体
conductor of electricity	导电体
cryolite ['kraiəulait] *n.*	冰晶石
electrolyte [i'lektrəlait] *n.*	电解，电解液，电解质
additive ['æditiv] *n.*	添加剂
cell [sel] *n.*	电解槽，小房间
probable ['prɔbəbl] *a.*	很可能的，大概的
ion ['aiən] *n.*	离子
dissociate [di'səuʃieit] *v.*	分离，游离，分裂
chloride ['klɔːraid] *n.*	[化] 氯化物
aluminium chloride	氯化铝
sodium ['səudiəm] *n.*	[化] 钠
anode ['ænəud] *n.*	[电] 阳极，正极
carbon anode	碳阳极
petroleum [pi'trəuliəm] *n.*	石油
tar [tɑː] *n.*	焦油，柏油
pitch [pitʃ] *n.*	沥青，柏油
coal tar pitch	煤焦油沥青
purification [ˌpjuərifi'keiʃən] *n.*	净化，提纯，精制（炼）
digestion [di'dʒestʃ(ə)n] *n.*	溶解，溶出，煮解
anhydrous [æn'haidrəs] *a.*	无水的
preferentially [ˌprefə'renʃəli] *ad.*	先取地，优先地
sedimentation [ˌsedimen'teiʃən] *n.*	沉淀，沉降
inherent [in'hiərənt] *a.*	固有的，内在的
supersaturate [sjuːpə'sætʃəreit] *v.*	[化] 使……过度饱和
exploit [iks'plɔit] *v.*	利用，使用，开发，开采
excess [ik'ses, 'ekses] *a.*	多余的，过度的，额外的
blow-off	排出，喷出
slurry ['sləːri] *n.*	不溶解物的悬浮液
clarify ['klærifai] *v.*	澄清，使（液体）纯净
sand trap	除沙槽

coarse [kɔːs] *a.* 粗的，粗糙的
fraction ['frækʃən] *n.* 小部分，微量
tank [tæŋk] *n.* 盛液体、气体的大容器（槽，桶，箱）
filter ['filtə] *n.* 过滤器，滤光器
pressure filter 压力过滤器
gibbsite ['gibzait] *n.* 水铝矿，铝土矿
green [griːn] *a.* 未加工（处理）的，湿的
socket ['sɔkit] *n.* 穴，孔，插座
modify ['mɔdifai] *v.* 调节，限制
crust [krʌst] *n.* 外壳，硬壳
interfacial [ˌintə(ː)'feiʃəl] *a.* 界面的，分界面的
tension ['tenʃən] *n.* 压力，张力，牵力
interfacial tension 界面张力
ionic [ai'ɔnik] *a.* 离子的
electrolysis [ilek'trɔlisis] *n.*【复数】-ses [-siːz] 电解（作用）
siphon ['saifən] *v.* 用虹吸管吸
holding furnace 保温炉
hooded ['hudid] *a.* 有罩盖的，戴头巾的
potline ['pɔtlaɪn] *n.* （制铝用的）电解槽系列
commensurate [kə'menʃərit] *a.* 相称的，相当的
be commensurate with 与……相称，与……相当
crane [krein] *n.* 起重机
assist [ə'sist] *n.* 帮助，辅助
cable ['keibl] *n.* 钢丝绳，缆，索
manipulate [mə'nipjuleit] *v.* （熟练地）操作，使用
robot ['rəubɔt, 'rɔbət] *n.* 机器人，遥控设备

Proper Names

Guyana [gai'aːnə, gai'ænə] *n.* 圭亚那（拉丁美洲）
Brazil [brə'zil] *n.* 巴西

Answer the following questions.

(1) What are the major areas where aluminium is used?
(2) Which process is used for production of aluminium worldwide today?
(3) What is the key to the successful production of aluminium?
(4) Where are the largest known bauxite reserves found?

(5) How many cells will a aluminium making plant consist of ?

(6) Can you simply describe the Bayer-Hall-Heroult process?

(1) Most of the alumina is extracted leaving an insoluble residue known as red mud, consisting largely of iron oxide and silica which is removed by filtration.

绝大部分氧化铝被萃取后只剩下被称为红泥的不可溶残渣，主要成分是铁氧化物和二氧化硅，可以通过过滤去除。

“known as red mud” 为过去分词短语作定语，修饰 “insoluble residue”。

(2) Since alumina precipitated during sedimentation is lost with the residue, the inherent stability of supersaturated sodium aluminate solution under pressure is exploited to hold the maximum amount of alumina in solution at the lowest possible temperature.

由于氧化铝发生沉淀会与残渣一起丢弃，这样在一定压力下，过饱和偏铝酸钠由于其自身的稳定性，会呈现出在最可能低的温度下溶解尽可能多的氧化铝。

(3) These additives permit operation at the desired temperature and adjust the hardness of the insulating crust that forms on the bath of the smelting cell, as well as the interfacial tension and activities of the ionic species present in the melt to reduce backreaction of the electrode products.

这些添加剂在给定的温度下可以调整熔炼室上面绝热壁的硬度。同时，这些添加剂可以调整界面张力和存在于熔融状态下的离子的行为，从而减少生成电极产物的逆反应。

“that forms on the bath of the smelting cell” 是定语从句，修饰 “insulating crust”。“to reduce backreaction of the electrode products” 是动词不定式短语，作目的状语。

(4) Oxides, dissolved hydrogen and unwanted trace elements are removed in a flowing molten stream by purification systems in which the metal is filtered and contacted with reactive gases as it is transported to the continuous casting units for ingot production.

氧化物、溶解的氢气和不需要的微量元素通过净化系统在流动的熔体中被去除，在净化系统中，金属过滤并与反应性气体接触，而后运输到连铸系统进行铸锭生产。

“in which the metal is filtered and contacted with reactive gases” 是定语从句，修饰 “purification systems”。

(5) The most sophisticated crane allows the operator to sit in an air-conditioned cab and go through these operations by manipulating robot arms.

而采用最精密的起重机时，只要求操作者坐在有空调的工作室中，通过操纵机器人手臂来完成上述操作。

Part Ⅱ Curriculum Ideological and Political

【The profiles of the enterprises】About CHALCO

Chalco is a leading company in China's non-ferrous industry. Chalco has made its mission to safeguard, develop and utilize national strategic resources, playing an essential role in the application of aluminum in the defense industry, aeronautics and space industry, rail transport, and the production of high-end alloys for civil purposes. The company has provided key aluminum profiles for China's construction, transportation and national defense industries to develop China's first man-made satellite, its first nuclear reactor and nuclear submarine, Long March rockets, Shenzhou spacecrafts, Chang'e lunar probes, commercial airplanes, aircraft carriers and high-speed trains. Chalco has been speeding up structural readjustment, transformation and upgrading, in order to be engaged in greater global cooperation and become a world-class company with international competitiveness and the ability to operate globally.

Part Ⅲ Further Reading

Nano aluminum powder

Nano aluminum powder is used in propellants and aluminum-water hydrogen production due to its advantages of high energy density, high heat release, low cost, and wide sources. Adding a certain amount of nano aluminum powder to the propellant can significantly increase the combustion rate and specific impulse. The high activity of nano-aluminum powder lead to the oxidation of nano-aluminum powder and the formation of the dense oxide layer on the surface, resulting in a decrease in active aluminum content. The thickness of the oxide layer limits the further increase of the burning rate in the propellant and also increases the induction time of the reaction in the aluminum-water hydrogen production process. Therefore, for the purpose of maintain the high activity of the nano aluminum powder (n-Al), the surface of the n-Al needs to be coated with modifier.

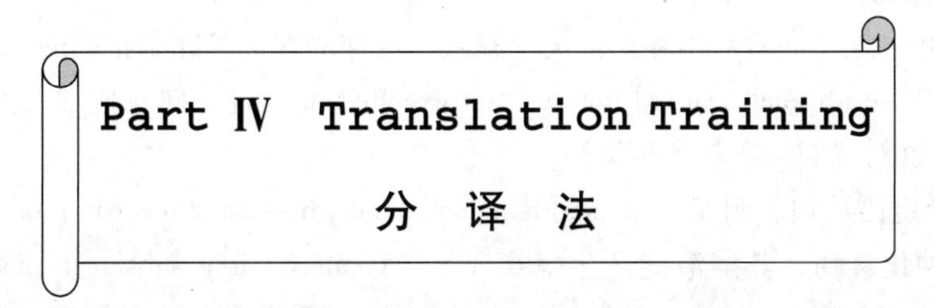

Part Ⅳ Translation Training

分 译 法

分译是指把英语中的一个句子分译成两个或两个以上的句子。总的来说，句子结构英语复杂、汉语简单。汉语和英语都有一词一义或多义的现象。但是同一个意思，英语常用长句表达，从句或短语环环相套，而汉语则用若干小句来表达。所以在翻译中，需要将原句的结构做较大的改变，以适合汉语的表达习惯，采用分译可使译文层次清楚、简洁明确。

1. 分译单句

单句分译为复句主要包括词和短语分化成小句，把原文的单句变为译文的复句。其原因是原文单句表达了丰富的思想，如按照原文语序译出不符合汉语习惯，这时需要用译文的复句来表达。

（1）分译副词。例如：The ancients tried unsuccessfully to explain how a rainbow is formed（古代人曾尝试解释彩虹是怎样形成的，但没有成功）；They，not surprisedly，did not respond at all（他们根本没有答复，这是不足为奇的）。

（2）分译名词结构。例如：The answer to many questions require facts not yet discovered（要对许多问题做出回答，就需要迄今尚未发现的事实）；Energy can neither be created nor destroyed，a universally accepted law（能量既不能创造也不能被消灭，这是一条普遍公认的规律）。

（3）分译动词不定式短语。例如：To find the hottest and coldest parts of the solar system on Mercury is quite strange（人们发现太阳系最热和最冷的地方都在水星上，这颇为奇怪）；The oxide of tungsten is reduced by hydrogen to obtain a pure products（通常用氢将钨的氧化物还原，以获得纯钨）。

（4）分译分词短语。例如：Exposed to the air，iron soon turns brown（铁暴露在空气中，很快变成褐色）；Radiowaves discovered a good many years ago are now being used for various purposes（许多年前发现的无线电波，现在正用于各种不同的目的）；He was lying on his side watching her（他侧身躺着，双目注视着她）。

（5）分译介词短语。例如：Their power increased with their number（随着人数的增加，他们的力量也随之加强）；Earlier scientists thought that during a man's lifetime the power of his brain decreased（从前科学家认为，人越老，脑子的技能就越衰弱）。

2. 分译复合句

把原文中含主语从句、定语从句等的复合句拆开，译成两个或两个以上的句子。

（1）分译主语从句。例如：It has been rightly stated that this situation is a threat to international security（这个局势对国际安全是个威胁，这样的说法是完全正确的）；It was a real challenge that those who had learned from us now excelled us（过去向我们学习的人，现在反而超过我们，这对我们确实是个鞭策）。

（2）分译定语从句。例如：An acid is a compound whose solutions can produce hydrogenions（酸是一种化合物，其溶液能产生氢离子）；A catalyst is a substance that speeds up a chemical reaction but does not enter into the reaction itself（催化剂是一种物质，它能加速化学反应，但并不参与化学反应）。

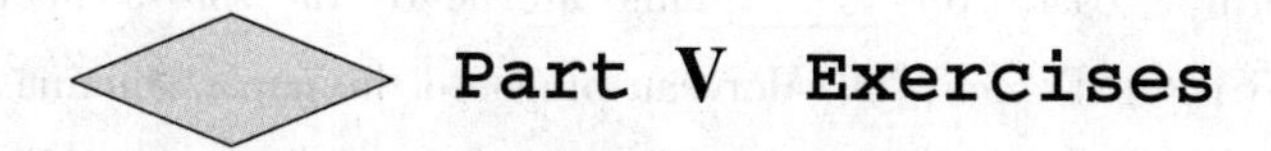

Part V Exercises

I . Translate the following expressions into English.

(1) 煤焦油沥青	(2) 保温炉	(3) 铁氧化物
(4) 压力过滤器	(5) 除沙槽	(6) 氢氧化钠
(7) 界面张力	(8) 电解槽	(9) 导电体

II. Fill in the blanks with the words from the text. The first letter of the word is given.

(1) A ______ is obtained from bauxite which is the name given to ores usually containing 40%~60% hydrated alumina together with impurities such as iron oxides, silica and titania.

(2) After cooling, the l ______ is seeded with crystals of alumina trihydrate to reverse the chemical reaction, the trihydrate being precipitated and the caustic soda recycled.

(3) The alumina trihydrate is then c ______ in a rotary kiln at 1200℃ to remove water of crystallization and alumina is produced as a fine powder.

(4) The exact mechanism for the electrolytic reaction in the cell is u ______.

(5) This ***blow-off*** s ______ is clarified in three steps to discard the undissolved residue.

(6) The third step is carried out in smelting c ______ where alumina is added to molten cryolite (Na_3AlF_6) modified by the addition of aluminium fluoride (AlF_3), calcium fluoride (CaF_2) and other additives.

(7) Molten aluminium is v ______ siphoned from the cells periodically and transferred to open-hearth holding furnaces to be melted with recycled aluminium scrap and alloyed.

(8) Fluoride values are recovered in wet or d ______ scrubbing systems and returned to the cell, the clean carbon oxides being discharged to the atmosphere.

(9) Smelting cells are arranged in long r ______ called potlines.

(10) The most sophisticated c ______ allows the operator to sit in an air-conditioned cab and go through these operations by manipulating robot arms.

III. Fill in the blanks with appropriate prepositions.

Alum has been known since ancient times, but not until the middle (1) ______ the eighteenth century did chemists learn that it contains two bases and not simply terra calcarea, a lime-like substance (2) ______ which alum had been identified since 1702. Pure alumina was first prepared (3) ______ alum in 1746 (4) ______ J. H. Pott and was first distinguished (5) ______ lime by Marggraff in 1754 when he prepared alumina from clay. Marggraff also showed that silica

(SiO_2) and the "earthy" base (6) ______ alum are nearly the sole components of pure white clay. (7) ______ 1761, in France, D. Morveau proposed the name alumina for this base in alum, and this name was accepted (8) ______ britain until (9) ______ 1820 when it was anglicized (10) ______ alumina.

Ⅳ. Decide whether the following statements are true or false (T/F).

(1) Alumina has a low melting point (2400℃) and is poor conductor of electricity. (　)

(2) The first commercial preparation of aluminium occurred in France in 1455, when H. Stainte-Claire Deville reduced aluminium chloride with sodium. (　)

(3) The alumina trihydrate is then calcined in a rotary kiln at 1200℃ to remove water of crystallization and alumina is produced as a fine powder. (　)

(4) In the preferred cell design carbon anodes are made by mixing and forming sized petroleum ore and coal tar pitch into green blocks, baking them and making poured cast iron electrical connections to a performed socket in the anode. (　)

(5) The carbon oxides containing traces of fluoride compounds, discharged at the anode both as particulate and gas, are diluted with 1000 parts of air and collected in a hooded exhaust system to meet environmental requirements. (　)

Ⅴ. Translate the following English into Chinese.

In 1913 Harry Brearley of Sheffield was experimenting with alloy steels for gun barrels, and among the samples which he threw aside as being unsuitable was one containing about 14 per cent of chromium. Some months later he noticed that most of the steels had rusted, but the chromium steel remained bright. This led to the development of stainless steels. Stainless steels are alloys having a high chromium content of 12 per cent, or more. They possess extremely high corrosion resistance. Most of these steels have much better mechanical properties at high temperatures. This group was first called stainless steel. With the emphasis on high temperature use, they are frequently referred to as heat and corrosion-resistant steels.

Unit 15 Zinc Metallurgy

Part Ⅰ Reading and Comprehension

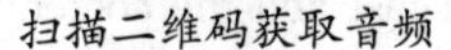

扫描二维码获取音频

扫描二维码获取 PPT

A large number of precious metals are produced mainly by hydrometallurgical techniques. Most of the copper, almost all of the aluminum, gold and platinum are involved as well as some of the base metals such as lead, zinc and nickel. Hydrometallurgy is the extraction and recovery of metals from their ores by processes in which aqueous solutions play a predominant role. Two distinct processes are involved in hydrometallurgy: putting the metal values in the ore into solution and extracting the metal values from solution via the operation known as leaching; and recovering the metal values from solution after a suitable solution purification or concentration step, or both. The advantages of hydrometallurgy are applicability to the treatment of low-grade ores (copper, titanium, gold, silver), amenability to the treatment of materials of quite different compositions and concentrations, adaptability to separation of highly similar materials (hafnium from zirconium), flexibility in terms of the scale of operations, simplified materials handling and good operational and environmental control as compared with pyrometallurgy.

Zinc is one of the softer metals, slightly less dense than iron and with a low melting point (419.5℃). The capability for resisting erosion and alloy performance of zinc are better. It is widely applied in industry because of their better physical and chemical properties. The main use of zinc is the protection of steel from rust. The making of brass and zinc alloy die castings account for 80 percent of the total consumption of zinc. The other uses of zinc include rolled zinc for storage batteries and buildings, and zinc dust used as protective paint. In addition, rolled zinc is also used for photoengraving plates, organ pipes and some trunk linings. The metal is contained in zinc sulfide ore deposits and widely distributed throughout the world. The important mines are in Canada, Australia, Peru and the U. S. A. Zinc metallurgy is developing rapidly in China, which has become the biggest country with zinc metallurgical production.

The original processes for the production of zinc was pyrometallurgy. Such methods consist of the retort process, electric-arc furnace zinc-making process and the imperial smelting process.

The zinc oxide formed by roasting the sulfide ore was heated, together with anthracite or a similar carbonaceous material in banks of retorts, to a temperature of about 1, 100℃. Zinc was formed as a vapor, which was caught as liquid metal in condensers adjoining the furnace. Now the process was being washed out because of more power consumption and serious pollution. After works experiments which started in the 1940s, Imperial Smelting Corporation, at Avonmouth in England announced in 1957 the successful development of a blast furnace for making zinc. In the Imperial Smelting process, zinc and lead sulphides are roasted to produce oxides which are charged in the blast furnace with proportioned amounts of coke. Preheated blast air enters the furnace through water-cooled tuyeres. Slag and molten lead containing precious metals and copper from the charge are tapped from the bottom of the furnace and separated. Zinc leaves the shaft as a vapour in the furnace gas containing carbon monoxide and carbon dioxide and is shock-cooled in a lead splash condenser, the zinc vapour being absorbed by the lead. Next the hot lead containing zinc in solution is pumped out of the condenser and cooled so that zinc is rejected from the solution and floats on the lead. The zinc layer is poured off and cast and the cooled lead is recycled to the condenser.

The characteristic of electric arc furnace zinc-making process is its ability to heat directly burden and retort continuously to produce zinc by using electric energy. Its raw materials is similar to retort's. The method can be used to deal with multiple metal zinc concentrate.

The electrolytic zinc process, developed during the First World War, now accounts for over four fifths of world zinc production. Zinc hydrometallurgy is to get zinc from zinc ore concentrate, which contains roasting-leaching-purifying-electrolytic deposit in sequence.

1. Roasting of zinc concentrates

Zinc spar (ZnS) and marmatite (mZnS · nFeS) are the main mineral raw materials of zinc producing. The most important roasting reactions are those concerning zinc sulfide concentrates and involve chemical combination with the roasting atmosphere.

Possible reactions include:

$$2ZnS + 3O_2 = 2ZnO + 2SO_2 \quad \text{(Dead roast)}$$

$$6ZnS + 11O_2 = 2(ZnO \cdot 2ZnSO_4) + 2SO_2 \quad \text{(Dead roast)}$$

$$ZnS + 2O_2 = ZnSO_4 \quad \text{(Sulfating roast)}$$

Other equilibria which need to be taken into account include:

$$S + O_2 = SO_2$$

$$2SO_2 + O_2 = 2SO_3$$

Zinc sulfide in concentrates can be converted into zinc oxide which can be soluble in dilute sulfuric acid by roasting, and it is also hoped that the roasted product contains a small amount of $ZnSO_4$, to compensate for the loss of sulfuric acid in the process of leaching and electrowinning.

2. Leaching of zinc calcine

Zinc calcine obtained by roasting the zinc concentrates in boiling furnace which consists of zinc oxide, other metal oxides and gangue. Leaching of zinc calcine is the process of extracting zinc and other metal values from a solid calcine by means of sulfuric acid solvent. It makes the metal values in solid calcine dissolve into the solution, separating them from the insoluble matters such as gangue.

Two purposes can be achieved:

(1) Making zinc in the material dissolve completely into the leaching solution as far as possible.

(2) Making the harmful impurities not dissolve into the solution as far as possible but go into slag.

3. Purification of leaching solution

The leaching solution often contains arsenic, antimony, copper, cadmium, cobalt, nickel and other impurities, and must be purified before electrolysis. The standard electrode potentials of the impurities are all higher than zinc, and they can be changed and purified by using zinc powder.

4. The electrolytic deposit of zinc

The electrolytic deposit is the last process step in hydrometallurgy of zinc, which extracts zinc from zinc sulfate solution with H_2SO_4 (electrolyte). The reaction is as follow:

$$ZnSO_4 + H_2O = Zn + H_2SO_4 + \frac{1}{2}O_2$$

Electrolytic tank is a rectangular bath. Pb-Ag alloys serve as anode (positive electrode), and cathodes (negative electrode) are made from rolling of aluminum sheets, both dipped into the electrolyte. The oxygen is liberated at the anodes, and zinc is deposited electrolytically on aluminum sheets from which it is stripped off, melted, and cast into slabs. The purity of the metal so formed is greater than 99.95 per cent and can be maintained above 99.99 per cent when desired. The energy consumption in electrolytic deposit process is very massive, and it has a great effect on capitalized cost in hydrometallurgy of zinc.

❋ Words and Expressions

扫描二维码获取音频

扫描二维码答题

hydrometallurgical [ˈhaidrəuˌmetəˈləːdʒikəl] *a.* 湿法冶金的

platinum [ˈplætinəm] *n.* 铂，白金

base [beis] *a.* 低劣的

hydrometallurgy [ˌhaidrəumeˈtælədʒi] *n.* 湿法冶金术

aqueous [ˈeikwiəs] *a.* 水的，水成的，水状的
predominant [priˈdɔminənt] *a.* 主要的，支配的
distinct [disˈtiŋkt] *a.* 独特的，有区别的
solution [səˈluːʃən] *n.* 溶液，溶解
extract [iksˈtrækt] *v.* 萃取，提取
recover [riˈkʌvə] *v.* 回收，恢复
recovery [riˈkʌvəri] *n.* 回收，恢复
leach [liːtʃ] *n.* 浸出，过滤
v. 浸出，过滤
applicability [ˌæplikəˈbiləti] *n.* 适用性，适应性
amenability [əmiːnəˈbiləti] *n.* 可处理性，可控制性
adaptability [əˌdæptəˈbiləti] *n.* 适应性，可变性，适合性
composition [kɔmpəˈziʃən] *n.* 构成，合成物
hafnium [ˈhæfniəm] *n.* 铪（元素符号 Hf）
zirconium [zəːˈkəuniəm] *n.* 锆（元素符号 Zr）
dense [dens] *a.* 稠密的，浓厚的
pyrometallurgy [ˌpairəumeˈtæləːdʒi] *n.* 火法冶金学，高温冶金学
melt [melt] *v.* 熔化，溶解
performance [pəˈfɔːməns] *n.* 性能
alloy [ˈælɔi, əˈlɔi] *n.* 合金
property [ˈprɔpəti] *n.* 性质，性能
die [dai] *n.* 冲模，钢模
casting [ˈkɑːstiŋ] *n.* 铸造，铸件
rolled [rəuld] *a.* 轧制的，滚制的
dust [dʌst] *n.* 灰尘，尘埃
zinc dust 锌粉
photoengraving [ˌfəutəuinˈgreiviŋ] *n.* 照相凸版，照相凸版印刷
organ [ˈɔːgən] *n.* 风琴，管风琴
pipe [paip] *n.* 管
trunk [trʌŋk] *n.* 汇集管
lining [ˈlainiŋ] *n.* 衬里，内衬
retort [riˈtɔːt] *n.* 曲颈瓶，蒸馏罐
v. 蒸馏，（在蒸馏罐中）提纯
retort process 反应罐直接还原法，蒸馏过程，干馏过程
imperial [imˈpiəriəl] *a.* 帝国的
smelt [smelt] *v.* 熔炼，冶炼
anthracite [ˈænθrəsait] *n.* 无烟煤
carbonaceous [ˌkɑːbəˈneiʃəs] *a.* 碳质的，碳的，含碳的

banks of retort	蒸馏罐
condenser [kən'densə] *n.*	冷凝器
adjoin [ə'dʒɔin] *v.*	邻近，毗连，邻接
corporation [ˌkɔːpə'reiʃən] *n.*	公司
proportioned [prə'pɔːʃənd] *a.*	相称的，成比例的
charge [tʃɑːdʒ] *n.*	炉料
molten ['məultən] *a.*	熔化的，铸造的
shaft [ʃæft] *n.*	竖炉
splash [splæʃ] *n.*	溅上的斑点，溅泼的量
v.	飞溅，泼
lead splash condenser	铅雨冷凝器
shock-cool	急速冷却
reject [ri'dʒekt] *v.*	拒绝，排斥
multiple ['mʌltipl] *a.*	多样的，许多的
electrolytic [iˌlektrəu'litik] *a.*	电解的，电解质的
electrolytic deposit	电解沉积
sequence ['siːkwəns] *n.*	连续，一连串
spar [spɑː] *n.*	晶石
marmatite ['mɑːmətait] *n.*	铁闪锌矿
zinc sulfide concentrate	硫化锌精矿
equilibria [iːkwi'libriə] *n.*	平衡，均势
【复数】equilibrium [iːkwi'libriəm]	
sulfate ['sʌlfeit] *n.*	硫酸盐
v.	使……成硫酸盐，用硫酸处理，硫酸盐化
sulfating roast	硫酸化焙烧
dilute [dai'ljuːt, di'ljuːt] *a.*	淡的，稀释的
v.	冲淡，稀释
sulfuric [sʌl'fjuːrik] *a.*	（正）硫的，含（六价）硫的
compensate ['kɔmpenseit] *v.*	补偿，赔偿
electrowinning [iˌlektrəu'winiŋ] *n.*	电解冶金法，电积金属法
arsenic ['ɑːs（ə）nik] *n.*	砷（元素符号 As）
antimony ['æntiməni] *n.*	锑（元素符号 Sb）
cadmium ['kædmiəm] *n.*	镉（元素符号 Cd）
cobalt ['kəubɔːlt] *n.*	钴（元素符号 Co）
electrode potential	电极电势
cathode ['kæθəud] *n.*	阴极
dip [dip] *v.*	浸，泡
strip [strip] *v.*	剥夺，剥去
slab [slæb] *n.*	平板，厚片

capitalize [ˈkæpitəlaiz] *v.*　使……资本化，积累资本

✲ Proper Names

Peru [pəˈruː] *n.*　秘鲁（拉丁美洲国家名）
Avonmouth [eivɔnˈmauθ] *n.*　埃文茅斯（在英国布里斯托尔附近）

✲ Answer the following questions.

(1) How many processes are involved in hydrometallurgy?
(2) In which countries are the important mines of zinc sulfide ore deposits?
(3) What can the electric arc furnace zinc-making process be used to deal with?
(4) What is leaching of zinc calcine ?
(5) What is the last process step in hydrometallurgy of zinc?

(1) Most of the copper, almost all of the aluminum, gold and platinum are involved as well as some of the base metals such as lead, zinc and nickel.

这包括了大部分的铜，几乎所有的铝、金和铂以及部分铅、锌和镍等贱金属。

“as well as” 意思是“除……之外（也），同……一样（也），和，也，还”。例如：A teacher should entertain as well as teach（教师不仅要教书，也要激发起学生的兴趣）；This is your responsibility as well as a survival necessity（这是你的职责，也是你赖以生存的必需品）。

(2) Hydrometallurgy is the extraction and recovery of metals from their ores by processes in which aqueous solutions play a predominant role.

湿法冶金是指从矿石中提取和回收金属的生产过程，其中水溶液起主要作用。

“which” 引导定语从句，代替“processes”。定语从句中，主语是“aqueous solutions”，“play a role in” 是动词短语，作谓语，意思是“在……中起作用”（也可以用“play a part in”）。介词“in”可以置于“which”前。

(3) The advantages of hydrometallurgy are applicability to the treatment of low-grade ores (copper, titanium, gold, silver), amenability to the treatment of materials of quite different compositions and concentrations, adaptability to separation of highly similar materials (hafnium from zirconium), flexibility in terms of the scale of operations, simplified materials handling and good operational and environmental control as compared with pyrometallurgy.

湿法冶金的优点为：适合处理低品位矿石（铜、钛、金、银），可处理成分和浓度完全不同的材料，适于分离高度相似的材料（从锆中分离铪）；与火法冶金相比，其运行规模灵活，物料处理简单，操作及环境控制良好。

此句为系表结构，"applicability to the treatment of low-grade ores（copper，titanium，gold，silver）""amenability to the treatment of materials of quite different compositions and concentrations""adaptability to separation of highly similar materials（hafnium form zirconium）""flexibility in terms of the scale of operations""simplified materials handling""good operational and environmental control"等短语为并列表语。

"applicability"是 applicable 的名词形式，意思是"适用性，可应用性"，与 to 连用。例如：The applicability of the rule to this case is obvious（这条规则显然适用于此情况）。

"amenability"是 amenable（/ə'miːnəbl/）的名词形式，意思是"可处理性，可控制性"，与 to 连用。例如：The technology has attracted a great deal of attention because of its ease in operation and amenability to automation（此技术因其操作简单、易于自动化而备受人们重视）。

"adaptability"是 adaptable 的名词形式，意思是"适应性"，与 to 连用。例如：The adaptability of wool to the manufacture of the cloth is one of its great attractions（羊毛对生产织物的适应性是其具有的巨大吸引力之一）。

"in terms of"意思是"在……方面，依据"。例如：They see all decisions in terms of victory or defeat（所有的决定在他们看来只有胜利和失败之分）。

（4）Zinc leaves the shaft as a vapour in the furnace gas containing carbon monoxide and carbon dioxide and is shock-cooled in a lead splash condenser，the zinc vapour being absorbed by the lead.

锌以蒸气形式混入含有一氧化碳和二氧化碳的炉气中排出炉外，在铅雨冷凝器中急速冷却，锌蒸气被铅吸收。

此句的谓语是"leaves"和"is shock-cooled"。"in the furnace gas containing carbon monoxide and carbon dioxide"作状语，修饰"a vapour"。其中，"containing carbon monoxide and carbon dioxide"是现在分词短语作后置定语，修饰"the furnace gas"。此外，"The zinc vapour being absorbed by the lead"是分词的独立主格形式，作状语，表示伴随状况。

（5）The characteristic of electric arc furnace zinc-making process is its ability to heat directly burden and retort continuously to produce zinc by using electric energy.

电弧炉炼锌法的特点是能够利用电能直接加热炉料并连续蒸馏生产锌。

（6）Zinc hydrometallurgy is to get zinc from zinc ore concentrate，which contains roasting-leaching-purifying-electrolytic deposit in sequence.

湿法炼锌是将锌精矿依次经焙烧-浸出-净化-电解沉积而获得金属锌。

"which contains roasting-leaching-purifying-electrolytic deposit in sequence"是非限制性定语从句，其中，关系代词"which"指代"zinc hydrometallurgy"，在定语从句中作主语。"in sequence"意思是"按顺序"，例如：The teacher requests us answering the questions in sequence（老师要求我们依次回答问题）。

（7）Pb-Ag alloys serve as anode（positive electrode），and cathodes（negative electrode）are made from rolling of aluminum sheets，both dipped into the electrolyte.

Pb-Ag 合金作为阳极（正极），阴极（负极）由压延铝板轧制而成，它们浸入电解液里。

“both dipped into the electrolyte”是分词独立主格结构作状语，表示伴随。分词独立主格结构的形式之一是“名词/代词+分词”的结构，例如：The rain having stopped，he went out for a walk（雨停了，他出去散步）。

（8） The oxygen is liberated at the anodes，and zinc is deposited electrolytically on aluminum sheets from which it is stripped off，melted，and cast into slabs.

氧气在阳极上释放出来，锌电解沉积到铝板上，从铝板上剥下，经熔化，铸成锌板。

“from which it is stripped off”是定语从句，其中“which”指代“aluminum sheets”。“strip off”意思是“剥去，除去”，例如：The paint will be difficult to strip off（这层漆很难除掉）。

Part Ⅱ　Curriculum Ideological and Political

【Engineering ethics】Sulfur dioxide escape accident

On 2019，sulfur dioxide escaping accident happened in a Zinc smelter. But the company did not report equipment failure and maintenance to the department of ecological environment，and did not take measures to decrease the load，repair flue leakage to reduce pollutants discharge. That made part of the flue gas from the weld escape to outside environment.

For an enterprise in modern society，it is certainly understandable to pursue profits. However，if an enterprise only pays attention to profits and even sacrifices environmental and social resources to obtain profits，it will have no value of existence and will not be able to promote the sustainable prosperity of social economy. Under the premise of protecting environment，resources and social benefits，the pursuit of corporate profits is the primary criterion for a modern enterprise to settle down.

Part Ⅲ　Further Reading

Extract precious zinc from fly ash

Fly ash contains zinc and other precious metals，which lead to the loss of precious metals directly in landfill. Researchers at Charms University of Technology in Sweden have developed a method to extract these precious metals，which can reduce pollution and recycle metals at the same time. This waste is treated by acid washing to separate zinc from ash. Then，zinc can be extracted，washed and processed into raw materials. In the experiment，they found that 70% of the zinc in the fly ash can be recycled. After zinc is extracted，the residual ash is burned again to decompose dioxin. 90% of the bottom ash can be used as building materials.

Part Ⅳ Translation Training

句子成分的转换

由于英语和汉语的表达方式不尽相同，翻译时往往需要改变原文的语法结构，所采用的主要方法是句子成分的转换。进行科技英语翻译时，在一定情况下适当改变原文中某些句子成分，可以达到译文逻辑正确、通顺流畅、重点突出等目的。

1. 宾语译成主语

在科技英语翻译中，为了符合汉语的表达方式，有时需要用原文中的宾语、介词宾语来更换原来的主语，使译文重点突出、行文流畅。例如：Hydrogen is found in many compounds. 在许多化合物中都有氢。A motor is similar to a generator in construction. 电动机的结构与发电机类似。

2. 谓语译成宾语

通过“把、将、使、得到、受到、获得、有了”等动词，可使英语的谓语译成汉语的宾语。例如：The protective device has been greatly improved（保护装置的性能得到了很大的改善）。

3. 定语译成谓语

有时科技英语中的定语也可以转译成汉语的谓语。例如：Water has a greater heat capacity than sand（水的热容比沙大）。

4. 主语译成定语

在句子成分的转换中，主语转译成定语也是科技英语翻译时常用的方法之一。例如：A large steam engine may have as much power capacity as 100000 horsepower（大型蒸汽机的功率可达 100000 马力）。

5. 状语译成补语

为了避免译句使读者感到生硬、别扭，还可以把状语译成补语。例如：Nylon is far stronger than rayon. 尼龙比人造丝牢固得多。

Part V　Exercises

Ⅰ. Translate the following expressions into English.

(1) 锌冶金	(2) 湿法炼锌	(3) 回收有价金属
(4) 熔点	(5) 物理化学性能	(6) 硫化锌矿床
(7) 氧化锌	(8) 矿物原料	(9) 死焙烧
(10) 硫酸化焙烧	(11) 锌焙砂	(12) 硫酸溶剂
(13) 浸出液	(14) 电极电势	(15) 锌的电解沉积

Ⅱ. Fill in the blanks with the words from the text. The first letter of the word is given.

(1) Hydrometallurgy is the e ______ and recovery of metals from their ores by processes in which aqueous solutions play a predominant role.

(2) The capability of resisting e ______ of zinc is better.

(3) The electrolytic zinc process now a ______ for over four fifths of world zinc production.

(4) Zinc hydrometallurgy is to get zinc from zinc ore c ______, which contains roasting-leaching-purifying-electrolytic deposit in sequence.

(5) Zinc oxide can be soluble in d ______ sulfuric acid by roasting.

(6) Zinc calcine consists of zinc oxide, other metal oxides and g ______.

(7) Leaching of zinc calcine makes the metal values in solid calcine d ______ into the solution, separating them from the insoluble matters.

(8) The leaching solution must be p ______ before electrolysis.

(9) The electrolytic deposit extracts zinc from zinc s ______ solution with H_2SO_4.

(10) The energy c ______ in electrolytic deposit process is very massive.

(11) The standard electrode potentials of the impurities are higher than zinc, and they can be changed and purified by using zinc p ______.

(12) Pb-Ag alloys serve as a ______ (positive electrode).

(13) The oxygen is liberated at the anodes, and zinc is d ______ electrolytically on aluminum sheets.

(14) From aluminum sheets zinc is stripped off, melted, and cast into s ______.

(15) The p ______ of the metal so formed is greater than 99.95 percent.

Ⅲ. Fill in the blanks by choosing the right words form given in the brackets.

In the Imperial Smelting process, zinc and lead sulphides are (1) ______ (roasting; roasted) to produce oxides which are charged in the blast furnace with (2) ______ (proportion; proportioned) amounts of coke. (3) ______ (Preheated; Heated) blast air enters the furnace (4) ____ (for; through) water-cooled tuyeres. Slag and molten lead containing precious metals and copper from the charge are tapped from the bottom of the furnace and (5) ______ (set; separated). Zinc leaves the shaft (6) ______ (for; as) a vapour in the furnace gas containing carbon monoxide and carbon dioxide and is shock-cooled in a lead splash condenser, the zinc vapour being absorbed by the lead. Next the hot lead (7) ______ (contained; containing) zinc in solution is pumped out of the condenser and cooled so that zinc is rejected from the solution and floats on the lead. The zinc layer is poured off and cast and the cooled lead is recycled to the condenser.

The characteristic of electric arc furnace zinc-making process is its ability to heat (8) ______ (correctly; directly) burden and retort continuously to produce zinc by using electric energy. Its raw materials is similar (9) ______ (to; for) retort's. The method can be used to deal (10) ______ (for; with) multiple metal zinc concentrate.

Ⅳ. Decide whether the following statements are true or false (T/F).

(1) Large numbers of precious metals are produced by hydrometallurgical techniques completely. ( )

(2) Hydrometallurgy is the extraction and recovery of metals from their ores by processes in which aqueous solutions play an important role. ( )

(3) Zinc is one of the hardest metals, slightly less dense than iron and with a low melting point (419.5℃). ( )

(4) Now the original process of zinc production was being washed out because of more power consumption and serious pollution. ( )

(5) Zinc hydrometallurgy is to get zinc from zinc ore concentrate, which contains roasting-purifying-electrolytic deposit in sequence. ( )

(6) Zinc sulfide in concentrates can be converted into zinc oxide which can be soluble in concentrated sulfuric acid by roasting. ( )

(7) Zinc calcine obtained by roasting the zinc concentrates in boiling furnace which consists of zinc oxide and gangue. ( )

(8) Leaching of zinc calcine is the process of extracting zinc from a liquid calcine by means of sulfuric acid solvent. ( )

(9) The leaching solution often contains arsenic, antimony, copper, cadmium, cobalt, nickel and other impurities, and must be purified before electrolysis. ( )

(10) Pb-Ag alloys serve as cathodes (negative electrode). ( )

Ⅴ. Translate the following English into Chinese.

Long before zinc was known as a metal, the Romans mixed calamine, which contains zinc carbonate, with copper ores; the smelting of the two materials produced brass. Zinc still could not be isolated for many hundreds of years after the discovery of brass. It was in China that zinc was first made. And it was exported to Europe in the early seventeenth century. England was the first European country to develop the manufacture of zinc. William Champion began smelting the metal on a commercial scale in 1738.

Unit 16 The Extraction of Gold

Part Ⅰ Reading and Comprehension

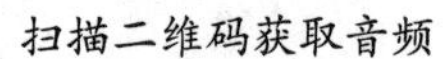

扫描二维码获取音频　　扫描二维码获取 PPT

Gold and then , silver , were called ***noble metals*** because they could be exposed to the atmosphere for a long time without tarnishing and could be melted repeatedly without much loss in weight. Owing to these characteristics, they were used originally for making jewelry and later for making coinage. Of all metals, gold has the best malleability and it is one of the most ductile metals. It can be forged into gold foil in a translucent state. On some airplanes, a sandwich of laminated glass containing a film of gold less than five over ten thousand of a millimeter thick is embedded in the windshield which has been heat treated so as to guard against the harmful rays of the sun. One ounce troy of gold can be drawn to a fine wire which is more than 80km long. Gold can be completely resistant to atmospheric attack and oxidation. The vital electronic systems that enabled man to take his first step on the moon were protected by coatings of gold from the take-off blast. Gold was also used in the circuits of computers on the moon voyage. The pure gold would be too soft for many uses, and therefore it is hardened by the addition of other alloying elements, copper being the most common one. In addition, silver, nickel, palladium and zinc are alloying elements in making jewelry.

Pellets or nuggets of pure gold are rarely found in nature; they are generally combined with other associated minerals. The economic value of a kind of ore depends on the price of the metal that is contained in the ore and its ease of extraction. The iron ore from which iron——less than 20 per cent of the weight of the ore——is extracted is reckoned as poor ore, while one part of gold in 100000 parts of ore is reckoned as rich ore. Gravity concentration is the earliest method of gold extraction; and it is based on a discriminating force, the magnitude of which varies with specific gravity. The other force that is usually operating in gravity methods is the resistance to relative motion exerted upon the particles by the fluid or semi-fluid medium in which separation takes place. Most gold ores are treated by cyanide process. Producing the chemical of potassium cyanide cheaply and utilizing it in the extraction of gold was a momentous discovery, which greatly influenced the development of gold mines and international trade in various parts of the world.

Cyanide gold leaching process can be grouped into three classes: lump leaching, percolation leaching, pulp agitation leaching.

1. Lump leaching

This method was first employed in the Harz Mountains in Germany during the sixteenth century. Depending upon the tonnage processed, an area of about 300 ft by 400 ft is leveled and then covered with an asphalt layer. Low grade ore is then dumped onto the site by dump trucks to a level of 3 ~ 6 meters high. Cyanide solution is then sprayed at the top of the dump, and leaching solution is collected in streams at the bottom of the heap. Sometimes, perforated vertical pipes are introduced at regular intervals inside the heap to facilitate the flow of solution and at the same time allow air circulation to facilitate the leaching process.

Heap leaching is used for low-grade gold ores (Au 1 ~ 2g/t). Heap leaching operation usually takes 30 ~ 90 days, and the recovery of gold is 50% ~ 70%.

2. Percolation leaching

The material to be leached is placed in a tank equipped with a false bottom covered with a filtering medium. The solvent is added at the top of the tank and is allowed to percolate through the material. In the tank a countercurrent system is usually employed. The new solids added to the last tank and the weak liquid added to the first are pumped successively from one tank to another till it reaches the last one, almost saturated.

The process is well suited to cases where the material is porous and sandy, but is inapplicable to the material which is impervious and tends to pile into a block.

Regularly the size of particles rather than the size of the matter is the chief factor governing good percolation. The method is, therefore, unsatisfactory if much residue is present. Its advantages are the minimum of solvent consumption, the production of high grade pregnant solution, and elimination of the use of expensive thickeners or filters. When leaching is finished, the tanks are emptied manually, and then a new batch is introduced.

3. Pulp agitation leaching

Leaching of the materials such as pulp of ores, concentrates and so on, with agitator, usually refers to grinding the material in water (to minimum dust) to produce particles with the optimum size. After being crushed into a fine powder, the gold ore is stirred in large tanks with a very dilute solution of cyanide, so that the gold is dissolved by the cyanide. Pulp densities vary from 40% to 70%. The leaching agent is added and the pulp is agitated continuously. Agitation leaching may be accomplished by:

(1) Mechanical agitation leaching: the propeller or impeller is commonly used here. This is usually for small leaching tanks.

(2) Air agitation leaching: the tank for leaching is a cylindrical tank of about 12ft in diameter and 45ft high, with a conical bottom of 60 degrees. A metal tube open at both ends is placed in the central part. Compressed air is introduced through the tube and cause the circulation of the pulp which is put in the tank.

(3) Combined air and mechanical agitation: for the leaching on a large scale, Dorr agitator are used extensively.

The pulp after leaching consists of gold-bearing solution and tailings. The solution is filtered and the gold precipitated from the solution.

Words and Expressions

扫描二维码获取音频

扫描二维码答题

tarnish ['tɑːniʃ] *v.*　失去光泽，变灰暗
malleability [ˌmæliə'biliti] *n.*　加工性，展延性
ductile ['dʌktail, 'dʌktil] *a.*　柔软的，易延展的
foil [fɔil] *n.*　箔，金属薄片
translucent [træns'lusənt, trænzlusənt] *a.*　半透明的
laminated ['læmineitid] *a.*　层压的，由薄片叠成的
embed [im'bed] *v.*　使……嵌入，使……插入
windshield ['windʃiːld] *n.*　挡风玻璃
ounce [auns] *n.*　盎司，英两
ounce troy [trɔi]　金衡盎司
coating ['kəutiŋ] *n.*　涂层
blast ['blæst] *n.*　冲击波
palladium [pə'leidiəm] *n.*　钯（元素符号 Pd）
pellet ['pelit] *n.*　小球
nugget ['nʌgit] *n.*　天然金块，矿块
reckon ['rekən] *v.*　估计，认为
discriminating [dis'krimineitiŋ] *a.*　识别的，有识别力的
magnitude ['mægnitjuːd] *n.*　大小，量级
specific gravity　相对密度
exert [ig'zəːt] *v.* (on or upon)　对…… 施加影响
cyanide ['saiəˌnaid] *n.*　氰化物
potassium [pə'tæsiəm] *n.*　钾（元素符号 K）

momentous [məu'mentəs] *a.* 重要的，重大的
percolate ['pɜːkəleit] *v.* 浸透，渗透
lump [lʌmp] *n.* 堆，块
pulp [pʌlp] *n.* 矿浆
asphalt ['æsfælt] *n.* 沥青
dump truck 自动倾卸（自卸）卡车
perforate ['pəːfəreit] *v.* 穿孔于，打孔穿透
interval ['intəvəl] *n.* 间隔，间距
facilitate [fə'siliteit] *v.* 促进，帮助
ditch [ditʃ] *n.* 沟渠
gravel ['grævəl] *n.* 碎石，砂砾
perpendicularly [ˌpəːpən'dikjuləli] *ad.* 垂直地，直立地
false [fɔːls] *a.* 假的
solvent ['sɔlvənt] *n.* 溶剂
saturated ['sætʃəreitid] *a.* 饱和的
manually ['mænjuəli] *ad.* 手动地，用手地
batch [bætʃ] *n.* 一批，一炉
minimum ['miniməm] *n.* 最小值，最低限度
propeller [prə'pelə] *n.* 螺旋桨，推进器
impeller [im'pelə] *n.* 叶轮，推进者
conical ['kɔnikəl] *a.* 圆锥形的
compressed [kəm'prest] *a.* 压缩的
precipitate [pri'sipiteit] *n.* 沉淀物
v. 使……沉淀

❊ Proper Names

Harz [hɑːrts] Mountains *n.* 哈尔茨山（德国中部山）
Germany ['dʒəːməni] *n.* 德国
Dorr agitator 多尔搅拌器

❊ Answer the following questions.

(1) Why were gold and silver called 'noble metals'?
(2) How are most gold ores treated?
(3) How many classes can cyanide gold leaching process be grouped? And what are they?
(4) Is heap leaching used for low-grade gold ores?
(5) Where is the material to be leached by percolation placed?

(1) On some airplanes, a sandwich of laminated glass containing a film of gold less than five over ten thousand of a millimeter thick is embedded in the windshield which has been heat treated so as to guard against the harmful rays of the sun.

在一些飞机上，把夹有一层不足万分之五毫米厚的金膜的夹层玻璃嵌入经过热处理的挡风玻璃中，以阻挡太阳的有害射线。

“containing a film of gold” 是现在分词作定语，修饰 “sandwich of laminated glass”。短语 “less than five over ten thousand of a millimeter thick” 是 “a film of gold” 的后置定语。“which has been heat treated” 是定语从句，修饰 “the windshield”。“so as to guard against the harmful rays of the sun” 是目的状语。

(2) The iron ore from which iron——less than 20 per cent of the weight of the ore——is extracted is reckoned as poor ore, while one part of gold in 100000 parts of ore is reckoned as rich ore.

铁质量分数小于20%的铁矿石即为贫矿，而含金1/100000的金矿则被看作富矿。

“which iron——less than 20 per cent of the weight of the ore——is extracted” 是定语从句，修饰 “the iron ore”。

“while” 是连词，用来表示对比或相反情况。例如：English is understood all over the world while Turkish is spoken by only a few people outside Turkey itself（英语在全世界通行，但土耳其语离开本国就很少有人说了）。

(3) The other force that is usually operating in gravity methods is the resistance to relative motion exerted upon the particles by the fluid or semi-fluid medium in which separation takes place.

另一种在重力法中起作用的力是通过在液体或半流质媒介中发生分离来阻止作用在矿石颗粒上的相对运动。

“that is usually operating in gravity methods” 是定语从句，修饰 “the other force”。“exerted upon the particles” 是过去分词作定语，修饰 “relative motion”。 “which separation takes place” 是定语从句，修饰 “medium”。

(4) After being crushed into a fine powder, the gold ore is stirred in large tanks with a very dilute solution of cyanide, so that the gold is dissolved by the cyanide.

金矿被破碎成细粉后，搅拌混入盛有很稀的氰化物溶液的大槽中，金即被氰化物溶解。

“being crushed into a fine powder” 是动名词的被动语态。“so that the gold is dissolved by the cyanide” 为目的状语从句。

(5) the tank for leaching is a cylindrical tank of about 12 ft in diameter and 45 ft high, with a conical bottom of 60 degrees.

浸出槽是一个直径约12英尺、高45英尺、带有60°圆锥底的圆柱形槽体。

"in diameter" 意思是"直径为"。例如：The tree is two feet in diameter（这棵树直径有2英尺）。

Part Ⅱ Curriculum Ideological and Political

【Craftsmen of the nation】Dripping water nuggets pouring "rare and precious" life—Pan Congming

"China's reserves of precious metals account for only about 0.39 percent of the world's total. Without advanced technologies, some precious metals are often merely byproducts of general commodity mining and can only be treated as waste — a huge loss for the country." Pan Congming said. His invention of "green and efficient extraction technology of platinum, palladium, rhodium and iridium in nickel anode slime" has greatly simplified the purification process of platinum group precious metals, which not only saves on costs, but also is more environmentally friendly. The technology can extract eight kinds of precious metals from nickel ore waste, and the purity of each precious metal can reach 99.99 percent.

"I'm never afraid of failure. Only through failure do I know that I need to take another approach to succeed," he said. Pan's hard work has won him national-level prizes including national model worker and national technical expert.

Part Ⅲ Further Reading

Microbiological metallurgy

Microbial metallurgy technology, also known as microbial ore leaching technology or microbial leaching technology, refers to that most microorganisms can act on different grades and different kinds of minerals through various ways, transform valuable elements in minerals into ions in solution, and use this characteristic of microorganisms to integrate hydrometallurgy process to form modern microbial metallurgy technology. Microbial metallurgy technology is especially suitable for processing lean ore, waste ore, and off surface ore, or for heap leaching and in-situ leaching of ore that is difficult to mine, concentrate, and smelt. At the same time, microbial metallurgy technology has the advantages of easy operation, low cost, low energy consumption, and small pollution, and has been widely used in practical production. In foreign countries, the biological extraction of copper and uranium, as well as the biological pre oxidation of arsenic bearing gold ore have been industrialized.

Part Ⅳ Translation Training

科技英语常用句型

科技文章中经常使用若干特定的句型，例如 It…that…句型结构、被动语态句型结构、分词短语句型结构、省略句型等，从而形成科技文体区别于其他文体的标志。举例如下：

(1) It is evident that a well lubricated bearing turns more easily than a dry one.

显然，润滑好的轴承比不润滑的轴承容易转动。

(2) It seems that these two branches of science are mutually dependent and interacting.

看来这两个科学分支是相互依存、相互作用的。

(3) It has been proved that induced voltage causes a current flow in opposition to the force producing it.

已经证明，感应电压产生一电流，其方向与产生此电流的磁场力方向相反。

(4) It was not until the 19th century that heat was considered as a form of energy.

直到 19 世纪人们才认识到热是能量的一种形式。

(5) Computers may be classified as analog and digital.

计算机可分为模拟计算机和数字计算机。

(6) The switching time of the new-type transistor is shortened three times.

新型晶体管的开关时间缩短了三分之二（或缩短为三分之一）。

(7) This steel alloy is believed to be the best available here.

人们认为这种合金钢是这里能提供的最好的合金钢。

(8) Electromagnetic waves travel at the same speed as light.

电磁波传送的速度和光速相同。

(9) Microcomputers are very small in size, as is shown in Fig. 5.

如图 5 所示，微型计算机体积很小。

(10) In water sound travels nearly five times as fast as in air.

声音在水中的传播速度几乎是在空气中传播速度的 5 倍。

(11) Compared with hydrogen, oxygen is nearly 16 times as heavy.

氧气与氢气相比，其重量大约是氢气的 16 倍。

(12) The resistance being very high, the current in the circuit was low.

由于电阻很大，电路中通过的电流就很小。

(13) Ice keeps the same temperature while melting.

冰在融化时，其温度保持不变。

(14) An object, once in motion, will keep on moving because of its inertia.

物体一旦运动，就会因惯性而持续运动。

(15) All substances, whether gaseous, liquid or solid, are made of atoms.

一切物质，不论是气体、液体还是固体，都是由原子构成的。

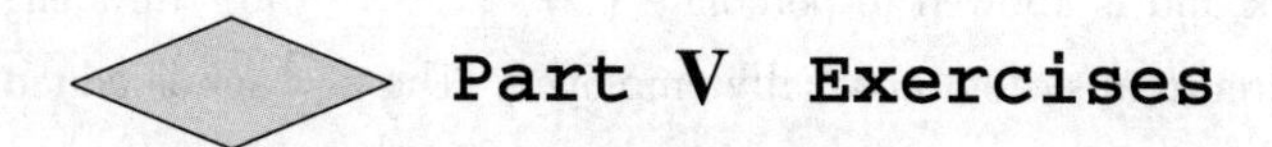

Ⅰ. Translate the following expressions into English.

（1）伴生矿物	（2）氰化法浸金	（3）堆浸法
（4）渗滤浸出法	（5）搅拌浸出法	（6）浸出槽
（7）金的回收率	（8）以一定间距	（9）假底
（10）逆流系统	（11）堆成块状	（12）母液
（13）矿浆密度		

Ⅱ. Fill in the blanks with the suitable words listed below. Change the form where necessary.

recovery	filtering	cylindrical	agent
cyanide	circulation	mechanical	propeller
expose	stir	pulp	

（1）Gold and silver were called ***noble metals*** because they could be ______ to the atmosphere for a long time without tarnishing.

（2）______ solution is then sprayed at the top of the dump, and leaching solution is collected in streams at the bottom of the heap.

（3）Heap leaching operation usually takes 30~90 days, and the ______ of gold is 50%~70%.

（4）The material to be leached is placed in a tank equipped with a false bottom covered with a ______ medium.

（5）After being crushed into a fine powder, the gold ore is ______ in large tanks with a very dilute solution of cyanide, so that the gold is dissolved by the cyanide.

（6）______ densities vary from 40% to 70%.

（7）The leaching ______ is added and the pulp is agitated continuously.

（8）Agitation leaching may be accomplished by ______ agitation leaching.

（9）The ______ or impeller is commonly used in mechanical agitation leaching.

（10）The tank for air agitation leaching is a ______ tank.

（11）Compressed air is introduced through the tube and cause the ______ of the pulp which is put in the tank.

Ⅲ. Fill in the blanks by choosing the right words form in the brackets.

Percolation leaching

The material to（1）______（leach; be leached; leaching;）is placed in a tank equipped with a false bottom covered with a filtering (2) ______ (media; medium). The solvent is added at the top of the tank and is allowed to percolate (3) ______ (to; through; for) the material. In the tank a countercurrent system is usually employed. The new solids added to the last tank and the weak liquid added to the first are pumped successively from one tank to another till it reaches the last one, almost saturated.

The process is well suited to cases where the material is porous and sandy, but is inapplicable (4) ______ (with; in; to) the material which is impervious and tends (5) ______ (pile; piled; to pile) into a block.

Regularly the size of particles rather than the size of the matter is the chief factor governing good (6) ______ (percolate; percolates; percolation). The method is, therefore, unsatisfactory if much residue is (7) ______ (presents; presence; present). (8) ______ (Their; It's; Its) advantages are the minimum of solvent consumption, the production of high grade pregnant solution, and elimination of the use of expensive thickeners or filters. When leaching is finished, the tanks are emptied (9) ______ (manual; manually), and then a new batch is (10) ______ (being introduced; introducing; introduced).

Ⅳ. Decide whether the following statements are true or false (T/F).

(1) Gold was called ***noble metals*** because it could be exposed to the atmosphere for a long time without tarnishing and could be melted repeatedly without any loss in weight. (　)

(2) The economic value of a kind of ore depends on the price of the metal that is contained in the ore only. (　)

(3) Cyanide gold leaching process can be grouped into three classes: lump leaching, percolation leaching, pulp agitation leaching. (　)

(4) Heap leaching operation usually takes 30 ~ 60 days, and the recovery of gold is 50% ~ 70%. (　)

(5) After being crushed into a fine powder, the gold ore is agitated in large tanks with a very dilute solution of cyanide, so that the gold is dissolved by the cyanide. (　)

(6) The pulp after leaching consists of solution of gold and tailings. (　)

Ⅴ. Translate the following English into Chinese.

To produce one troy ounce (31.3 grams) of gold from a typical rich mine, about three tons of ore must be blasted from the reefs, conveyed to the shaft, hoisted to the surface, crushed to the consistency of powder, and then through the processes of agitation, filtration and treatment, the metal is separated with potassium cyanide.

Unit 17 Metal Forming Processes

Part Ⅰ Reading and Comprehension

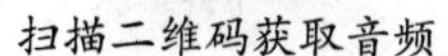

扫描二维码获取音频

扫描二维码获取 PPT

Almost every industrial product has metal parts or is made by machines with metal parts. All of these metal parts must be formed or shaped. As an important content of parts design, metal forming method and process is an issue makers extremely concerned, and the key factors in the process of material processing, Metal forming methods and processes can be grouped into following classes: casting, plastic molding, mechanical processing molding, joint molding, powder metallurgy, injection molding, semi-solid molding and 3D printing molding.

1. Casting

Casting is the earliest known form of liquid metal processing and is still an important process today. Casting is the introduction of molten metal into a cavity or mold where, upon solidification, it becomes an object whose shape is determined by mold configuration. The technological process is as follows: liquid metal → mold filling → solidification → casting.

The characteristics of casting process are mainly shown in the following aspects:

(1) It can produce castings with complex shapes, especially those with complex inner cavities.

(2) Strong adaptability. Alloy type and casting size are not limited.

(3) Wide material sources, remodeling waste products, low equipment investment.

(4) High rejection rate, low surface quality and poor working conditions.

The main casting processes include sand casting, investment casting, pressure casting, low pressure casting, centrifugal casting, permanent-mold casting, vacuum casting, die casting, lost foam casting and continuous casting etc.

2. Plastic Molding

Plastic forming is the technology of processing work pieces under the force of tool and die by using plastic. There are many kinds of plastic forming, including forging, rolling, drawing, extrusion, stamping and so on. For a better understanding of the mechanics of various forming operations, we shall briefly discuss each of these processes.

Forging techniques are subdivided into: hammer forging, and die forging. Hammer forging is one of the oldest metalworking processes known to man. As early as 2000 B. C. , forging was used to produce weapons, implements, and jewelry. The process was performed by hand using simple hammers. Today, the material is squeezed between two or more dies to alter its shape and size in forging. Depending on the situation, the dies may be open [shown in Fig. 17-1(a)] or closed [shown in Fig. 17-1(b)].

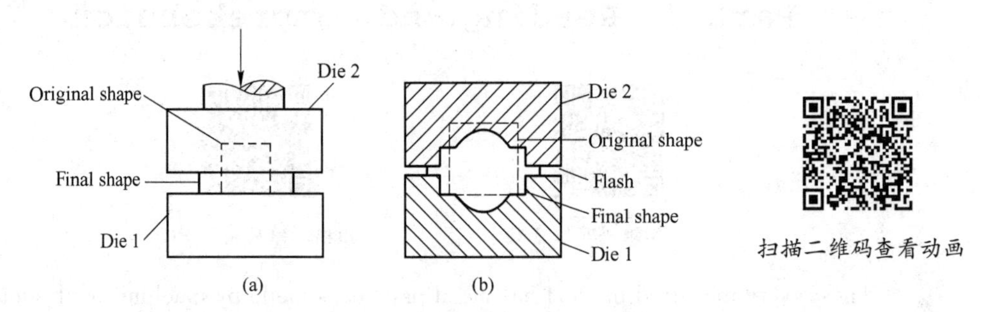

Fig. 17-1　Forging operation

(a) open die forging; (b) closed die forging

Rolling is the most widely used process of metal plastic forming. As early as the sixteenth century soft metals such as lead, gold and silver had been rolled into sheets, but it was not until the eighteenth century that the process was employed on any large scale for the rolling of harder metals. There are records of a rolling mill in Birmingham in 1755, capable of hot-rolling bars of metal 7.5 cm wide and reducing them to a quarter of their former thickness, with corresponding extension of length. The contents about rolling are discussed in detail later.

In drawing process, the cross-section of a wire or that of a bar or tube is reduced by pulling the workpiece through the conical orifice of a die. When high reduction is required, it may be necessary to perform the operation in several passes.

The process of extrusion involves a similar principle to that of squeezing toothpaste from the tube or making cake decorations by squirting icing sugar from a bag. Extrusion is one of the most potential and useful metal working processes and has a large number of variations in the mode of application. It can be performed under both hot and cold conditions. A prodigious pressure would be necessary to extrude most of the common metals while cold, and the plant is generally designed to extrude metals hot. The process is so economical that is a normal production procedure for such products as rods, bars, tubes, and strips.

Stamping is a forming processing method of pressing on the plate, strip, pipe and profile by pressing machine and die to produce plastic deformation or separation, so as to obtain the required shape and size of the work piece (stamping). 60 to 70 percent of the world's steel are plate and most of them are stampings after stamping. Automobile body, chassis, fuel tank, radiator, boiler drum, container shell, electric motor, electrical core silicon steel sheet are all made by stamping. There are also a large number of stamping partsin instruments, household appliances, bicycles, office machinery, household utensils and other products.

3. Mechanical processing molding

Mechanical processing molding is a machining process in which the excess metal layer thickness is directly cut off from the blank with a tool in the process of part production to meet the technical requirements of dimension accuracy, shape and position mutual accuracy and surface quality as required by the drawing. Common methods are: turning, milling, shaping and planing, drilling, boring, grinding, gear machining and CNC machining.

4. Joint molding

The attention which has been paid in recent years to joining processes, such as **soldering**, welding, has made built-up construction both rapid and efficient, and it is often cheaper than one-piece manufacture. The main types of joining processes are riveting, soldering, welding by electrical, fusion welding, plasma-jet welding, etc.

A weld has been described as the whole of metallurgy in miniature, for it calls for the knowledge of the melting, casing, and forging of metals.

5. Powder metallurgy

Powder metallurgy is a process for making a wide range of components and shapes from a variety of metals and alloys in the form of powder. In powder metallurgy, the three main steps are: (1) the preparation of the metal powders, (2) the compacting or pressing of the basic form to shape, and (3) the sintering or heating of the metal powders to a nearly liquid state in an inert atmosphere. Powder metallurgy is used primarily in the making of tools. In addition to the obvious advantages of producing a finished component which does not require further machining, powder metallurgy offers great possibilities for recycling.

6. Injection molding

Injection molding is a molding method in which metal powder and its binder plasticized mixture is injected into the model. The selected powder is mixed with the binder, then the mixture is pelleted and injected into the desired shape. The injection molding process is divided into four unique processing steps (mixing, forming, degreasing and sintering) to achieve the production of parts and determine whether surface treatment is required for product characteristics.

7. Semi-solid molding

Semi-solid forming process takes advantage of the characteristics of metal transition from liquid to solid (coexistence of solid and liquid). It combines the advantages of solidification forming and plastic forming with low processing temperature, small deformation resistance, complex shapes and high precision requirements. At present, it has been successfully used in the manufacturing of master cylinder, steering system parts, rocker arm, engine piston, wheel hub, transmission system parts, fuel system parts, air conditioning parts and so on.

8. 3D printing molding

3D printing is a rapid prototyping technology based on digital model files. This technology uses materials such as powdered metals or resins to construct objects by printing them layer by layer. The comparison of several 3D printing molding technologies is shown in Table 17-1.

Table 17-1 Printing molding technology introduction table

Technology	FDM (Fused Deposition Modelling)	SLA (Stereo lithography appearance)	SLS (Selective Laser Sintering)	3DP (3D Printing)
Materials	thermoplastic fed in the form of filiform, such as wax, ABS, PC, PPSF, nylon, PLA, etc.	liquid photosensitive resin	wide range of materials: thermoplastic, resin coated sand, Polycarbonate, metal powder, Ceramic powder	ceramic powder, metal powder
Methods	fused deposition modelling	UV irradiation (wavelength $x = 325$mm, strength $\omega = 30$mW) achieves photopolymerization and curing	high intensity CO_2 laser sintering	micro droplet spray binder (such as silicagel) molding
Layer height/mm	0.15~0.4	0.016~0.15	0.08~0.15	0.013~0.1
Holder	need as necessary	need	no need	no need
Characteristics	Simple maintenance and low cost, desktop application is the most widely used	The highest precision molding method at present	It is widely used in large-scale structural design, such as aerospace	High speed and colorful molding

Words and Expressions

扫描二维码获取音频

扫描二维码答题

plastic molding 塑性成型
mechanical processing molding 机械加工成型
joint molding 接合成型
powder metallurgy 粉末冶金
injection molding 注射成型
semi-solid molding 半固态成型
3D printing molding 3D 打印成型
cavity [ˈkæviti] *n.* （铸造）型腔

configuration [kən͵figju'reiʃən] *n.*	轮廓，外形
intricate [intrikit] *a.*	复杂的；难懂的
sand casting	砂型铸造
investment casting	熔模铸造
pressure casting	压力铸造
low pressure casting	低压铸造
centrifugal casting	离心铸造
permanent-mold casting	金属型铸造
vacuum casting	真空铸造
die casting	挤压铸造
lost foam casting	消失模铸造
continuous casting	连续铸造
drawing	拉延
extrusion [eks'truːʒən] *n.*	挤压；推出
stamping ['stæmpɪŋ]	冲压
mechanics [mi'kæniks] *n.*	机械学；力学
hammer forging	锤锻
die forging	模锻
weapon ['wepən] *n.*	武器；兵器
implement ['implimənt] *n.*	工具；器具
jewelry ['dʒuːəlrɪ] *n.*	珠宝；珠宝饰物
squeeze [skwiːz] *v.*	挤出，挤压
alter ['ɔːltə] *v.*	改变，变更
strain [strein] *n.*	张力；应变
compression [kəm'preʃ(ə)n] *n.*	压缩，压紧
soft metal	软金属
lead [liːd] *n.*	铅
sheet [ʃiːt] *n.*	（一）片，（一）张，薄片
rolling mill	轧机
in detail	详细地
extension [iks'tenʃən] *n.*	延长；扩充
workpiece ['wəːkpiːs] *n.*	工件
conical ['kɔnikəl] *a.*	圆锥的；圆锥形的
orifice ['ɔrifis] *n.*	孔；口
conical orifice	锥形孔
pass [pɑːs] *n.*	通道；轧道
squirt [skwəːt] *v.*	喷出
metalworking ['metəl͵ wəːkɪŋ] *n.*	金属加工
prodigious [prə'didʒəs] *a.*	巨大的，庞大的
extrude [eks'truːd] *v.*	挤压出

plasticized mixture　增塑的混合物
sintering ['sɪntərɪŋ] *n.*　烧结
coexistence [ˌkəuɪg'zɪstəns] *n.*　共存
rapid prototyping　快速成型

Answer the following questions.

(1) How many classes can metal forming method and processes be grouped into?
(2) What are the characteristics of casting process?
(3) What do casting processes include ?
(4) When was forging used to produce weapons, implements, and jewelry ?
(5) When were soft metals such as lead, gold and silver rolled into sheets ?
(6) What do Forging techniques include?
(7) What are the main types of joining processes?

(1) Casting is the introduction of molten metal into a cavity or mold where, upon solidification, it becomes an object whose shape is determined by mold configuration.

铸造是将熔化的金属导入型腔或铸模中，在那里，金属一旦凝固就会变成一个形状由铸模轮廓确定的物件。

"upon" 在此是介词，意思是"一……就，在……时"。例如：Rivets contract upon cooling（铆钉在冷却时会收缩）。

(2) If the working temperature is higher than the recrystallization temperature of the material, then the process is called hot forming. Otherwise the process is termed as cold forming.

如果加工温度高于材料的再结晶温度，那么这一过程就称为热成形。否则，这一过程就称为冷成形。

(3) The process was performed by hand using simple hammers. Today, the material is squeezed between two or more dies to alter its shape and size in forging. Depending on the situation, the dies may be open [shown in Fig. 17-1(a)] or closed [shown in Fig. 17-1(b)].

这一工艺过去是手工使用简单的锤子完成的。今天的锻造材料是在两个或多个模具间受到挤压，以改变其形状和尺寸。根据情况不同，模具可以是开式的［见图 17-1(a)］，也可以是闭式的［见图 17-1(b)］。

(4) …, but it was not until the eighteenth century that the process was employed on any large scale for the rolling of harder metals.

……，但是直到 18 世纪，轧制法才大规模地用来轧制较硬的金属。

"It was…that" 是强调句，这个句型可以强调主语、宾语和状语。如强调含 "not…until" 的句子，要注意把主句中的 "not" 与后面状语一起放在被强调的位置。例如：直到午夜，雨才停止，正常语序可这样写：It did not stop raining until midnight。如果用上述强调句，可写成：It was not until midnight that it stop raining.

（5）There are records of a rolling mill in Birmingham in 1755, capable of hot-rolling bars of metal 7.5cm wide and reducing them to a quarter of their former thickness, with corresponding extension of length.

根据记载，1755 年在伯明翰开始有轧机，当时可以热轧出 7.5 厘米宽的条材，其厚度减小到原来的四分之一，相应地延展了长度。

（6）In drawing process, the cross-section of a wire or that of a bar or tube is reduced by pulling the workpiece through the conical orifice of a die.

在拉延工艺中，金属丝的截面或者是条钢或钢管的截面，由于工件被拉过模具的锥形孔而减小。

（7）The process of extrusion involves a similar principle to that of squeezing toothpaste from the tube or making cake decorations by squirting icing sugar from a bag.

挤压法的原理同从牙膏筒里挤牙膏和从口袋里往糕点上挤糖衣花饰一样。

句中“that”代表“principle”。

（8）The attention which has been paid in recent years to joining processes, such as soldering, welding, has made built-up construction both rapid and efficient, and it is often cheaper than one-piece manufacture.

近年来，金属的接合方法如钎焊、焊接等，受到了重视。这些方法可以使“装配”结构既快又效率高，而且往往比“整体制造”便宜。

（9）A weld has been described as the whole of metallurgy in miniature, for it calls for the knowledge of the melting, casing, and forging of metals.

焊接一向被描述为整个冶金业的缩影，因为它要求具有金属熔炼、铸造和锻造方面的综合知识。

“in miniature”意思是“小型，小规模，小比例”。例如：She is just like her mother in miniature（她简直是她母亲的缩影）。

（10）Powder metallurgy is a process for making a wide range of components and shapes from a variety of metals and alloys in the form of powder.

粉末冶金是以粉末形式的各种金属和合金制造各种各样零件和形状的方法。

“a wide range of”与“a variety of”的意思都是“各种各样的，种种的”。例如：He left for a variety of reasons（他因种种原因而离开了）；The new model comes in a wide range of colors（这种新式样有各种各样的颜色）。

Part Ⅱ Curriculum Ideological and Political

【The innovation-driven development】The Report of the 20th National Congress of the Communist Party of China | Accelerating the implementation of the innovation-driven development strategy

Setting our sights on the global frontiers of science and technology, national economic development, the major needs of the country, and the health and safety of the people, we should speed up efforts to achieve greater self-reliance and strength in science and technology.

To meet China's strategic needs, we will concentrate resources on original and pioneering scientific and technological research to achieve breakthroughs in core technologies in key fields. In order to enhance China's innovation capacity, we will move faster to launch a number of major national projects that are of strategic, big-picture, and long-term importance. We will strengthen basic research, prioritize original innovation, and encourage researchers to engage in free exploration.

Part Ⅲ　Further Reading

Hand tear steel

Taiyuan Iron and Steel Group (TISCO), a leading stainless steelmaker, has achieved a major technological breakthrough by making an extremely thin stainless steel sheet called hand tear steel.

Wang Tianxiang, general manager of Shanxi Taigang Stainless Steel Precision Strip Co., Ltd, a subsidiary of TISCO, said in the past only a few countries such as Germany and Japan could produce thin steel sheets, in a narrow shape and width of 350mm to 400mm. However, the product, that TISCO developed after over 700 failed attempts, was just 0.02mm thick-equivalent to one fourth of a piece of standard A4 paper, but also 600mm wide, the world's first such product. Such hand tear steel can be applied in a wide range of areas such as aviation, new energy and flexible foldable screen, said Wang.

"High-quality development means to do things that others cannot do or cannot do well and to use specialized products to satisfy the demand, create the need and lead the market. In order to do that, innovation is the core, talent is the base and the mechanism is the key." said Gao Xiangming, president of TISCO.

Part Ⅳ　Translation Training

定语从句的翻译

英语中定语从句很常见，一般分为限制性定语从句和非限制性定语从句。限制性定语从句与先行词关系密切，非限制性定语从句与先行词关系松弛。进行英译汉时，要根据定语从句的结构和含义灵活处理。有的定语从句比较短而且结构简单，可以译成汉语带“的”的定语词组；有的定语从句比较长而且结构复杂，可译成汉语的并列分句；有的定语从句是整个句子的重点，可以和主句融合成一个句子；有的定语从句对主句起解释说明作用，可以分译成一个独立的句子；有的定语从句兼有状语的职能，可译成汉语的偏正复句。

（1）限制性定语从句与先行词关系密切，没有它主句的意思就不完整，一般可按汉语的习惯译成带“的”的定语词组，放在修饰词的前面。还有一些较短的描述性的非限制性定语从句，也可以用上述方法翻译。例如：Oxygen is a gas which unites with many substances（氧气是一种能和许多物质化合的气体）；The earth contains a large number of metals which are useful to man（地球蕴藏着大量对人类有用的金属）；The furnace that is used for separating iron from the other elements combined with it in the iron is called a blast furnace（用于把铁从铁矿石内与之化合的其他元素中分离出来的炉子称为高炉）；The iron ore we find in the earth is not pure（我们在地球上发现的铁矿石不是纯的）；The sun，which had hidden all day，now came out in all its splendor（那个整天躲在云层里的太阳，现在又光芒四射地露面了）。

（2）非限制性定语从句对先行词不起限制作用，只对它附加说明。其在形式上与先行词以逗号分开，与先行词关系比较松弛。非限制性定语从句在语义结构上与主句接近并列结构，翻译时可译成并列分句。结构复杂的限制性定语从句，如译成前置定语，则显得太长，不符合汉语的表达习惯，往往也译成后置的并列分句。例如：Metal has many useful properties，of which strength is the most important（金属有许多性能，其中强度最重要）；By mixing aluminium with other metals，scientists have been able to produce a variety of alloys，some of which have the strength of steel but weigh only one third as much（科学家把铝同其他金属混合已能生产许多金属，其中有一些强度与钢相同，但重量只有钢的三分之一）；Daylight comes from the sun，which is a mass of hot，glowing gas（日光来自太阳，太阳是一团炽热、发光的气体）；A fuel is a material which will burn at a reasonable temperature and produce heat（燃料是一种物质，在适当的温度下它能够燃烧并放出热量）。

（3）在英语中有一些限制性定语从句与主句关系密切，并突出了全句的重点，而主句仅起结构上的作用；也有些定语从句是作主句的一个修饰成分。对于这些复合句，可以把主句和定语从句融合在一起译成一个独立的句子，这样符合汉语的习惯。例如：There are some metals which possess the power to conduct electricity（某些金属具有导电的能力）；This is the steelworks that they set up in 1980（这个钢厂是他们在 1980 年建造的）；There are many people who want to visit the rolling mill（许多人想参观这个轧钢厂）；Science plays an important role in the society in which we live（科学在我们生活的社会里起着重要作用）。

（4）有些非限制性定语从句对先行词不起限制作用，只是对先行词加以解释和补充说明。翻译时可将其与主句分开，译成一个独立的汉语句子，这样译文的表达简洁明了。例如：Galileo，who made the first telescope，died in 1642（伽利略死于 1642 年，他制造了第一架望远镜）；The most important form of energy is electric energy，which is widely used in our daily life（电能是一种重要的能量形式，它广泛用于我们的日常生活中）；Nevertheless the problem was solved successfully，which showed that the computation were accurate（不过问题还是圆满地解决了，这说明计算得准确）。

（5）在英语中有些定语从句在形式上是定语从句，但在意义上却起着状语的作用，说明原因、结果、目的、让步等关系，翻译时应根据逻辑关系，译成汉语相应的各种偏正复句。例如：Metals，which possess strength，ductility，malleability，and other remarkable properties，have supplanted other classes of engineering materials to a considerable extent（由于金属具有强度、可延性、韧性以及其他显著的特性，因而在很大程度上已代替了其他种类的工程材料）；Matter has certain features or properties that enables us to recognize it easily（物质具有一定的特性或特征，所以我们能很容易地识别各种物质）；He wishes to write an article that will attract public attention to the matter（他想写一篇文章，以便引起公众对这件事的注意）；The problem，which is very complicated，have been solved（这个问题虽然复杂，但已经解决了）。

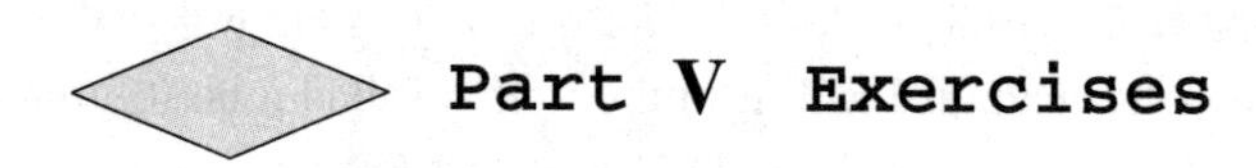

Part V　Exercises

Ⅰ. Translate the following expressions into English.

（1）金属成型工艺	（2）液态金属成型	（3）铸模轮廓
（4）复杂形状	（5）金属铸造工艺	（6）离心式铸造
（7）批量生产	（8）锤锻	（9）粉末冶金
（10）锥形孔	（11）“整体”制造	（12）等离子喷射焊接

Ⅱ. Fill in the blanks with the suitable words listed below. Change the form where necessary.

conical	forge	tool	impact	industrial	wide
form	strength	technique	potential	produce	liquid

(1) Almost every ______ product has metal parts or is made by machines with metal parts.

(2) Casting is the earliest known form of ______ metal processing and is still an important process today.

(3) Ingot castings are ______ by pouring molten metal into a permanent or reusable mold.

(4) Hot and cold ______ of metals were once done by forging with a hammer.

(5) Forging ______ are subdivided into: 1) hammer forging; 2) die forging.

(6) Rolling is the most ______ used process of metal plastic forming.

(7) In drawing process, the cross-section of a wire or that of a bar or tube is reduced by pulling the workpiece through the ______ orifice of a die.

(8) Extrusion is one of the most ______ and useful metal working processes and has a large number of variations in the mode of application.

(9) Powder metallurgy is used primarily in the making of ______.

(10) Today, forging is done mostly by the plastic deformation of a metal through the ______ of compression forces.

(11) All of these metal parts must be ______ or shaped.

(12) Now one valuable feature of forging is that it improves the ______ of the metal by refining the structure and making it uniform.

Ⅲ. Fill in the blanks by choosing the right words form given in the brackets.

In order to (1) ______ (obtain; obtaining) a metal in a certain form, it may be hot-worked or cold-worked in special machines which (2) ______ (shapes; shape) the metal by means of pressure. The most important parts of these processes are hammering or forging, pressing and rolling. Forging is the simplest method of (3) ______ (shape; shaping) metal to the required form by deformation. Forging is usually (4) ______ (doing; done) by hammers. The most common type of hammers (5) ______ (using; used) in industry are steam hammers and air hammers. Metal is forged or pressed hot, but for pressing it should be heated to a (6) ______ (low; lower) temperature than for forging. Rolling is the process of shaping metal in a machine (7) ______ (calls; called) rolling mill. Ingots of metal are rolled by forcing them between two rollers rotating in opposite directions, thus pressing the (8) ______ (requiring; required) shape. There are two kinds of rolling: hot rolling and cold rolling. Cold rolling produces a higher surface finishing sheet and gives it a very exact size. Stamping or cold forging means (9) ______ (to press; pressing) cold-rolled or hot-rolled sheets between two dies. One of them has a hollow of certain shape, while the other die forces the sheet metal into this hollow, thus forcing the metal sheet (10) ______ (following; to follow) the shape of the hollow in the die. This operation is very useful because by means of this process, objects with various shapes can be produced in mass production more economically and efficiently.

Ⅳ. Decide whether the following statements are true or false (T/F).

(1) Two broad categories of metal-casting processes exist: ingot casting (which includes continuous casting) and casting to shape. ( )

(2) Now one valuable feature of forging is that it improves the brittleness of the metal by refining the structure and making it uniform. ( )

(3) Powder metallurgy is a process for making a wide range of components and shapes from a variety of metals and alloys in the form of liquid. ( )

(4) Almost every industrial product has metal parts or is made by machines with metal parts. ( )

(5) Not all of these metal parts must be formed or shaped. ( )

Ⅴ. Translate the following English into Chinese.

Modern use of powder metallurgy began in the 18th century and it has become extensively employed during the first half of the 20th century. Powder metallurgy is the art of producing metal powders and utilizing them to make serviceable objects. The processes involve the heating (termed sintering) of a compact formed by compression. Basically, powder metallurgical methods are used in preference to conventional methods for only two reasons: to secure properties not otherwise obtainable, or to eliminate machining and subsequent operations.

Unit 18 Rolling

Part Ⅰ Reading and Comprehension

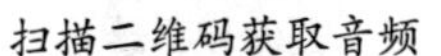

扫描二维码获取音频

扫描二维码获取PPT

1. Introduction to steel rolling

Steel rolling is a continuous or stepwise forming with the aid of morethan two rotating rolls. The rolls act on the metal primarily through pressure (shown in Fig. 18-1. Therefore rolling is classified among the pressure-forming methods.

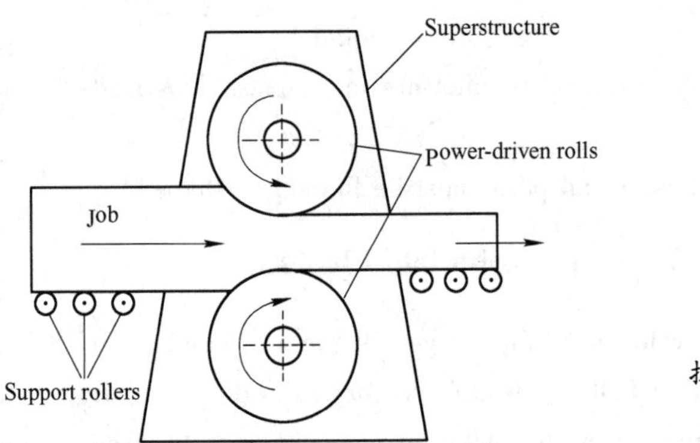

扫描二维码查看动画

Fig. 18-1 Rolling operation

From the perspective of metallography, steel rolling is divided into hot rolling and cold rolling according to the rolling temperature. Hot rolling refers to the process above the recrystallization temperature, while the plastic process below the recrystallization temperature is called cold rolling.

Steels may be rolled hot or cold. Hot rolling is carried out at an elevated rolling-stock temperature, and cold-rolling is performed without heating.

Hot-rolling mills are generally divided into the following zones: furnace area (for heat supply prior to deformation), rolling area, and finishing area. Rolling stocks will be homogeneously preheated to defined temperature by socking pits in the furnace area. The rolling area completes the rolling process. Rolling trains are normally adapted to the products to be rolled. Hence, there are many very different types of train, such as roughing trains, intermediate and finishing trains. The most important tasks of the finishing shop are: cutting, straightening, surface protection stacking and retrieving, inspection, checking, sorting, marking, collecting, bundling, packing. The sequence and extent of these jobs will depend on the nature of the product.

Cold-rolling mills generally include the following areas: pickling area, rolling train, heat treatment, and finishing area. Steel is cold rolled mainly for producing flat products such as deep-drawing sheet, tin sheet and stainless sheet. Sectional steel products and tubes are also cold rolled. The most wide-spread process is the cold rolling of strip. Strip is cold rolled on two-high, four-high or multiple-roll mills. In order to eliminate work hardening after cold rolling, heat-treatment by annealing is frequently applied. A combination of cold forming and heat treatment permits specified technological properties to be obtained.

The advantage of hot rolling is that it can be reduced in thickness much more easily. But the surface finish and accuracy are not so good as those obtainable by cold rolling. The most general use of hot rolling, therefore, is for breaking down large ingots; cold rolling is used to make smooth and accurate thin sheets of metal. As cold steel is harder than hot steel, cold steel mills need harder rolls and the power required for cold rolling is greater than that for hot rolling.

Two main groups of rolled steel products are semi-finished products and finished products. Semis are the blooming and slabbing as well as the hot wide strip. Normally, they are deformed in the steel plants into finished products. Finished products are those whose hot forming has been completed in the rolling mills. Finished rolled steel products are classified according to the shape of their cross-sections as: section product , flat product, tube, and wire rod.

Section products are subdivided into groups each with differing individual profiles. They include:

(1) **Sectional steel.** Sectional steel include circular, square, rectangular, octagonal or semi-circular sections, and I, H, U, profiled sections and wide-flanged beams etc.

(2) **Reinforcing steel.** With a diameter measuring 6 to 8 mm, reinforcing steel is also a bar with a circular cross-section. Its surface is usually ribbed.

(3) **Rail accessories.** This category covers all the parts needed for building railway tracks.

(4) **Flat.** Flats are rectangular in cross-section. The width is much larger than the thickness. The surface is mostly smooth but may also be patterned. Flat products are classified by thickness into: heavy flat steel (>60mm), medium steel plat (4~60mm) and steel sheet (<4mm).

(5) **Tube.** Tubes are hollow sections. Although normally circular in cross-section, other forms are also produced.

(6) **Wire rod.** Wire rod's surface is mostly smooth. Its cross-section may be circular, oval, square, rectangular, etc. Thicknesses range from 5 to 40 mm. Most wire rod undergoes further treatment by cold drawing or cold rolling.

2. Controlled rolling and cooling of steel

Controlled rolling and controlled cooling technology is an advanced steel rolling technology which can save metal, simplify working procedure and energy consumption. It can fully tap the potential of steel through technological means and greatly improve the comprehensive properties of steel.

Controlled rolling is a new rolling process that combines thermoplastic deformation with solid phase transition through reasonable control of metal heating system, deformation system and temperature system in the hot rolling process, so as to obtain fine grain structure and provide steel with excellent comprehensive mechanical properties.

Controlled cooling is a new process for controlling the cooling rate of rolled steel to improve the microstructure and properties of rolled steel. The combination of controlled rolling and controlled cooling can combine the two strengthening effects of hot rolled steel (fine grain strengthening and precipitation strengthening) to further improve the steel's strength and toughness and obtain excellent comprehensive mechanical properties.

Words and Expressions

扫描二维码获取音频

扫描二维码答题

stepwise [stepwaiz] *a.* 逐步的，逐渐的
pressure ['preʃə(r)] *n.* 压，压力
prior to 居先，在前
deformation [ˌdiːfɔː'meiʃən] *n.* 变形
finishing area 精整区
rolling-stock 轧制材料
rotating rolls 旋转轧辊
socking pits 均热炉
train [trein] *n.* 滚道，轧机组，机列
roughing trains 粗轧机组
intermediate [ˌintə'miːdjət] *a.* 中间的
finishing shop 成品车间
stacking ['stækiŋ] *n.* 堆垛
retrieve [ri'triːv] *v.* 保持，保存
marking ['mɑːkiŋ] *n.* 做记号，做标记
bundling ['bʌndliŋ] *n.* 捆扎
packing ['pækiŋ] *n.* 包装
pickling ['pikliŋ] *n.* 酸洗

deep-drawing	深冲，深压
flat [flæt] *n.*	板
annealing [ə'niːliŋ] *n.*	退火
surface finish	表面粗糙度
breaking down	粗轧，开坯
semis ['semis] *a.*	半成品的
semi-finished	半制成品
bloom [bluːm] *n.*	初坯，大方坯，毛坯
flat product	板材
profile ['prəufail] *n.*	剖面，外形
sectional steel	型钢
circular ['səːkjulə] *a.*	圆形的，环形的
rectangular [rek'tæŋgjulə] *a.*	矩形的
octagonal [ɔk'tægənl] *a.*	八边形的，八角形的
profiled section	异型断面
wide-flanged beams	宽缘钢梁
reinforce [ˌriːin'fɔːs] *v.*	增强，加强
rib [rib] *v.*	加肋于，加肋材于
accessories [æk'sesəriz] *n.*	辅助设备，附件
track [træk] *n.*	轨道，铁轨
hollow ['hɔləu] *a.*	空的
oval ['əuvəl] *a.*	椭圆的
range [reindʒ] *v.*	在……范围内变化
carry out	实现，完成，实行

✻ Answer the following questions.

(1) What zones are hot-rolling mills generally divided into?

(2) What are the most important tasks of the finishing shop?

(3) What areas do cold-rolling mills generally include?

(4) What is the advantage of hot rolling?

(5) What are finished rolled steel products classified as according to the shape of their cross-sections?

(6) What types are flat products classified by thickness?

(7) What products undergo further treatment by cold drawing or cold-rolling?

(8) What are two main groups of rolled steel products?

(9) What products is steel cold rolled mainly for producing?

(1) Hot rolling is carried out at an elevated rolling-stock temperature, and cold-rolling is performed without heating.

热轧在轧件高温时进行，而冷轧不需要加热。

"carry out/perform or conduct (a survey, etc)" 意思是"进行（勘察等）"。例如：carry out an investigation, an enquiry（进行调查、查询）。

(2) Rolling trains are normally adapted to the products to be rolled.

轧制道次要与轧制的产品相适应。

"(be) adapted to" 意思是"适应"。例如：He adapted to the cold weather.（他适应了寒冷的天气）；When he moved to Canada, the children adapted to the change very well（他移居加拿大后孩子们很能适应变化）。

(3) A combination of cold forming and heat treatment permits specified technological properties to be obtained.

冷轧和热处理的结合可以使钢材获得规定的技术性能。

"permit/make (sth.) possible" 意思是"使……有可能"。例如：The new road system permits the free flow of traffic at all times（新道路系统可使车辆在任何时候都畅通无阻）；The windows permit light and air to enter（这些窗户采光及通风性能良好）。

(4) The surface is mostly smooth but may also be patterned.

表面基本是光滑的，也可以用图案装饰。

"patterned/decorated with a pattern" 为形容词，意思是"有图案装饰的，有花样的"。例如 patterned china, wallpaper（有图案的瓷器、壁纸）。

(5) Thicknesses range from 5 to 40mm. Most wire rod undergoes further treatment by cold drawing or cold rolling.

厚度从 5 毫米到 40 毫米不等。多数线材需要经过冷拔和冷轧进一步处理。

"range" 意思是"（在一定范围内）变化，变动"。例如：Prices ranged from 5 dollars to 10 dollars（价格自 5 美元至 10 美元不等）；The increase ranged from several to several dozen times（增长几倍到几十倍不等）。

Part Ⅱ　Curriculum Ideological and Political

【Create dreams with ingenuity】 Rolling giant—Niu Guodong

Niu Guodong, born in 1975, member of CPC, monitor of continuous rolling area of Stainless Cold Rolling Plant of Taiyuan Iron & Steel (Group) Co., Ltd. Delegate of 18th and 19th CPC National Congress. He is the leader of "National model worker Innovation Studio" and "National skill Master Studio". For 22 years, he stuck to his original aspiration. During the commissioning of

No. 12 rolling mill, he created the record of the shortest commissioning time of single stand rolling mill, the earliest commissioning and the highest starting point of product quality. He bravely undertook the mission and was the first man who successfully finished the rolling of martensitic steel with large deformation. Relying on the platform of "Niu Guodong Innovation Studio", he has trained 92 senior rolling mill workers and 6 senior technicians, with a cumulative effect of more than 90 million yuan. He has won the "May 1 Labor Medals", "May Fourth Medals", "Super Model Workers of Shanxi Province", " San Jin Outstanding talent" and other honors.

Part Ⅲ Further Reading

New process of high quality special steel, high efficiency and low cost special metallurgy

1. Tertiary refining technology

After the conventional electric furnace or converter process, add three times of refining, such as vacuum consumable furnace or electroslag remelting smelting, to obtain special steel with high cleanliness, which can be used to efficiently and low-cost prepare special steel materials and other high-performance metal materials for aerospace and other applications.

2. New generation of special steel purification and homogenization refining technology

Research and develop a new generation of special steel ladle clean refining technology that is pollution-free to molten steel and characterized by heating and deoxidation, a new technology of high-end stainless steel pressurized nitrogen increasing metallurgy, and electroslag remelting technology based on conductive crystallizer to produce high-end alloy steel.

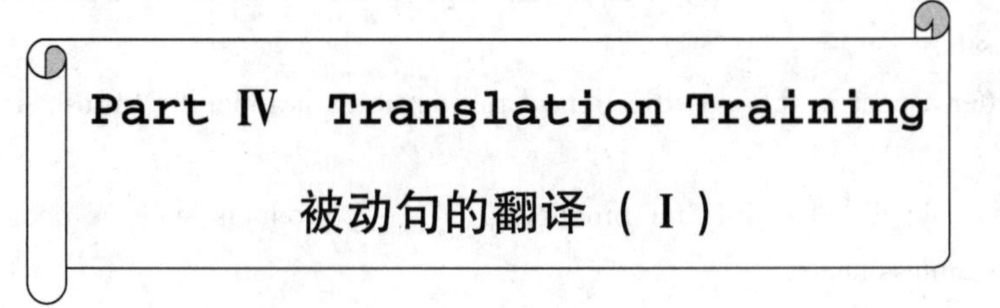

Part Ⅳ Translation Training

被动句的翻译（Ⅰ）

被动语态是英语动词的一种变化形式，被动结构的句子是以动作的承受者作主语，也就是说，英语的被动句把要说明的人和事物放在主语的位置上。在英语中被动语态使用的范围很广泛，汉语中虽然也有被动语态，但使用的范围狭窄得多。汉语中许多被动句是用无主句来表达的，英语的被动句译成汉语时，要根据句子的整体结构，采用合适的方法来处理。一般来讲，英语被动句翻译成汉语的主动句、判断句、无主句、被动句。

（1）把英语的被动句译成汉语的主动句，即英语被动句的主语仍作译文的主语，或用由 by 引出的逻辑主语作译文的主语。例如：The production plan has already been drawn up（生产计划已经制订出来了）；These substances can be dissolved in water（这些物质可以溶解于水）；Large quantities of fuel are used by modern industry（现代工业耗用大量的燃料）；Scanning was first used in the transmission of pictures by telegraph（扫描最初用于电报的图像传真）；The heat energy can be changed into electrical energy（热能可以转变为电能）。

（2）把英语的被动句译成汉语的判断句，就是将英语的被动句译成汉语的“……是……的”，用此句型来说明人和事物的客观情况。它与英语被动句在意义上是相通的，具有被动的含义。例如：History is made by people（历史是人民创造的）；Some plastics have been discovered by accident（某些塑料是偶然发现的）；These materials are obtained by a new method（这些材料是用一种新方法获得的）；Concrete is made of cement，sand，stones and water（混凝土是用水泥、沙子、石子和水制成的）；The mail is sorted out by hand（邮件是人工分拣的）。

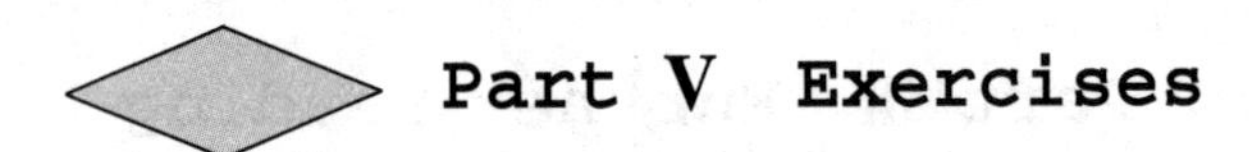

Part V Exercises

Ⅰ. Translate the following expressions into English.

（1）无缝管	（2）成品区	（3）冷拉拔
（4）拉力	（5）薄金属板	（6）连铸厂
（7）热成形	（8）空心断面	（9）钢筋

Ⅱ. Fill in the blanks with the words from the text. The first letter of the word is given.

（1）Steel rolling is a continuous or stepwise forming with the aid of several r ______ rolls.

（2）Rolling mill installations usually consist of main motor，d ______ systems and roll stands.

（3）The roll s ______ are the central elements of a rolling mill installation.

（4）If the metal is to be deformed by hot rolling，it must first be homogeneously p ______ to defined temperature.

（5）A further aspect in the classification of rolling trains is roughing trains，i ______ and finishing trains.

（6）Steel is cold rolled mainly for producing f ______ products such as deep-drawing sheet，tin sheet and stainless sheet.

（7）The most wide-spread process is the cold rolling of s ______.

（8）The advantage of working a metal hot is that it can be reduced in t ______ much more easily than when cold rolled.

（9）Finished products are those whose hot forming has been c ______ in the steel or rolling mills.

（10）Flats are r ______ in cross-section.

（11）Most wire rod undergoes further treatment by cold d ______ or cold rolling.

（12）Pipes and tubes are h ______ sections.

Ⅲ. Fill in the blanks by choosing the right words form given in the brackets.

Rolling is the most（1）______（wide；widely）used process of metal plastic forming. In the rolling process，as the metal passes between the power-driven rolls，it is pressed（2）______ (thin; thinner) and longer, although it changes very little in (3) ______ (wide; width). Metal which deforms (4) ______ (on; along) the roll surface behaves as though the roll were an inclined plane. That is, the smaller the roll, the (5) ______ (steep; steeper) the incline. Roll diameter is a major factor in the rolling process. The reduction (6) ______ (at; in) metal thickness cannot exceed $1/8^{th}$ the diameter of the roll in plastic deformation (hot work). In cold work, it cannot exceed $1/200^{th}$ the diameter of the roll. In mill practice, the metal is rolled in three or four passes, or (7) ______ (in; through) several sets (stands) of rolls. The first rolls in a series are usually (8) ______ (large; larger) than the other, because the first passes make the greatest reduction in thickness of the metal.

Ⅳ. Decide whether the following statements are true or false (T/F).

(1) Depending upon the reciprocal movements of tool and workpiece, we distinguish between: longitudinal rolling, cross rolling, and skew rolling. (   )

(2) The disadvantage of working a metal hot is that it can be reduced in thickness much more easily than when cold rolled. (   )

(3) Flats are rectangular in cross-section. The width is much smaller than the thickness. (   )

(4) Most flat and sectional products at the steel mills are produced by cross rolling. (   )

(5) Cold drawing is also a most important cold-forming technique. (   )

V. Translate the following English into Chinese.

The reheating furnaces employed to heat steel for rolling employ water quite extensively for cooling furnace doors, skid pipes, skewbacks, and so on. Water also is used extensively as a coolant for rolling-mill rolls to maintain their contours, minimize fire checking, and lengthen their service lives. On hot rolling mills, water in the form of high-pressure jets is used to remove scale from the hot steel before rolling and to keep the surface clean between certain passes. Hot-strip mills also use cooling sprays over the run out table to cool the strip to the proper temperature for coiling.

Unit 19 Tube and Wire Rod Making

Part Ⅰ Reading and Comprehension

扫描二维码获取音频　　扫描二维码获取 PPT

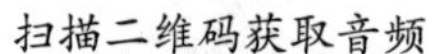

Finished products produced by rolling are so diverse that it is difficult to give a general description, but a large works might produce steel tubes and steel wires in the following ways.

1. Tube-making

Just over a hundred years ago, steel tubes began to be used in making bicycles. They possess the correct combination of strength, flexibility and lightness, so that nowadays tubes are used widely at home, in traffic, in industries, etc.

One of the world's largest manufacturers has modestly described the two basic ways of making tubes as either wrapping an accurate hole in strip steel and pushing a very strong hole through a steel bar.

The first method begins with coils of steel strip which are fed automatically through a series of forming rolls which gradually curl the flat strip into a tube shape. The edges of the tube are heated by electric induction and pressed together so that they join by welding. Then the tube travels through a series of rolls which size and straighten it. Although many such tubes are supplied in straight lengths, a great number of tubes are manipulated bent, tapered, tapped and slotted, into the shapes required by manufacturers.

The second type of process, where the ***strong hole is pushed through a steel bar***, has many variations but the products are generally described as seamless tubes. The Mannesmann process for making seamless steel tubes is illustrated in Fig. 19-1. Firstly, a piece of rolled steel bar is cut to billet length by torch cutter. Secondly, each piece, called a billet, is passed to a heat furnace by a conveyor. Next, the hot billets are passed to a hydraulic press which makes a coneshaped indentation in the center of the billet's end, making it ready for the piercing operation which will follow. Then a solid rod of hot steel is spun between two mutually inclined rolls which rotate in the same direction, so that the rod is pulled forward between them, and passes over the mandrel which is shown in the drawing. In this way a thickwalled tube is produced, the dimensions of which can be varied by the setting of the rolls and the size of the nose-piece of the mandrel. After that the tube-shape can be fabricated into a thinwalled tube by further processes. If a tube of optimum surface appearance and strength is required for, say, aircraft structures or hypodermic needles, a further process of cold drawing is applied.

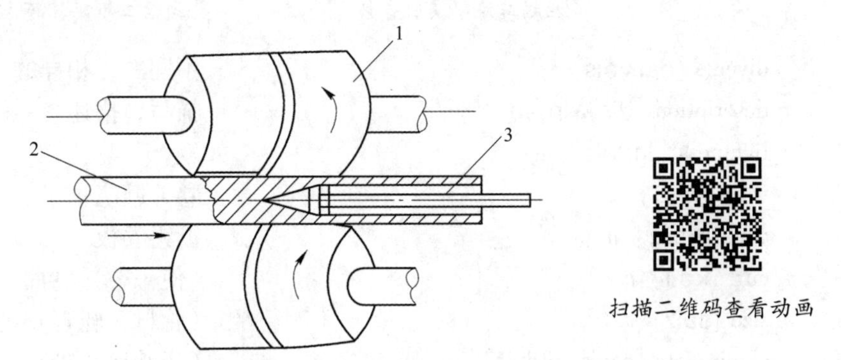

Fig. 19-1 Mannesmann tube-making process
1—Roller; 2—A steel billet; 3—Tube-shape

2. The drawing of wires and rods

Wire has been in use for many hundreds of years. It was made then and for many centuries afterward by beating metal into plates which were then cut into strips and rounded by further beating. Wire drawing was known in the 14th century, but the machinery used in this process was not perfected until the 19th century. In modern rolling steel mill the raw material is received in the form of small bars or billets. The billets are heated and conveyed to a set of rolls to be reduced in size. Finally, for ordinary sizes of wire, the billet is rolled down to a rod smaller than a lead pencil. The heated rod is carried through a pipe to a coiling device which coils the rod. The coiled metal is cooled and taken to the drawing plant where it drawn into wire of all sizes. First the scale which has accumulated is removed by an acid bath and the acid removed in an alkali bath. Next the rod with its point made small enough enters a bell-mouthed hole in a draw plate or die made of hard steel, and emerges from the smaller end of the hole reduced in size. The process of drawing the wire through smaller and smaller holes continues until the desired size is reached. As the metal is drawn finer it becomes harder and more brittle, so that from time to time it must be annealed to make it soft and tough and it must be constantly oiled as it is drawn through the dies or perforated plates. Wires are produced as shown in Fig. 19-2.

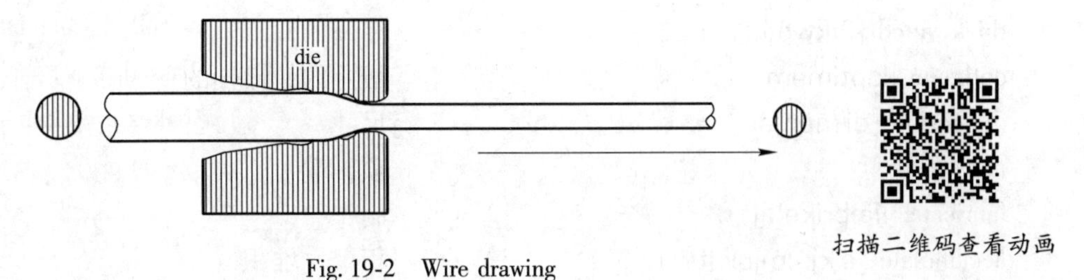

Fig. 19-2 Wire drawing

Words and Expressions

扫描二维码获取音频

扫描二维码答题

diverse [dai'və:s] *a.* 不同的，相异的
description [dis'kripʃən] *n.* 描写，描述
lightness ['laitnis] *n.* 轻
wrap [ræp] *v.* 卷，缠绕
coils [kɔilz] *n.* 镀锡卷板
curl [kə:l] *v.* （使……）卷曲
size [saiz] *v.* （管材，轧管）定径
manipulate [mə'nipjuleit] *v.* （巧妙地）处理
bend [bend] *v.*【过去式/过去分词】bent 弯曲
tapered ['teipəd] *a.* 锥形的
tapped ['tæpd] *a.* 内螺纹的
slotted [slɔtid] *a.* 有槽的
variation [ˌvεəri'eiʃən] *n.* 变更方法，变异
seamless ['si:mlɪs] *a.* 无缝的
seamless tubes 无缝管
illustrate ['iləstreit] *v.* 加插图于
torch cutter 火焰切割机
conveyor [kən'veiə] *n.* 传送带，输送机
hydraulic [hai'drɔ:lik] *a.* 水力的，水压的
hydraulic press 水压机，液压机
cone-shaped ['kəun'ʃeɪpt] *a* 锥形的
indentation [ˌinden'teiʃən] *n.* 凹痕，压痕
piercing ['piəsiŋ] *a.* 刺穿的
mutually ['mju:tʃuəlɪ, 'mju:tjuəlɪ] *ad.* 互相地
spin [spin] *v.*【过去式/过去分词】spun 自旋（转），旋转
rotate [rəu'teit] *v.* 旋转，自转
mandrel ['mændrɪl] *n.* 顶杆，芯棒
thickwalled [θɪkwɔ:ld] *a.* 厚壁的
optimum ['ɔptiməm] *a.* 最好的，最佳的
dimension [di'menʃən] *n.* 尺寸
nose-piece 顶，端
fabricate ['fæbrikeit] *v.* 制作
accumulate [ə'kju:mjuleit] *v.* 积聚，堆积

bell-mouthed	喇叭口的
emerge [i'məːdʒ] *v.*	排（涌，射，冒）出
oil [ɔil] *v.*	给……加油，涂油
perforated ['pəːfəreitid] *a.*	多孔的，穿孔的
hypodermic [ˌhaipəu'dəːmik] *a.*	皮下的，皮下注射用的
grip [grip] *v.*	紧夹，紧握
comparatively [kəm'pærətɪvlɪ] *ad.*	比较地，相当地
annealing [æ'niːliŋ] *n.*	退火

✻ Answer the following questions.

(1) What are the two basic ways of making tubes that one of the world's largest manufacturers has modestly described?

(2) When did steel tubes begin to be used in making bicycles?

(3) How was wire made in the early times?

(4) When was wire drawing known to people and when was the machinery used in this process perfected?

(5) What is the raw material for making wire?

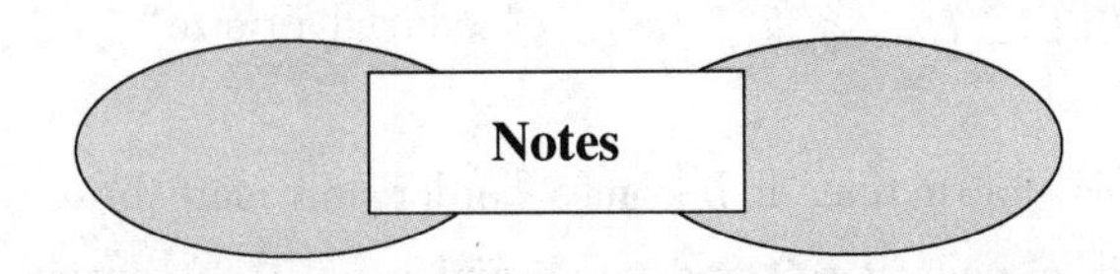

Notes

(1) The first method begins with coils of steel strip which are fed automatically through a series of forming rolls which gradually curl the flat strip into a tube shape.

第一种方法首先从带钢卷开始，带钢自动喂入一系列成形辊，这些辊子逐渐把钢带卷成管状。

句子中有两个由“which”引导的定语从句，分别修饰“coils of steel strip”和“forming rolls”。

(2) Although many such tubes are supplied in straight lengths, a great number of tubes are manipulated bent, tapered, tapped and slotted, into the shapes required by manufacturers.

虽然许多这样的钢管是以直管方式定尺交货的，但也有许多管子是按用户的要求加工成弯形、锥形、带外螺纹和内螺纹等多种形状。

“a number of”意思是“若干，一些，许多”。例如：A number of new products have been successfully trial-produced（许多新产品已试制成功）。

(3) Next, the hot billets are passed to a hydraulic press which makes a coneshaped indentation in the center of the billet's end, making it ready for the piercing operation which will follow.

接下来，热管坯经过一台水压机，水压机将管坯一端的中心压出一个锥形凹坑，以便下一步进行穿孔。

由“making”引出的分词短语作状语，起进一步解释说明的作用。

(4) In this way a thickwalled tube is produced, the dimensions of which can be varied by the setting of the rolls and the size of the nose-piece of the mandrel.

通过这种方法可生产出厚壁管，管子的尺寸根据辊缝的大小和芯棒顶头的尺寸变化而不同。

"the dimensions of which…of the mandrel" 是非限制性定语从句，"which" 代替 "a thick-walled tube"。

(5) If a tube of optimum surface appearance and strength is required for, say, aircraft structures or hypodermic needles, a further process of cold drawing is applied.

如果要求生产最佳表面和高强度的管子，如飞机构件或注射针头，则需要进一步进行冷拔。

"say" 在这里的意思是 "比如说"。例如：You could learn to play chess in, say, three months（学下国际象棋要用，比如说，三个月吧）。

(6) Next the rod with its point made small enough enters a bell-mouthed hole in a draw plate or die made of hard steel, and emerges from the smaller end of the hole reduced in size.

接着，将端部已压缩得很小的线材送入拉模板的喇叭口或进入用硬钢制成的拉模。大孔进，小孔出，断面便缩小了。

"with its point made small enough" 是介词短语作定语，修饰 "rod"。"made of hard steel" 是过去分词短语作定语，修饰 "die"。"reduced in size" 是过去分词短语作定语，修饰 "hole"。

(7) As the metal is drawn finer it becomes harder and more brittle, so that from time to time it must be annealed to make it soft and tough and it must be constantly oiled as it is drawn through the dies or perforated plates.

金属在被拉得越来越细的同时，也会变得越来越硬、越来越脆，因此必须进行中间退火以使其变软并增加其韧性。在材料通过数个拉模或板孔时，需不断上油加以润滑。

Part Ⅱ　Curriculum Ideological and Political

【The steel and iron bones】National stadium—the Bird's Nest

The National Stadium Bird's Nest is the home stadium of the 2008 Beijing Olympic Games and the venues for the opening and closing ceremonies of the Beijing 2022 Winter Olympics and the Winter Paralympics. During the construction of the Bird's Nest, the exterior steel structure alone weighed 42000 tons, supported by 12 pairs of V-shaped columns. Experts believe that the column must use Q460E-Z35 steel whose strength, low temperature toughness and shockproof 3 indicators reach the highest limit of the national standard. The Q460E-Z35, 110mm thick, has not been produced by any manufacturer in the world. Hegang and Wugang bravely undertook the development of Q460E-Z35 steel plate for the honor of the country. After more than half a year of trial production, in October 2005, all 3600 tons of Bird's Nest steel was delivered, of which 680 tons of Q460E-Z35 steel plate became the strongest "steel bar and iron bone" of Bird's Nest.

Part Ⅲ Further Reading

The quality control of high speed wire rod production

As an important product in steel production, wire rod has the characteristics of repeatability, continuity and timeliness. Its quality is directly related to the economic benefits of enterprises. The application of advanced technology can improve the quality of products.

At present, the advanced technologies for high-speed wire rod production include:

（1）Regenerative combustion technology. This technology has the advantages of high preheating temperature, low calorific value fuel, shortened heating time and energy saving.

（2）Endless rolling technology. Endless rolling technology mainly refers to welding two billets end to end for rolling when the billets are released from the furnace.

（3）Precision rolling mill. It has been greatly improved in terms of rolling accuracy and surface quality, and is mainly divided into reducing and sizing rolling mills and dual-module rolling mills.

（4）Controlled cooling system. The controlled cooling system mainly realizes the production of high-speed wire rods through water cooling lines and air cooling lines.

Part Ⅳ Translation Training

被动句的翻译（Ⅱ）

（3）将英语的被动句译成汉语的无主句，即翻译时把原文的主语译成译文的宾语，也可以加“把、将”等词，将宾语放在动词的前面。英语中不带 by 短语或含有情态动词的被动句，都可以用这种方法译成汉语。有些特殊的被动句是由短语动词中的名词作主语构成的，翻译时可将原句的主语和谓语合起来译成汉语的无主句。例如：Computers are now being widely used in different branches of science（在不同的科学部门，正在广泛使用计算机）；Many scientific research institutes have been set up（近几年建立了许多科学研究所）；Great attention has been paid to the development of industry for many years（多年以来一直非常注重发展工业）；Use is being made of advanced techniques and technical processes（正在利用先进技术和工艺）；A conclusion has been arrived at（已得出一个结论）。

（4）将英语的被动句译成汉语的被动句，即翻译时原句的主语仍作译文的主语，再添加“被、用、（遭）受、把、由”等字译成汉语的被动句。当原句中含有动作的执行者时，则在动作执行者前加“用、把、由”等表示被动意义。例如：Atomic energy has been made used of in production（原子能已被用于生产）；Last year the region was visited by the worst drought in 60 years（去年这个地区遭受到 60 年来最严重的旱灾）；When water is heated, its temperature grows higher and higher until it boils（当水受热时，其温度越来越高，直到沸腾为止）；A chemical element is represented by a symbol（化学元素可用符号来表示）；The mechanical properties of this material will be studied by us（这种材料的力学性能将由我们来研究）。

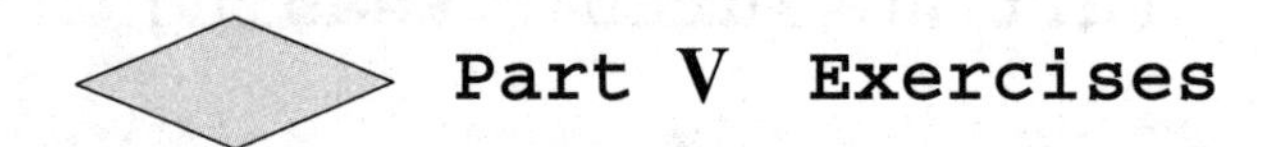

Part V　Exercises

Ⅰ. Translate the following expressions into English.

（1）管材	（2）线材	（3）钢棒
（4）带钢卷	（5）成形辊	（6）电感应
（7）火焰切割机	（8）加热炉	（9）水压机
（10）注射针头	（11）高强度管	（12）赛车
（13）钢条	（14）尖端	（15）加工温度

Ⅱ. Fill in the blanks with the words from the text. The first letter of the word is given.

（1）Just over a hundred years ago, steel tubes began to be used in making b ______.

（2）The edges of the tube are heated by electric i ______ and pressed together so that they join by welding.

（3）Then the tube travels through a series of rolls which s ______ and straighten it.

（4）Nowadays tubes are used widely at home, in t ______, in industries etc.

（5）The first method begins with coils of steel strip which are fed automatically through a series of forming rolls which gradually c ______ the flat strip into a tube shape.

（6）Firstly, a piece of rolled steel bar is cut to b ______ length by torch cutter.

（7）Secondly, each piece, called a billet, is passed to a heat furnace by a c ______.

（8）If a tube of optimum surface appearance and strength is required for, say, aircraft structures or hypodermic needles, a further process of cold d ______ is applied.

（9）This has the effect of reducing the d ______ and the wall thickness of the tube and increasing its length.

（10）The raw material for making w ______ is hot-rolled rod about 6mm diameter.

Ⅲ. Fill in the blanks by choosing the right words form given in the brackets.

When you wake up in the morning rested after a good night's（1）______（sleeping; sleep; sleeps）you look at your watch to see the time: the hairspring of the mechanism is made of the finest tempered and flattened wire. If it is dark, the electric light which you（2）______（turn; turning; turned）on is charged with power transmitted to you over steel and copper wires. The street car which（3）______（take; takes; taking）you to school or to work gets its power from the heavily（4）______（charged; charges; charging）trolley wires overhead. Or if you ride in（5）______（the; an; a）automobile or a bicycle, wire entered into the composition of the spokes and springs, and the bridge you（6）______（crossed; across; cross）may suspend from cables or wire. Every time you use a pin or needle or drive a nail, the wire industry supplies you (7) ______ (with; to; for) the instrument. So you conclude that the improvement in the manufacture of wire is one of the great (8) ______ (step; steps; stepping) forward that industry has taken in the last century. The commonest wires used are made of steel or copper or alloys of both. Several other metals are used, the commonest ones (9) ______ (be; being; been) nickle, platinum, silver, iron, alumium and gold. Wire has been in (10) ______ (use; using; usage) for many hundreds of years.

Ⅳ. Decide whether the following statements are true or false (T/F).

(1) Just over five hundred years ago, steel tubes began to be used in making bicycles. ( )

(2) One of the world's largest manufacturers has modestly described the two basic ways of making tubes as either wrapping an accurate hole in strip steel or pushing a very strong hole through a steel bar. ( )

(3) The raw material for making wire is hot-rolled rod about 60mm diameter. ( )

(4) The second type of process, where the ***strong hole is pushed through a steel bar***, has many variations but the products are generally described as seamless tubes. ( )

(5) Wires are produced as shown in Fig. 19-2 and are always made at a comparatively high working temperature. ( )

Ⅴ. Translate the following English into Chinese.

The raw material for making wire is hot-rolled rod about 6mm diameter. After annealing, colling, and removal of scale, the steel rod is pointed at one end and then inserted through a hole in the die slightly smaller than the size of the rod. The pointed end is gripped from the other side and the rod pulled through, thus reducing its diameter. By the time the wire enters the last die it is much reduced in diameter and is traveling at high speed in modern continuous methods.

Unit 20 The Heat Treatment of Steel

Part Ⅰ Reading and Comprehension

扫描二维码获取音频

扫描二维码获取 PPT

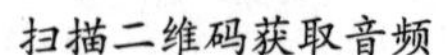

The heat treatment technology of steel is a heat treatment technology that changes the structure of steel by means of heating, heat preservation and cooling to obtain the required properties of the work piece. Mechanical properties such as hardness, tensile strength and toughness of metals can be improved by proper heat treatment. The procedures of heat treatment of steel include hardening, tempering, annealing, and surface heat treatment.

1. Hardening

Hardening is a process of heating and cooling steel to increase its hardness and tensile strength, to reduce its ductility, and to obtain a fine grain structure. The simple experiments described below may be enlightening. Two old-fashioned steel knitting needles, a gas ring, a bowl of water, a piece of sandpaper and a pair of pliers are needed.

Experiment 1: Bend one needle slightly to feel how tough and springy it is. Now hold the needle in the flame, using the pliers, the bowl of water being close at hand. When it is bright red, dip the end of the needle as quickly as possible into the water. The needle should still be red hot as it is being quenched. Then try to bend the quenched end; it is hard and brittle and will snap off.

2. Tempering

Tempering is done by reheating the metal to low or moderate temperature, followed by quenching or by cooling in air. It is a process that follows the hardening procedure and makes the metal as hard and tough as possible and removes the brittleness from a hardened piece.

Two important points are to be noted in connection with the quenching and tempering of steel:

(1) To get a fully hardened structure, the steel should be heated to the appropriate temperature. and then rapidly cooled. The appropriate temperature before quenching depends on the carbon content of the steel concerned.

(2) Tempering does not restore the pearlitic structure. Before that can be done, the steel has to be heated to the appropriate temperature and then slowly cooled.

3. Annealing

Annealing is the process of softening steel to relieve internal strain. This makes the steel easier to machine. The metal is heated above the critical temperature and cooled slowly. The most common method is to place the steel in the furnace and heat it thoroughly. Then turn off the furnace, allowing the metal to cool slowly. Another method is to pack the metal in clay, heat it to the critical temperature, remove it from the furnace, and allow it to cool slowly.

Experiment 2: Take the second needle and heat it until it is red-hot; maintain it at this temperature for about a quarter of a minute. Then withdraw it very slowly, so that it cools gradually. If you now test this end (which you have just annealed) it will bend, like a piece of soft wire, and furthermore it will remain bent.

4. Surface heat treatment

Surface heat treatment is a process of hardening the outer surface or case of ferrous metal. It is mainly used for gear, CAM, crankshaft and various shaft parts to work under the alternating load of torsion, bending and so on, while bearing the friction and impact, and its surface bears higher stress than the heart. The main technologies are as following:

(1) Surface quenching. Surface quenching is a heat treatment method that rapidly heats the work piece to the quenching temperature and then cools it quickly so that only the surface layer can obtain the quenching structure. Its purpose is to obtain the surface with high hardness and high wear resistance, while the core still retains the original good toughness. According to the different heat sources on the surface of the work piece, there are many kinds of surface quenching processes for steel, such as induction heating, flame heating, electric contact heating, electron beam heating, electrolyte heating and laser heating. One of the most advanced surface treatment technologies is laser hardening, which has the greatest advantage of enabling selective area hardening without hardening the entire component.

(2) Chemical heat treatment. The heat treatment process in which the metal work piece is put into a chemical medium containing some active atoms, and the atoms in the medium are diffused into the surface layer of the work piece at a certain depth by heating, and the chemical composition, structure and properties of the metal work piece are changed is called chemical heat treatment. There are various chemical heat treatments. According to the different of infiltration element, it can be divided into carburizing, nitriding, nitrocarburizing, multicomponent permeation, carburizing, carburizing metal etc.

(3) Deformation heat treatment. Deformation heat treatment is a new heat treatment technology, which combines plastic deformation and heat treatment. The process can not only improve the strength of the steel, but also improve the plasticity and toughness of the steel, as well as simplify the process and save energy.

At present, deformation heat treatment is one of the important means to improve the strength and toughness of steel.

As with many other aspects of metallurgy, the current trends of heat-treatment are towards higher throughputs, lower energy use and reduced environmental pollution.

Words and Expressions

扫描二维码获取音频

扫描二维码答题

tempering ['tempəriŋ] *n.* 回火
annealing [æ'ni:liŋ] *n.* 退火
case hardening 表面硬化
tensile ['tensail] *a.* 可拉长的
enlightening [in'laitəniŋ] *a.* 有启发作用的
knitting needle 编织针
sandpaper ['sændpeipə(r)] *n.* 砂纸
pliers ['plaiəz] *n.* 钳子
tough [tʌf] *a.* 有韧性的
springy ['spri:ŋi] *a.* 有弹性的
quench [kwentʃ] *v.* 淬火，水冷（钢铁）
snap off 突然折断
hardened structure 硬化组织
pearlitic structure 珠光体组织
remove [ri'mu:v] *v.* 消除，除去
hardened ['hɑ:dənd] *a.* 变硬的，淬火的
piece [pi:s] *n.* 部件，构件
restore [ris'tɔ:] *v.* 恢复，使……回复
represent [ˌri:pri'zent] *v.* 代表
relieve [ri'li:v] *v.* 减轻，解除
strain [strein] *n.* 张力，应变
thoroughly ['θʌrəli] *ad.* 十分地，彻底地
clay [klei] *n.* 黏土，泥土
pack [pæk] *v.* 包装，把……包起来
withdraw [wɪð'drɔ:] *v.* 取回，提取
crankshaft ['kræŋkʃɑ:ft] *n.* 曲轴；曲柄轴
torsion ['tɔ:ʃn] *n.* （物体等一端固定的）扭转
friction ['frɪkʃn] *n.* 摩擦；摩擦力
resistance [rɪ'zɪstəns] *n.* 抵抗；抗力；抵抗力
laser heating ['leɪzə(r)'hi:tɪŋ] 激光加热

atom [ˈætəm] *n.* 原子
infiltration [ˌɪnfɪlˈtreɪʃən] *n.* 渗入；渗透
carburizing [ˈkɑːbjəraɪzɪŋ] *n.* 渗碳
v. 渗碳于（carburize 的 ing 形式）
nitriding [ˈnaɪtraɪdɪŋ] *n.* 渗氮法
nitrocarburizing *n.* 气体碳氮共渗；氮碳共渗
multicomponent permeation 多元共渗

✻ Answer the following questions.

(1) What do the procedures of heat treatment of steel include?
(2) Can tempering remove the brittleness from a hardened piece?
(3) What is the purpose of annealing?
(4) What is surface quenching about?
(5) What is the very latest surface treatment techniques mentioned in this text?
(6) Does tempering restore the pearlitic structure?

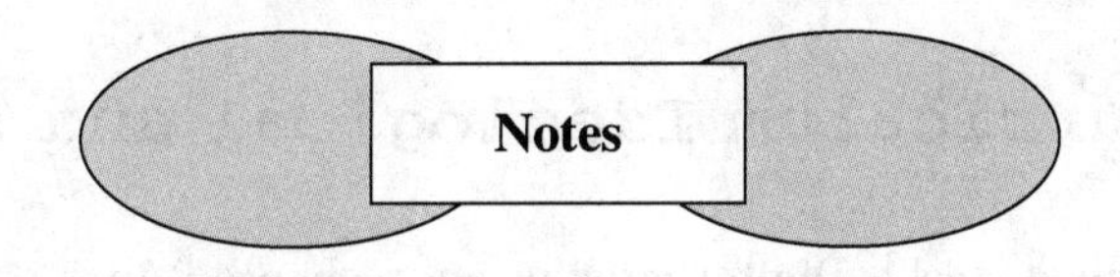

(1) Bend one needle slightly to feel how tough and springy it is. Now hold the needle in the flame, using the pliers, the bowl of water being close at hand.

轻轻弯曲一根毛衣针，试一试它的韧性和弹性，然后用钳子夹住一端放在火上烧，手旁放一碗水。

"at hand" 意思是"近在手边；在附近；即将到来"。

(2) It is a process that follows the hardening procedure and makes the metal as hard and tough as possible and removes the brittleness from a hardened piece.

回火是在淬火之后采用的使金属尽可能变硬、变韧的方法，它能消除淬火工件的脆性。

"follows" "makes" 和 "removes" 是定语从句的并列谓语。

(3) Two important points are to be noted in connection with the quenching and tempering of steel.

关于钢的淬火和回火有两个要点需要注意。

"in connection with" 意思是"关于……，与……有关"。例如：I am waiting to you in connection with your job application（我等你是有关你求职的事情）。

(4) According to the different heat sources on the surface of the work piece, there are many kinds of surface quenching processes for steel, such as induction heating, flame heating, electric contact heating, electron beam heating, electrolyte heating and laser heating.

根据工件表面加热热源的不同，钢的表面淬火有很多种，例如感应加热、火焰加热、电接触加热、电子束加热、电解液加热以及激光加热等表面淬火工艺。

"according to" 意思是"按照，根据"。例如：According to the arrangement, you need to arrive at the hotel at 11 o'clock（根据安排，你应该在 11 点到达酒店）。

（5）The heat treatment process in which the metal work piece is put into a chemical medium containing some active atoms, and the atoms in the medium are diffused into the surface layer of the work piece at a certain depth by heating, and the chemical composition, structure and properties of the metal work piece are changed is called chemical heat treatment.

将金属工件放入含有某种活性原子的化学介质中，通过加热使介质中的原子扩散渗入工件一定深度的表层，改变其化学成分、组织和性能的热处理工艺叫做化学热处理。

"Which" 是关系代词，引导定语从句。

（6）As with many other aspects of metallurgy, the current trends of heat-treatment are towards higher throughputs, lower energy use and reduced environmental pollution.

与许多其他冶金部门一样，热处理工艺的发展趋势是提高产量、降低能源消耗、减少环境污染。

"as with" 意思是"如同……一样，正如……的情况一样"。例如：As with our hands, the friction of a tire with the road makes heat which warms the tyre（如同摩擦双手那样，轮胎与道路的摩擦产生热从而使轮胎变热）。

Part Ⅱ　Curriculum Ideological and Political

【Smart steel】Digital intelligence reshapes future steel

Digital technologies are changing the world and dramatically improving the operating way of organizations. Today, steel and metals manufacturers face a huge opportunity. To transform their operational model by implementing digital technology can enable them to improve operational efficiency, customer service, inventory levels and profit margin.

The era of digitization in the steel and metal industries has arrived. Cost of data acquisition, storage and analysis has dropped dramatically in the past years. As a digital technology, predictive analytics has already demonstrated its potential to revolutionize the operational model, in terms of speed, cost, and ease of implementation. It adapts advanced, self-learning algorithms to sift through large volumes of data generating insights and identifying patterns. The traditional prediction method depends on people's experience, where limited data is collected and analyzed, and not builds thought through mechanism. With sensors and machine learning algorithms, digital solutions can greatly improve prediction accuracy and allow extra time before unplanned shutdowns to fix potential issues.

Part Ⅲ　Further Reading

The normalizing process of steel

Normalizing is a heat treatment process in which the steel is heated to an appropriate temperature above AC_3 (or AC_{cm}), and then cooled in air to obtain a pearlite structure (usually sorbite).

Normalizing can be used as a preparatory heat treatment to provide suitable hardness for machining, and can also refine grains, eliminate stress, eliminate Widmancer's structure and banded structure, and provide a suitable structure state for final heat treatment. Normalizing can also be used as a final heat treatment to provide suitable mechanical properties for some carbon steel structural parts with low stress and low performance requirements.

Part Ⅳ Translation Training

英语长句的翻译

在进行英汉翻译的过程中，常常遇到很长的英语句子。首先要正确判断句子是并列句还是复合句，找出句子的主要成分和修饰成分，确定各成分之间的内在联系、逻辑关系，再按汉语的特点和表达方式，正确译出原文的意思。长句的翻译方法一般有四种，即顺译法、倒译法、分译法和综合法。

1. 顺译法

英语与汉语在表达上有相同之处，有些句子是按动作发生的先后顺序叙述的，也有的是按长句内容的逻辑关系排列的，因此翻译时可按原文顺序译出。例如：With the more complex shapes and when less ductile and malleable materials are used, it is necessary to go through a series of stages before the final shape can be produced from the raw material（如果形状比较复杂，而所用材料的可延性和可锻性较差，就需要经过一系列过程，才能使原材料成型)。(按句子顺序翻译)

2. 倒译法

英、汉是两种不同的语言，在表达上相同之处少，而不同之处多。例如：英语的定语从句、同位语从句、介词短语和分词短语常常被放在被修饰词的后面，其他从句如时间状语从句、目的状语从句以及条件状语从句，放在主句后面的情况也不少。当遇到这种情况时，必须采用倒序进行翻译。例如：The rusting of iron is only one example of corrosion, which may be described as the destructive chemical attack of a metal by media with which it comes in contact, such as moisture, air and water（可以认为，腐蚀是金属在接触湿气、空气和水等介质时所受到的破坏性化学侵蚀，铁生锈只是其中的一个例子)。(倒译定语从句)

3. 分译法

英语中长句多，汉语句子一般比较短。翻译时可按汉语表达习惯，把长句中的从句和短语切断来译，转化为句子分开叙述，然后再按时间先后、逻辑顺序重新组合。为了使语意连贯，还可增加词语。例如：Other students of the brain, noticing that disease and physical damage can change personally and distort the mind, believe the brain to be nothing more than a fantastically complex computer（另外，有一些脑学家注意到疾病和身体损伤能改变人的秉

性，歪曲人的心灵。他们认为人脑仅仅是一个神奇复杂的计算机而已）。（分译分词短语）

4. 综合法

有些英语长句子顺译、倒译和分译都感到不合适，这时就应该仔细推敲，或按时间先后，或按逻辑顺序，有顺有逆、有主有次地全面进行综合处理。例如：The iron, as it were, breathes air as we do, "and as it breathes, softening from its merciless hardness, it falls into fruitful and beneficent dust, gathering itself again into the earths from which we feed and the stones with which we build, into the rocks that frame the mountains and the sands that bound the sea"（铁像人一样也能呼吸空气，"而且它一面呼吸空气，一面由铁矿石般坚硬变得松软，成为肥沃富饶的有益的尘土，进而成为我们赖以获取食物的土壤，成为我们用以建造房屋的石头，成为构成山脉骨架的岩石，成为环绕大海的沙滩"）。（分译分词短语和定语从句，倒译定语从句）

Part Ⅴ Exercises

Ⅰ. Translate the following expressions into English.

（1）热处理	（2）抗拉强度	（3）表面硬化
（4）细粒结构	（5）回火	（6）退火
（7）硬化组织	（8）珠光体组织	（9）内应力
（10）临界温度	（11）软金属丝	（12）外表面
（13）淬火	（14）活塞销	（15）激光硬化

Ⅱ. Fill in the blanks with the suitable words listed below. Change the form where necessary.

laser	harden	dip	nitrocarburizing	critical	describe
nitriding	depend	treat	springy	cyaniding	moderate

（1）Heat ______ is a term applied to a variety of procedures for changing the characteristics of metal by heating and cooling.

（2）The simple experiments ______ below may be enlightening.

（3）When it is bright red, ______ the end of the needle as quickly as possible into the water.

（4）Tempering is done by reheating the metal to low or ______ temperature, followed by quenching or by cooling in air.

（5）It will be seen, therefore, that the appropriate temperature before quenching ______ on the carbon content of the steel concerned.

（6）The metal is heated above the ______ temperature and cooled slowly.

（7）Case ______ is a process of hardening the outer surface or case of ferrous metal.

（8）In industry, ______ is also used for case hardening, because a thinner carbon case can be applied by immersing the steel in a bath of molten sodium cyanide.

(9) Another method of hardening the surface of steels is known as ______.

(10) In addition, other hardening processes have been developed, under the general term ______; with these methods carbon and nitrogen are diffused into the surface of ferrous components at relatively low temperatures.

(11) One of the very latest surface treatment techniques is ______ hardening.

(12) Bend one needle slightly to feel how tough and ______ it is.

Ⅲ. Fill in the blanks by choosing the right words form given in the brackets.

Heat treatment is a method by which the physical properties of a metal can be changed. There are three main operations in the heat treatment of steel: hardening, (1) ______ (temper; tempering; to temper), and annealing. The hardening operation (2) ______ (consist; consists; consisting) of heating the steel above its critical range and then quenching it, that is, rapidly (3) ______ (cool; cooling; cools) it in a suitable medium such as water, brine, or some other liquid. Having been (4) ______ (harden; hardened; hardening), the metal must be given a tempering treatment which consists of (5) ______ (reheat; reheating; reheated) the hardened steel to a temperature below the critical range, thus producing the required physics properties. Tempering is also (6) ______ (call; called; calling) drawing the temper, because this operation (7) ______ (give; gives; giving) a steel object the temper being required. Annealing is the uniform heating of a metal above usual hardening temperatures, followed by very slow cooling. Annealing may be (8) ______ (carry; carried; carrying) out either to soften a piece that is (9) ______ (to; too; very) hard to machine or to remachine a piece (10) ______ (having; have; has) been hardened.

Ⅳ. Decide whether the following statements are true or false (T/F).

(1) Annealing is a process of heating and cooling steel to increase its hardness and tensile strength, to reduce its ductility, and to obtain a fine grain structure. ( )

(2) Case hardening is the process of softening steel to relieve internal strain. ( )

(3) Hardening is a process of hardening the outer surface or case of ferrous metal. ( )

(4) Tempering is done by reheating the metal to high or moderate temperature, followed by quenching or by cooling in air. ( )

Ⅴ. Translate the following English into Chinese.

Metal products can thus be specified as hot rolled or cold rolled, and their surface finish is quite different. In the hot rolled process, a metal is in a red to orange glowing heated state. Under this condition of heat, and in the presence of air and moisture, oxidation or corrosion begins at a rapid pace. Even as the metal is rolled and formed, the oxide coating, sometimes called scale, is ever reforming. As the rolls pass over the metal and scale, the very hard layer of scale is pressed into the metal surface below it and forms a somewhat pitted and rough surface. When metal products must be accurate in size and must posses a smooth flawless surface, the cold rolling process is used. The absence of heat has the effect of not producing an oxide coating during the rolling; in some cases, a fine spray of oil is used to protect the cold rolled finish.

Development Trend of Iron and Steel Metallurgy

Part Ⅰ Reading and Comprehension

扫描二维码获取音频

扫描二维码获取 PPT

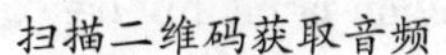

As the traditional basic material with the largest consumption, iron and steel material plays an important supporting role in the construction of national economy and the development of national defense. Since the founding of the People's Republic of China in 1949, China's iron and steel industry has risen rapidly and made **brilliant** achievements. The rapid development of iron and steel industry leads to increasingly **prominent** problems such as the contradiction between supply and demand and environmental pollution. The development of China's iron and steel industry will be **constrained** by resources-energy, environment-**ecology**, market-brand and so on. The future development of the iron and steel industry is based on improving the technological **innovation** ability and **core** competitiveness of the steel industry, developing green and intelligent steel process technology, meeting the needs of national economic construction, major projects and high-end equipment manufacturing, and supporting the transformation and **upgrading** of the iron and steel industry and **sustainable development**.

1. Green metallurgy development

To solve the difficulties and pain points in the **transformation** and development of metallurgical industry, it is the only way to **persevere in** green development and promote industrial **energy-saving** and **low-carbon** technologies.

(1) Make full use of resources. In metallurgical processes, **fuel combustion** and raw material consumption produce a large number of solid wastes. Besides iron elements, these solid wastes consist of different types of other metal elements, which have high secondary utilization value. Green technology can be used to properly process them and turn them into treasure. For example, **blast furnace slag** is **granulated** by extreme cold water and used as **cement additive**.

(2) Conserve energy and reduce emission. The total energy consumption and carbon emissions of the metallurgical industry are huge. The energy consumption of **ferrous metal** smelting and roll-

ing processing **accounts for** 14% of the national energy consumption, among which the energy consumption of the iron and steel industry accounts for about 11% of the national energy consumption, and the carbon emissions account for about 15% of China's total carbon emissions. It is an **inevitable** trend to develop new green technology and establish **resource-saving** and **environment-friendly** business model.

2. Intelligent metallurgy development

The iron and steel industry has certain advantages in carrying out intelligent and information development. Firstly, the iron and steel industry is a traditional industry that needs to be transformed and upgraded urgently and has urgent internal drive. Secondly, the iron and steel industry is a typical industry for the generation and application of big data, which is feasible to realize. Thirdly, the production process of iron and steel industry belongs to the manufacturing industry with long process and model, which is widely **implemented** and **reproducible**.

The intelligence of iron and steel metallurgical production is mainly based on PLC, DCS and industrial computer technology, and systematically manage the production line in the process of metallurgical production **simulation** control. Comprehensive process control is the main trends of metallurgical intelligent development, namely the integrated use of new type of sensing technology and data **fusion** and processing technology in metallurgical automation to improve the comprehensive control of the production process. Comprehensive process control can effectively improve production efficiency and quality and increase the productivity of metallurgical enterprises.

Words and Expressions

扫描二维码获取音频

扫描二维码答题

brilliant [ˈbrɪliənt] *a.* 巧妙的；使人印象深的；明显的
prominent [ˈprɒmɪnənt] *a.* 重要的；著名的；杰出的；显眼的；显著的
constrain [kənˈstreɪn] *v.* 强迫；强制；迫使；限制
ecology [iˈkɒlədʒi] *n.* 生态；生态学
innovation [ˌɪnəˈveɪʃn] *n.* （新事物、思想或方法的）创造；创新
core [kɔː(r)] *n.* 最重要的部分；核心；要点
a. 最重要的；主要的；基本的
intelligent [ɪnˈtelɪdʒənt] *a.* 聪明的；有智力的；智能的
upgrade [ˌʌpˈgreɪd, ˈʌpgreɪd] 升级；提高；改进；提升
sustainable [səˈsteɪnəbl] *a.* 合理利用的；可持续的
sustainable development 可持续发展
transformation [ˌtrænsfəˈmeɪʃn] *n.* 变化，改观，转变，改革

persevere in　坚持
energy-saving *a.*　节省能源的
blast furnace slag　高炉渣
fuel combustion　燃料燃烧
low-carbon *a.*　低碳（的），含碳低的
granulate [ˈgrænjʊleɪt] *v.*　（使）成颗粒；粒化
cement [sɪˈment] *n.*　水泥
v.　粘结，胶合；加强
additive [ˈædətɪv] *n.*　（尤指食品的）添加剂，添加物
ferrous metal [ˈferəsˈmetl] *n.*　黑色金属；含铁金属
account for　（数量或比例上）占；解释，说明
inevitable [ɪnˈevɪtəbl] *a.*　不可避免的
resource-saving　资源节约
environment-friendly *a.*　有利环境的，环保的
implement [ˈɪmplɪment] *v.*　使生效；贯彻；执行；实施
reproducible [ˌriːprəˈdjuːsəbl]　可再生的；可复制的
simulation [ˌsɪmjuˈleɪʃn] *n.*　模拟；仿真
fusion [ˈfjuːʒn] *n.*　融合；熔接

✻ Answer the following questions.

(1) What is the future development of the iron and steel industry?

(2) How to solve the difficulties and pain points of metallurgical industry?

(3) How can we make full use of resources in metallurgical processes?

(4) What are the advantages of carrying out intelligent and information development in the iron and steel industry?

(5) What is the comprehensive process control?

(6) Can you describe the function of comprehensive process control?

(1) As the traditional basic material with the largest consumption, iron and steel material plays an important supporting role in the construction of national economy and the development of national defense.

钢铁材料作为用量最大的传统基础材料，对国民经济建设和国防军工发展起着重要支撑作用。

"play an important role in" 意思为"在……中其重要作用"。例如：Nowadays, computers play an important role in our daily life（如今，计算机在我们的日常生活中起着重要的作用）。

(2) The future development of the iron and steel industry is based on improving the technological innovation ability and core competitiveness of the steel industry, developing green and intelligent steel process technology, meeting the needs of national economic construction, major projects and high-end equipment manufacturing, and supporting the transformation and upgrading of the iron and steel industry and sustainable development.

未来钢铁工业的发展是以提升钢铁产业科技创新能力和核心竞争力为出发点，开发绿色化、智能化钢铁流程技术，满足国民经济建设、重大工程及高端装备制造等需求，支撑钢铁工业转型升级和可持续发展。

(3) To solve the difficulties and pain points in the transformation and development of metallurgical industry, it is the only way to persevere in green development and promote industrial energy-saving and low-carbon technologies.

破解冶金行业转型发展的难点和痛点，坚持绿色发展，推广工业节能低碳技术，是冶金工业今后发展的必经之路。

“be the only way to” 意思是“是……的唯一之路，必经之路”。例如：Study is the only way to change your fate（学习是改变你命运的唯一之路）。

“persevere in” 意思是“坚持”。例如：She failed to persevere in dancing（她未能坚持跳舞）。

(4) Besides iron elements, these solid wastes consist of different types of other metal elements, which have high secondary utilization value.

这些固体废弃物除了铁元素之外还有不同类型的其他金属元素，具有较高的二次利用价值。

“consist of” 意思是“由……组成（构成）；包括”。例如：A healthy diet should consist of vegetables（健康饮食包括蔬菜）。

“different types of” 意思是“不同种类的，不同类型的”。例如：There are different types of entertainments in China（在中国有不同类型的娱乐活动）。

(5) It is an inevitable trend to develop new green technology and establish resource-saving and environment-friendly business model.

积极开发新型绿色技术，建立资源节约型与环境友好型的企业模式，是冶金工业的必然趋势。

“inevitable” 意思是“不可避免的”。例如：It was inevitable that there would be unemployment（失业已是不可避免的事）。

Part Ⅱ Curriculum Ideological and Political

【The workforce development strategy】The Report of the 20th National Congress of the Communist Party of China | Cultivating a large workforce of high-quality talent who have both integrity and professional competence

Cultivating a large workforce of high-quality talent who have both integrity and professional competence is of critical importance to the long-term development of China and the Chinese nation. We will move faster to build world hubs for talent and innovation, promote better distribution and balanced development of talent across regions, and strive to build up our comparative strengths in global competition for talent. We will speed up efforts to build a contingent of personnel with expertise of strategic importance and cultivate greater numbers of master scholars, science strategists, first-class scientists and innovation teams, young scientists, outstanding engineers, master craftsmen, and highly-skilled workers. We will increase international personnel exchanges and make the best use of talent of all types to fully harness their potential.

Young people: You should steadfastly follow the Party and its guidance, aim high but stay grounded, and dare to think big and take action but make sure you can deliver. You should strive to be the new era's great young generation, a generation with ideals, a sense of responsibility, grit, and dedication.

Part Ⅲ Further Reading

Extensively applying the core socialist values—The Report of the 20th National Congress of the Communist Party of China

The core socialist values have immense power to rally the people's support and pool their strength. We will carry forward the long line of inspiring principles for the Chinese Communists that originated with the great founding spirit of the Party; Put resources related to the Party's heritage to great use; Conduct extensive public awareness activities to promote the core socialist values; Enhance commitment to patriotism, collectivism, and socialism; And foster a new generation of young people to shoulder the mission of realizing national rejuvenation.

We will develop and institutionalize regular activities to foster ideals and convictions, and we will carry out public awareness initiatives on the history of the Party, the People's Republic of China, reform and opening up, and the development of socialism, in order to foster love for the Party and the country. These efforts will help strengthen the people's commitment to our common ideal of socialism with Chinese characteristics. We will uphold both the rule of law and the rule of virtue and see that the core socialist values are incorporated into efforts to advance the rule of law, into social development, and into the people's daily lives.

Part Ⅳ Translation Training

there be 句型的译法

there be 句型在英语中一般表示“某处存在某物”，后接时间或者地点状语。但在科技英语中，由于专业特点不同，很多时候不带状语，而是带有不同形式的定语，这就增加了翻译的难度，通常不直接翻译成“有……”。下面介绍科技英语中“there be”常见译法。

一、译成有主语的“有”

例如：There can be considerable overlap in the metallurgical functions that various secondary steelmaking processes achieve（许多炉外精炼工艺所达到的冶金功能会相互重叠）；There are two main stages in the pyrometallurgical extraction of copper（火法提取铜主要有两个阶段）；There are different ways of classifying materials（工程材料有许多不同的分类方法）。

二、译成无主语的“有”

例如：There are several ways of turning sunshine directly into electricity（有几种方法可以把太阳能直接转换成电）；There are many people who do not carefully observe the difference between temperatures and heat（有很多人不太注意温度与热之间的区别）。

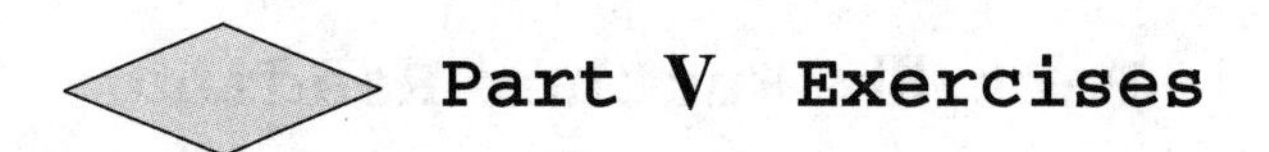

Part V Exercises

Ⅰ. Translate the following expressions into English.

(1) 钢铁材料	(2) 低碳（的）	(3) 生态
(4) 黑色金属	(5) 高炉渣	(6) 水泥添加剂
(7) 可持续发展	(8) 燃料燃烧	(9) 资源节约
(10) 固体废弃物	(11) 数据融合	(12) 过程控制全面化

Ⅱ. Fill in the blanks with the words from the text. The first letter of the word is given.

(1) The rapid development of iron and steel industry leads to increasingly p______ problems such as the contradiction between supply and demand and environmental pollution.

(2) The development of China's iron and steel industry will be c______ by resources-energy, environment-ecology, market-brand and so on.

(3) In m______ processes, fuel c______ and raw material consumption produce a large number of solid wastes.

(4) Green technology can be used to properly p______ them and turn them into treasure.

(5) B______ furnace slag is g______ by extreme cold water and used as cement additive.

(6) The energy c______ of ferrous metal smelting and rolling processing accounts for 14% of the national energy consumption.

(7) Firstly, the iron and steel industry is a traditional industry that needs to be transformed and u______ urgently and has urgent internal drive.

(8) Comprehensive process control is the main trends of metallurgical i______ development.

Ⅲ. Fill in the blanks with appropriate prepositions.

The intelligence (1) ______ iron and steel metallurgical production is mainly based

(2) ______ PLC, DCS and industrial computer technology, and systematically manage the production line (3) ______ the process (4) ______ metallurgical production simulation control. Comprehensive process control is the main trends (5) ______ metallurgical intelligent development, namely the integrated use (6) ______ new type of sensing technology and data fusion and processing technology (7) ______ metallurgical automation (8) ______ improve the comprehensive control (9) ______ the production process. Comprehensive process control can effectively improve production efficiency and quality and increase the productivity (10) ______ metallurgical enterprises.

Ⅳ. Decide whether the following statements are true or false (T/F).

(1) Iron and steel material is of little use in the construction of national economy and the development of national defense. ( )

(2) Green technology can be used to properly process solid wastes and turn them into treasure. ( )

(3) The iron and steel industry is a traditional industry that doesn't need to be transformed and upgraded. ( )

(4) The production process of iron and steel industry can be widely implemented and reproducible. ( )

(5) The technological innovation ability and core competitiveness of the steel industry play an important role in development of the iron and steel industry. ( )

Ⅴ. Translate the following English into Chinese.

In metallurgical processes, fuel combustion and raw material consumption produce a large number of solid wastes. Besides iron elements, these solid wastes consist of different types of other metal elements, which have high secondary utilization value. Green technology can be used to properly process them and turn them into treasure. For example, blast furnace slag is granulated by extreme cold water and used as cement additive.

第1单元　钢铁冶炼的历史

☞ 第一部分　课文参考译文

钢铁工业是世界上最重要的工业之一，也是传统史上最古老的工业之一。早在3000年前，铁就是人类文化与文明的基础。

20世纪50~60年代　钢铁时代开始

在20世纪初，钢铁行业成为主要产业，科学正逐步揭开钢铁的神秘面纱。

在20世纪50年代和60年代，钢铁工艺取得重大发展，这使得钢铁生产从军事和航运转移到汽车和家用电器，给消费者带来大量的钢铁家用电器。战后欧盟贸易也是制成品寻找资源和销售的重要因素。

20世纪50年代　连铸工艺

在连铸过程中凝固速度越快，合金的均匀性越高，质量越好。

1951年　团体的形成

欧洲煤钢共同体（ECSC）是由内部六国，即法国、意大利、比荷卢经济联盟国家（比利时、荷兰、卢森堡）和西德在巴黎条约（Treaty of Paris）之后成立的。

20世纪50年代　电弧炉（EAF）发展

在19世纪，许多人利用电弧熔化铁。

20世纪50年代　模铸

连续铸钢也称为连铸，是将熔融的金属凝固成半成品钢坯、坯或板坯，然后在铸锭机中轧制的过程。

1959年　从高炉到小型轧机

小型钢厂的小型设备提供私营企业可以负担得起的最新技术（电弧、连铸、水冷）。小型钢厂的兴起伴随着废料供应的增加。

1967年　钢铁世界

世界钢铁协会1967年10月19日在比利时布鲁塞尔成立。

1969年　小型轧机改造

1969年，当现在美国最大的钢铁生产商之一的纽科尔（Nucor）决定进入长条材市场时，他们选择了创办一家以电弧炉为炼钢核心的小型钢厂。这一举动很快在20世纪70年代被其他制造商效仿。

20世纪70年代　东日本的创新

日本在20世纪60年代和70年代追求快速增长，紧随其后的是韩国，开发了大量最先进的综合设备。

20世纪90年代　发展中的行业

钢铁工业的重心已经转向新兴经济体，城市化和工业化需要大量的钢铁。

21世纪前十年　钢铁巨龙

截至2011年底，中国是迄今为止全球最大的钢铁生产国，产量高于6.8亿吨。

21世纪前十年　安赛乐米塔尔来临

安赛乐米塔尔是2006年由米塔尔钢铁公司收购并兼并阿塞洛公司的第一家全球钢铁公司，在创立之初，它是全球最大的钢铁生产商。

21世纪一十年代　大规模合并发生

2011年，新日铁公司与住友金属合并成为新日铁住友金属公司。2016年，宝钢集团与武汉钢铁集团合并组建中国宝武钢铁集团，成为全球第二大钢铁企业。

☞ 第二部分　课程思政

【钢联世界】中国钢铁，用行动让世界更美好

1949年，新中国的钢铁工业在中国解放战争的灰烬中诞生，从年产15.8万吨钢开始。

中国的钢铁企业和中华人民共和国一起白手起家，但他们开辟了一条新的道路，从犁头到桥梁，从铁轨到“两弹一星”。钢铁工业一直致力于推进新中国建设的进程。1996年，中国钢铁产量首次突破1亿吨的门槛。自那时起，28年来，中国一直保持着世界最大钢铁生产国的地位。

钢铁行业一直在幕后努力，支持中国各行各业的腾飞，为中华民族走向繁荣奠定了坚实的基础。与此同时，中国最先进的钢铁产品质量上乘、技术先进、品种繁多，也为世界各地的建设和发展提供了优质选择，从非洲到欧洲，从东亚到北美。中国钢铁公司，作为工业文明的支柱，通过覆盖各种优质材料，满足了世界一半以上的钢铁需求。我们用钢铁来振兴城市。我们用钢来连接世界。我们用钢材来发现宇宙的奥秘。我们用钢板来保护我们的生活。我们用钢来照亮人类文明。

☞ 第三部分　延伸阅读

钢铁行业面临的挑战与今后的发展重点

我国当前社会主要矛盾已经转化为人民日益增长的美好生活需要和不平衡不充分的发展之间的矛盾，发展中的矛盾和问题集中体现在发展质量上。在这一大背景下，对钢铁产业的要求不再是数量规模为主，而是品种质量、绿色低碳、创新发展。重点从以下几个方面开展工作：

（1）继续推进供给侧结构性改革，维护行业平稳运行；

（2）加强产业链供应链建设，促进上下游协调发展；

（3）坚持绿色低碳方向，加大科技创新力度；

（4）加快数字化建设，推进行业智能制造发展。

☞ 第四部分　课文问题参考答案

Answer the following questions.

（1）France, Italy, the Benelux countries（Belgium, Netherlands and Luxembourg）and west Germany.

（2）During the 19th century.

（3）The International Iron and Steel Institute（IISI）in Brussels, Belgium on 19 October 1967.

（4）ArcelorMittal.

（5）In 2010s.

☞ 第五部分　练习题参考答案

Ⅰ. Translate the following expressions into English.

（1）the iron and steel industry　（2）civilization　（3）blast furnace
（4）die casting　（5）Electric arc furnace　（6）urbanisation
（7）industrialisation　（8）Continuous casting　（9）solidification

Ⅱ. Fill in the blanks with the words from the text. The first letter of the word is given.

（1）dawn, increasingly　（2）continuous, semifinished　（3）mills
（4）massive　（5）urbanization, industrialization　（6）producer

Ⅲ. Fill in the blanks by choosing the right words form given in the brackets.

（1）produced　（2）made　（3）improving　（4）absorbs　（5）smelting

Ⅳ. Decide whether the following statements are true or false（T/F）.

（1）T　（2）F　（3）T　（4）F　（5）F

Ⅴ. Translate the following English into Chinese.

1969 年，当现在美国最大的钢铁生产商之一的纽科尔（Nucor）决定进入长条材市场时，他们选择了创办一家以电弧炉为炼钢核心的小型钢厂，这一举动很快在 20 世纪 70 年代被其他制造商效仿。

第 2 单元　炼铁原料

☞ 第一部分　课文参考译文

高炉炼铁的原料可以分为含铁原料、燃料以及熔剂。

1. 含铁原料

高炉主要的含铁原料是铁矿石、烧结矿和球团矿。它们的作用是为炼铁生产提供93%~94%的铁元素。

按照化学成分铁矿石分为氧化物、硫化物、碳酸盐等矿石（见表 2-1）。

表 2-1　铁矿石分类表

分类和矿物学名称	纯物质的化学成分	通用名称	分类和矿物学名称	纯物质的化学成分	通用名称
氧化物			**碳酸盐**		
磁铁矿	Fe_3O_4	二价—三价铁氧化物	菱铁矿	$FeCO_3$	碳酸铁
赤铁矿	Fe_2O_3	三价铁氧化物	**硫化物**		
钛铁矿	$FeTiO_3$	钛铁氧化物	黄铁矿	FeS_2	—
褐铁矿	$n Fe_2O_3 \cdot m H_2O$	含水铁氧化物	磁黄铁矿	FeS	铁硫化物

赤铁矿是最广泛使用的铁矿之一，它的理论含铁量（质量分数）是 70%。这种矿石的红颜色是由于含有三价铁造成的。在这类红铁矿中，铁和氧混合物的结合不是很“紧密”，因此，赤铁矿是“容易还原”的。磁铁矿，一种磁性矿石，因为两方面的原因其使用量在（不断）增加：第一，它可以通过磁选法从岩石中分离出来；第二，它含铁量高。磁铁矿中铁原子和氧原子的结合很紧密，这样使得磁铁矿“很难还原”。褐铁矿是一种棕褐色矿石，它含有水，这就意味着铁氧化物和水（结合水）能形成很稳定的化合物。菱铁矿含有 30%~40%的铁，它是一种相对易还原的矿石。由于大多数铁矿石中（质量分数）含有 10%~20% 的由氧化铝和二氧化硅组成的脉石，因此铁的含量（质量分数）仅为 50%~60%。如果脉石主要含有石灰，则矿石是“碱性”的；如果脉石中 SiO_2 占多数，则矿石是“酸性”的。

矿石中过细的颗粒不能直接装入高炉，通常需要烧结成块。重要的造块工艺是：烧结生产和球团生产。

烧结生产工艺由 5 个阶段组成：（1）把烧结原料与煤粉或焦粉混合；（2）将混合料铺在（烧结台车）箅条上；（3）点火燃烧和烧结。当空气穿过该混合料时，混合料中的燃料进行燃烧，形成的高温足以使这些细小的颗粒烧结成块，以便满足高炉冶炼的要求；（4）冷却；（5）装入高炉前进行破碎和筛分。要获得更好的效果，可以加入粉状熔剂，使之在烧结过程与矿石中的脉石结合。通常烧结矿的含铁量（质量分数）为 50%～60%。

球团生产期间，由极细的铁精矿粉（小于 0.074mm）以及粒度远小于 1mm 的黏结剂组成的混合物，首先制成直径略大于 6mm 但小于 15mm 的“生”球，然后将生球装入竖炉、回转窑或是移动床（带式焙烧机）上焙烧使之变硬。通常球团矿含铁量（质量分数）为 60%～67%。

与天然块矿和烧结矿相比，球团矿的优点是：有一个窄小的粒度范围、稳定的质量和良好的还原性，此外，球团矿适合于输送和储存。但还原阶段球团矿的膨胀和黏结必须避免。

2. 燃料

高炉燃料可以是焦炭、煤粉、重油或天然气。它们被用于产生熔化过程中所需要的热量、将铁矿石还原为金属铁并进行渗碳（每吨铁中大约含有 40～45kg 的碳）。此外，由于焦炭在高温环境下仍保持其强度，因此，它还具有防止未熔化物料落入炉缸的结构支撑作用。

目前，高炉中的部分焦炭通常被煤粉所取代。高炉可以喷吹无烟煤，也可以喷吹烟煤和混合煤。高炉喷煤可以大幅度降低入炉焦比，减少对日益匮乏的焦煤资源的依赖，是炼铁降低成本的最有效手段，已成为高炉炼铁技术进步的一项重要内容。

3. 熔剂

高炉熔剂主要由石灰石、白云石和生石灰组成。它们的主要作用是将焦炭中的灰分和矿石中的脉石结合成可被顺利排出炉缸的液态炉渣。为了提高炉渣的脱硫能力及流动性，需要精心控制渣中碱性氧化物和酸性氧化物的比例。

☞ 第二部分　课程思政

【穿越时空】中国古代的钢铁冶炼技术成就

中国古代钢铁发展的特点与其他各国不同。世界上长期采用固态还原的块炼铁和固体渗碳钢，而中国铸铁和生铁炼钢一直是主要方法。由于铸铁和生铁炼钢法的发明与发展，中国的冶金技术在明代中叶以前一直居世界先进水平。

在中国，钢铁总产量在唐代就已达到年产 1200t 的水平，宋代为 4700t，明代最多达到 4 万吨。在 13 世纪，中国是世界上最大的铁的生产国和消费国，直到 17 世纪仍保持领先地位。从汉代到明朝，中国人不仅在数量上处于领先地位，而且拥有世界上最先进的钢铁冶炼技术。

☞ 第三部分　延伸阅读

微波烧结预还原技术

微波烧结即通过微波来实现烧结造块矿，而不是过去的焦粉煤粉加热烧结。经过微波烧结后，烧结矿还有高温余热，通过氢跟氧化铁的还原吸热效应，既起到冷却的效果，又实现了烧结矿的预还原，从而提高烧结矿的金属化率，降低烧结矿进入高炉以后对高炉还原剂的消耗，达到减碳的目的。

☞ 第四部分　课文问题参考答案

Answer the following questions.

（1）The raw materials for the production of iron in the blast furnace can be grouped as follows：iron-bearing materials，fuels and fluxes.

（2）Magnetite is increasing in use for two reasons. Firstly，it can be separated from the rock by magnetic means；secondly，it has high iron content.

（3）Yes，I can. The sintering process is in five stages：1）mixing of the raw materials and fine coal or coke；2）placing the mixture on a grate；3）igniting and sintering，air drawn through the mixture burns the fuel at a temperature high enough to frit the small particles together into a cake so that they can be charged into the blast furnace satisfactorily；4）cooling；5）crushing and screening before charging to the furnaces.

（4）Coke is used for producing the heat required for smelting，and reducing the iron oxides into metallic iron and carburizing the iron（about 40 to 50 kilograms per ton of iron）. In addition，because the coke retains its strength at high temperature，it provides the structural support that keeps the unmelted burden materials from falling into the hearth.

（5）The blast furnace can inject hard coal，soft coal and mixed coal.

（6）Because the sulphur-holding power of the slag as the fluidity must be preserved.

（7）Compared with lump ores，the advantages of pellets are：a narrow size range，constant quality and good permeability during reduction. Furthermore，pellets are well suited for transport and storage.

（8）At present，some of the coke in the blast furnace is usually replaced by coal.

（9）Fluxes mainly include limestone，dolomite and lime.

☞ 第五部分　练习题参考答案

Ⅰ. Translate the following expressions into English.

（1）raw material for the production of iron　（2）easily reducible ore　（3）acid gangue
（4）basic gangue　（5）pellet　（6）iron-bearing material
（7）injecting pulverized coal　（8）magnetite　（9）hematite
（10）ash in the coke　（11）sinter　（12）flux

Ⅱ. Fill in the blanks with the words from the text. The first letter of the word is given.

（1）bearing　（2）gangue　（3）basic　（4）Hematite　（5）Limonite
（6）agglomerated　（7）pellets　（8）fuel　（9）limestone，dolomite

Ⅲ. Fill in the blanks by choosing the right words form given in the brackets.

（1）one　　（2）is　　（3）hematite　　（4）combined　　（5）is

Ⅳ. Decide whether the following statements are true or false（T/F）.

（1）T　　（2）F　　（3）F　　（4）T　　（5）F

Ⅴ. Translate the following English into Chinese.

钢铁冶炼最重要的原料是铁矿石。目前对铁矿石的质量要求是：含铁量要高，冶金性能和还原性要好，杂质元素的含量要低，粉末要少（小于5mm的粉末要低于5%）。要达到这些要求，通常应对矿石进行处理。

第3单元　高炉车间

☞ 第一部分　课文参考译文

一个完整的高炉车间是由很多部分组成的（见图3-1）。最重要的组成部分是高炉、料仓、炉顶装料设备、出铁场、热风炉、煤气逸出和净化设备。

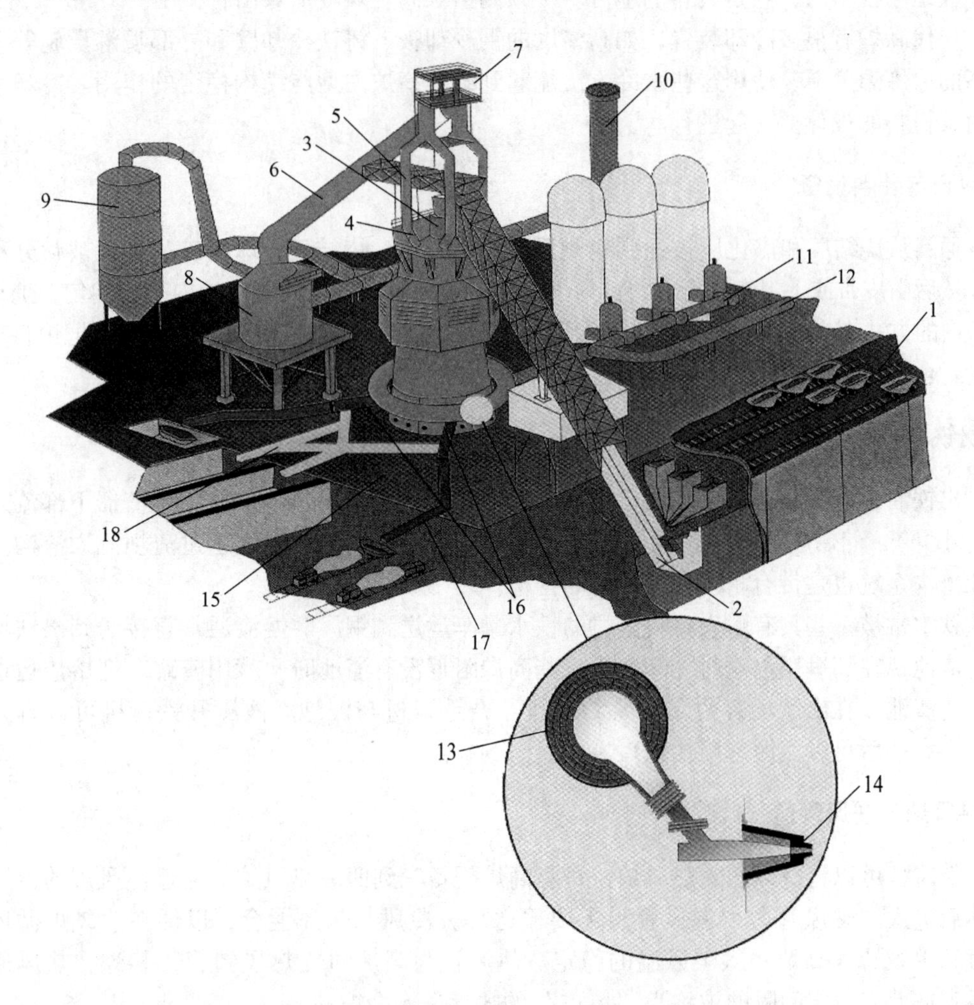

图3-1　高炉车间

1—料仓；2—称量车/料车上料系统；3—“料钟”系统或无（料）钟炉顶系统；
4—排气管，导出管；5—上升管；6—下降管；7—分压器；8—集尘器；9—煤气净化系统；
10—净煤气管；11—热风主管；12—混风管；13—热风围管；14—风口；
15—出铁场；16—主沟；17—铁沟；18—渣沟

1. 高炉

高炉是连续工作的竖炉，它的早期形状是细长喇叭形的。现在，它的轮廓是截短的圆柱体和锥体。高炉是具有耐火材料炉衬的大型钢制容器，并且整个高炉要求进行冷却。

下面介绍现代高炉的组成部分及其作用。高炉建造在能够支撑高炉及其炉料的打桩的混凝土地基上。高炉内型（自上而下）可以描述为炉喉、炉身、炉腰、炉腹、炉缸。较低的部分称为炉缸，炉缸四周由炭砖砌筑，底部中心复合砌筑。炉缸底部有一个或多个铁口，铁水和炉渣从这些铁口排出。铁口的上部是渣口，在现代低渣量操作的情况下，渣口仅在开炉和紧急条件下使用。炉缸上部是炉腹，炉腹角在高炉设计中是一个关键的参数。最宽的部分是炉腰，可以通过各种方法冷却。一般情况下，炉腹和炉身采用陶瓷内衬。高炉炉喉是圆柱形的，它是装料的地方。炉顶则引导高炉煤气从导出管排出。

现代高炉有很多冷却装置，如高密度的铜冷却板、铸铁冷却壁和外部喷淋设施等。在较高的冶炼温度下，使用这些水冷方法能起到维持高炉内型与结构稳定的作用。冷却系统与耐火材料的设计要一起进行。

2. 料仓与装料设备

原料从烧结厂和焦化厂被送到料槽系统，在这里，烧结矿、球团矿和焦炭被筛分和称量，然后由皮带或上料车进行上料。为了避免浪费高炉煤气，同时使炉内布料均匀，炉料装入炉内的方法有点复杂。高炉最近的发展是无料钟炉顶，它可以借助于旋转溜槽把炉料从顶部密封仓布入高炉炉内。

3. 出铁场

出铁后，炉渣浮在铁水上面，通过“撇渣器”与铁水分离。铁水由撇渣器下部流入铁槽（小井），然后越过“铁沟坝”进入铁沟。撇渣器前端的炉渣则越过渣坝流入渣沟。铁沟和渣沟系统也是带有耐火材料里衬的沟槽。

铁水常被一些具有耐火材料里衬的铁水罐车运送到钢厂。熔渣通常直接送到出铁场附近的渣池，在那里用水喷射冷却。当一些高炉附近没有渣池时，就用渣罐车把熔渣运送到较远的渣池。在出铁场有两个重要的设备：开铁口机与泥炮。液压开铁口机可以打开铁口。出完铁后可以用泥炮堵住铁口。

4. 热风炉与送风管道

鼓风机可以将冷风送入热风炉。每座高炉配备三到四座热风炉。通过热风炉的冷风被加热后进入“热风管”。“混风管”需要输送部分冷风与热风混合，以便在热风炉循环使用过程中维持（高炉）入炉风温的稳定。热风管与高炉的“热风围管”相连，热风通过一系列彼此有一定间隔的被称为“风口”的喷嘴送入高炉。

5. 煤气排出与净化设备

炼铁过程中产生的煤气，从高炉顶部排出，经过一系列具有耐火材料内衬的煤气“导出管”“上升管”“下降管”被送入煤气净化系统。在上升管的顶部有一系列被称为“放散阀”的煤气减压阀。下降管把含尘煤气送入“集尘器”中，它能除去约60%的灰尘颗粒，然后煤气被送到“煤气净化系统”，通过净化，煤气中含有的99.9%的灰尘颗粒可以被除去。从这点来看，煤气是一种适合于热风炉和其他地方使用的燃料。

☞ 第二部分　课程思政

【炉火丹心】高炉卫士——孟泰

1948年11月，孟泰重回鞍山钢铁厂，而此时的鞍钢在饱经战乱之后，已经残破不全。但是他丝毫没有退缩，爱厂如家，艰苦创业。他冒着严寒，刨冻雪抠备件，迎着臭气，扒废铁堆找原材料。每天泥一把、油一身、汗一脸，拣回一根根铁线、一颗颗螺丝钉、一件件备品。在他的带动下，全厂工人在短短的数月内回收了上千种材料、上万个零备件，建成了当时著名的“孟泰仓库”，并为恢复生产起了重要的作用。而后他又勇于攻克技术难关，先后解决了十几项技术难题。

☞ 第三部分　延伸阅读

富氢碳循环高炉

富氢碳循环高炉是指在传统冶金工艺中以氢代碳，大幅度减少钢铁冶金流程的温室气体排放，直至实现钢铁冶金生产过程的碳中和。富氢碳循环高炉技术的特点是全氧。如果高炉实现全氧鼓风，那么高炉炉顶煤气中的大量氮气就不会出现，高炉煤气就可以很容易实现CO和CO_2的分离。CO_2回收利用后，剩下的高浓度CO再通过管道输送到风口和炉身，实现CO和H_2重新富集成高还原势的煤气，重新回用至高炉，用于还原铁矿石，这就是碳循环。通过这样的碳循环，实现碳化学能的完全利用。有了煤气循环，大量使用富氢物质就不会浪费氢的化学能了，氢也可以在高炉循环，从而降低高炉流程对化石能源的消耗。

☞ 第四部分　课文问题参考答案

Answer the following questions.

(1) The most important components of the blast furnace are: the blast furnace, bunkers, charging equipment, the cast house, hot blast stoves, top gas removal and cleaning equipment.

(2) The lower portion of the furnace is called the hearth.

(3) Yes, it does. The modern blast furnace has much cooling equipment, such as, high density copper cooling plates, cast-iron stave coolers, external shower cooling equipment.

(4) There are two critical pieces of equipment in the casthouse: the taphole drill and the mudgun.

(5) Liquid slag will float on the iron and is separated from the iron by the skimmer after tapping.

(6) Each blast furnace has three or four hot blast stoves.

☞ 第五部分　练习题参考答案

Ⅰ. Translate the following expressions into English.

(1) blast furnace plant　(2) cast house　(3) blast pipe
(4) shaft　(5) bosh angle　(6) skimmer
(7) offtake　(8) iron runner　(9) slag pit
(10) mudgun　(11) dustcatcher　(12) bell-less top
(13) iron notch　(14) rotating chute　(15) belt system

Ⅱ. Fill in the blanks with the words from the text. The first letter of the word is given.

(1) plant　(2) hearth　(3) belly　(4) bell　(5) casthouse
(6) taphole, mudgun　(7) cooling　(8) bustle　(9) cleaning　(10) vessel

Ⅲ. Fill in the blanks with the words listed below.

(1) support　(2) notch　(3) ceramic　(4) undertaken
(5) double　(6) skimmer　(7) plug　(8) bleeders

Ⅳ. Decide whether the following statements are true or false (T/F).

(1) F　(2) T　(3) T　(4) F

Ⅴ. Translate the following English into Chinese.

钢铁工业中水的总耗量是很大的。下面讨论水在高炉中的专门用途。

为了有效地操作，高炉需要大量的水来冷却炉子的各个部分及其辅助设备。例如，风口、渣口、炉身、炉腹以及内壁冷却器，炉子阀门等地方必须有冷却水不断循环流通。清洗煤气，即除去由炉顶排出的废气中的灰尘，需要相当数量的水。此外，炉渣形成颗粒及其他用途也需要一定数量的水。

第 4 单元　热风炉

☞ 第一部分　课文参考译文

每座热风炉都是一个用于预热高炉用风的大换热器，它能利用高炉煤气的热值把空气加热到 1000~1350℃。当设计合理、操作得当时，热风炉的热效率将达到 80%~85%。

热风炉由炉壳、燃烧室、格子砖蓄热室以及能够调节和传输各种气体的控制阀和多根管道组成（见图 4-1）。炉壳是直径为 6~9m、高为 20~40m 的焊接圆钢筒，它的内部是耐火材料炉衬。燃烧室可以布置在内部、顶部或外部，据此，热风炉有内燃式、顶燃式和外燃式三种类型，它们的工作原理是一样的。燃烧室的容积应最小化，以便使蓄热室尽可能大些。在保证良好燃烧和有效稳定操作方面，热风炉燃烧器的设计很关键，用煤气和助燃空气混合效果好的陶瓷燃烧器能够满足这一要求。蓄热室是用格子砖堆砌而成的，它能提供许多直径比较小的高温气体直通管。当格子砖的表面积与体积比增加时，热风炉的效率就会提高。

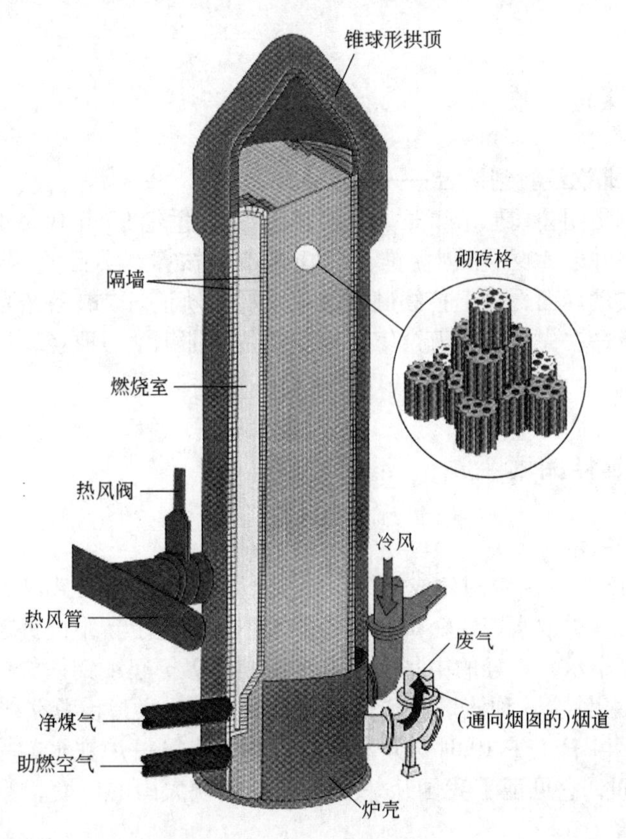

图 4-1　热风炉

三座热风炉工作时，通常是一座送风，而另两座在燃烧。燃烧时，助燃空气与高炉净煤气被送入燃烧室（进行燃烧），必要时，高炉煤气中可以通入天然气或焦炉煤气进行浓化。点火后，煤气与助燃空气涡旋混合后将产生较短而密的火焰，火焰温度通常高达 1200～1400℃。这时，热废气通过格子砖向下运动，在排到烟道阀以前，废气的温度下降到 300～400℃。

目前，热风炉的燃烧根据顶温和废气温度的目标采用 PLC（可编程控制器）自动控制。在燃烧的最初阶段，热风炉的拱顶温度比预设温度要低，大量的煤气与空气以合适的空燃比快速燃烧。当达到顶温的设定值以后，把顶温当作自动控制目标，也就是保持恒定的煤气流量和较大的空气流量。当燃烧废气温度达到设定值以后，操作模式自动转换到把废气温度当作自动控制目标上。

送风时，压缩的冷风进入蓄热室向上运动（逐渐地加热），而后通过热风阀排出热风炉。一部分冷风从热风炉周围通过并（经过混风管）再次进入热风系统。冷、热风混合能保证热风温度恒定。混风阀在每一循环的开始打开，在热风炉送出的风温与所需的风温相等时逐渐关闭。炉温的进一步损失则标志着另一座热风炉送风和下一个循环的开始。

由于高风温能够降低焦炭的消耗，所以热风炉耐火材料的设计也是变化很快的。高风温热风炉的耐火材料显然承受更高的操作温度。使用高铝质耐火材料可安全承受炉顶 1315℃的温度，而温度进一步升高时则需要使用抗蠕变的硅砖，炉内的耐火材料必须能很好地抵抗热循环和碱金属、氧化铁的影响。炉内耐火材料的寿命可以很长，但它主要取决于炉内气体中杂质对炉子污染的程度，这种污染是由含有铁和碱金属元素的气体逆流通过炉子时所产生的。

☞ 第二部分　课程思政

【改革先锋】邯钢经验创造者——刘汉章

刘汉章，邯钢集团公司原董事长、总经理。在 20 世纪 90 年代，他把市场机制引入企业内部经营管理，并于 1991 年创立推行“模拟市场核算、实行成本否决”的经营机制，使企业迅速扭转被动局面，走上持续健康快速发展的轨道。“邯钢经验”在全国掀起了一场企业管理模式革命，先后有 2 万余家企事业单位到邯钢学习取经，被誉为“我国工业战线上的一面红旗”。

☞ 第三部分　延伸阅读

武钢的第一炉铁水

1958 年 9 月 13 日，新中国建成的第一个大型钢铁企业——武汉钢铁公司的第一座大型高炉，流出了第一炉铁水。这一炉炙热的铁水不仅托起了新中国民族工业发展的向往和希冀，也向全世界昭示，年轻的中华人民共和国向世界挺起了钢铁脊梁。当时的人民日报发表社论《贺武钢出铁》，称该事件“标志着我国社会主义建设总路线的伟大胜利”。

2019 年 10 月 14 日上午 10 时 9 分，曾为武钢炼出第一炉铁水的一号高炉停炉，永久退出生产。61 年间，它见证了我国从一穷二白到钢铁大国的巨变，为国民经济的建设发展作出了巨大贡献。

☞ 第四部分 课文问题参考答案

Answer the following questions.

（1） There are three types of hot blast stoves—external-combustion，internal-combustion and top-combustion stoves nowadays.

（2） The stove consists of several parts：the shell，the combustion chamber，the checker work，and control valves and lines that regulate and deliver the various gasses.

（3） The combustion chamber can be arranged at the inside，top or outside.

（4） When the surface area to volume ratio for the checker mass is increased，the efficiency of the stove is improved.

（5） The common flame temperatures range from 1200 to 1400℃.

（6） The mixer valve is open at the start of each cycle.

（7） In combustion initial stage，the dome temperature is lower than the preset value.

☞ 第五部分 练习题参考答案

Ⅰ. Translate the following expressions into English.

（1） internal-combustion hot stove （2） combustion chamber （3） checker brick
（4） ceramic burner （5） chimney （6） cold blast valve
（7） checker chamber （8） high temperature （9） natural gas

Ⅱ. Fill in the blanks with the words from the text. The first letter of the word is given.

（1） stove （2） combustion （3） large （4） blast （5） life

Ⅲ. Fill in the blanks with appropriate prepositions.

（1） on （2） into （3） by （4） of （5） in
（6） after （7） of （8） by （9） through （10） to

Ⅳ. Decide whether the following statements are true or false（T/F）.

（1） F （2） F （3） T （4） T （5） F

Ⅴ. Translate the following English into Chinese.

耐火材料是应用于钢铁工业中的重要材料，它主要用作炼钢炉和炼铁炉的内衬、承装和运输金属的钢包内衬、下道工序加热钢坯的炉子内衬以及传导热气的烟道的内衬。历史证明，坚持不懈地寻求和开发更合理的冶金工艺，极大地推动了耐火材料的发展，这些耐火材料问题的迅速解决又成为近代钢铁工业不断发展的重要因素。

第 5 单元　炼铁的新发展

☞ 第一部分　课文参考译文

在过去的一个多世纪里，为了找到能够取代传统高炉的炼铁技术，人们做了许多努力。本章将介绍成功应用于工业化炼铁生产的一些新工艺和新技术。

1. 直接还原技术

在钢铁冶炼技术的历史上，最早使用的就是直接还原法，高炉取代了直接还原法炼铁是冶金技术的重大进步。随着钢铁工业的发展，200 多年以来高炉一直使用的焦炭价格每年都在增加，合适的焦煤越来越少。在 18 世纪末，直接还原法又得以恢复。

现在，那些在低于铁的熔点温度下用还原铁矿石的方法来炼铁的技术通常称为直接还原法，其产品称为直接还原铁（DRI）。

根据还原剂的使用种类，直接还原法一般可分为气体还原法与固体还原法两类。按使用的还原反应器类型，直接还原法主要分为流化床、竖炉、反应罐和回转窑法四大类。

（1）气体还原法。在气体还原法中，还原剂是一氧化碳、氢气或这两种气体的混合物。它们通常从天然气中获得。天然气中甲烷转化为一氧化碳和氢气的反应如下：

$$CH_4 + H_2O = CO + 3H_2$$

竖炉法在气体还原法中起着重要的作用。50%以上的还原过程是在竖炉中进行的，例如米德列法就是一个代表（见图 5-1）。

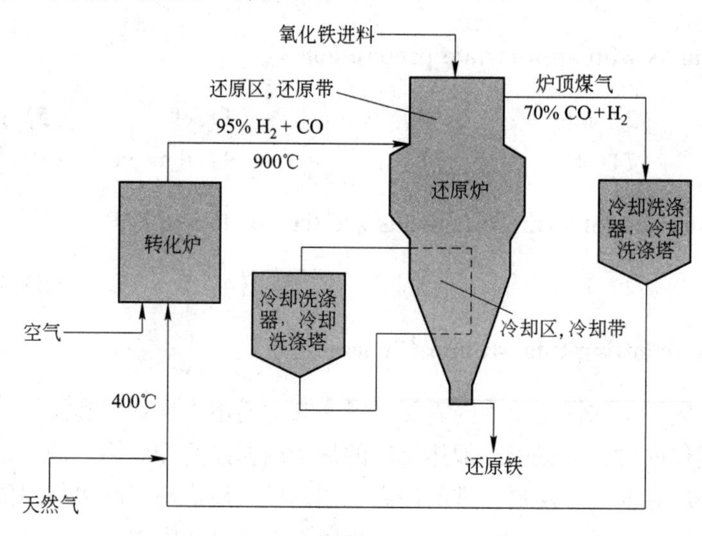

图 5-1　米德列法

米德列竖炉自1969年最早建成并投产以来，已发展成为直接还原法的主要形式。最大的米德列竖炉，年生产能力达到了80万吨。

这种工艺的主要部分包括直接还原炉、气体转化炉和气体冷却系统。生产时，还原气通过位于还原带底部的环形风管进入还原炉内，迎着下降炉料向上运动。固体炉料通过料封管不断从竖炉顶部进入。还原炉的设计使得炉料能够依靠自身的重力作用均匀地通过预热带、还原带、冷却带。冷却后的直接还原铁可以不断地通过炉底料封管排出。冷却气体的流向与竖炉内冷却带炉料的流向相反。从炉顶逸出的气体流向洗涤器，在那里冷却，洗涤除去炉尘。大部分炉顶气体被压缩后与天然气混合预热到约400℃，并导入重整管，被重整成一氧化碳和氢气，然后循环到直接还原炉内。

（2）固体还原法。固体还原法使用的还原剂是呈粒化状态的煤和合成的煤。回转窑是使用固体还原剂的重要设备，这个工艺是连续生产的。目前，煤基直接还原流程成熟的工艺是德国的SL/RN法、英国的DRC法和法国的Codir法。

直接还原法的产品是海绵铁，这名字的产生是源于它表面的多孔性。如果不采取相应的措施，海绵铁在处理和储存过程中就面临重新氧化的危险。采用热压成块、外层喷涂“石灰壳”或水玻璃钝化，可以阻止再氧化。

2. 熔融还原技术

熔融还原法是指不用焦炭从矿石中还原液态铁水的方法。熔融还原法主要有两种工艺：熔融气化炉法和铁浴反应炉法。熔融气化炉工艺中由奥钢联（Voest Alpine）与德国科尔夫（Korf）公司共同开发的COREX法已经用于工业生产。

熔融还原和直接还原一起代表了传统高炉工艺的重要发展趋势。

3. 低碳炼铁技术

（1）高炉煤气的循环作用。高炉炉顶煤气循环利用是将炉顶煤气除尘净化和脱除CO_2后，将其中的还原成分（CO和H_2）通过风口或者炉身适当位置喷入高炉，从而重新回到炉内参与铁氧化物的还原，加强C和H元素的利用。该工艺被认为是改善高炉性能、降低能耗和减少CO_2排放量的有效措施之一。

（2）高炉喷吹含氢物质。高炉喷吹含氢物质强化氢还原已成为当今研究的热点。高炉喷吹含氢物质主要包括废弃塑料、天然气、焦炉煤气等。

（3）高炉富氧鼓风技术。高炉富氧鼓风是指向高炉中加入工业氧，使鼓风中的氧含量超过大气中的含氧量。高炉使用富氧鼓风可以加速碳燃烧，在燃料比不变的情况下使产量增加。但富氧鼓风使进入高炉的风量减少，带入高炉的热量也减少。

4. 氢冶金技术

（1）氢气直接还原。随着CO_2减排压力的增大，氢还原技术将会越来越受到钢铁行业的重视，迎来蓬勃发展的机会。开展氢还原技术的研究，例如，生物质还原技术、核氢还原工艺等，应是主要发展方向。因此，低成本氢的来源成为了重要问题。与核能行业合作，开展核氢还原工艺研究，也是一个重要的方向。

（2）基于氢冶金的熔融还原直接炼钢。该方案以冷态除杂的超纯铁精矿为原料，实现源头减排。通过 1200℃的飞速氢还原和 1600℃的高能量密度铁浴熔融还原，得到超纯净的钢水。再经过连铸连轧得到高品质、高洁净度的钢铁材料。这一过程完全取消了炼铁，实现连续装料、连续炼钢、连铸连轧的全连续、一体化的生产方式，工艺简化，生产效率提高。

☞ 第二部分　课程思政

【绿色发展】习近平论积极稳妥推进碳达峰碳中和

实现碳达峰碳中和是一场广泛而深刻的经济社会系统性变革。立足我国能源资源禀赋，坚持先立后破，有计划分步骤实施碳达峰行动。完善能源消耗总量和强度调控，重点控制化石能源消费，逐步转向碳排放总量和强度“双控”制度。推动能源清洁低碳高效利用，推进工业、建筑、交通等领域清洁低碳转型。

☞ 第三部分　延伸阅读

推进文化自信自强——二十大报告

提高全社会文明程度。实施公民道德建设工程，弘扬中华传统美德，加强家庭家教家风建设，加强和改进未成年人思想道德建设，推动明大德、守公德、严私德，提高人民道德水准和文明素养。统筹推动文明培育、文明实践、文明创建，推进城乡精神文明建设融合发展，在全社会弘扬劳动精神、奋斗精神、奉献精神、创造精神、勤俭节约精神，培育时代新风新貌。加强国家科普能力建设，深化全民阅读活动。完善志愿服务制度和工作体系。弘扬诚信文化，健全诚信建设长效机制。发挥党和国家功勋荣誉表彰的精神引领、典型示范作用，推动全社会见贤思齐、崇尚英雄、争做先锋。

☞ 第四部分　课文问题参考答案

Answer the following questions.

(1) Because with the development of iron and steel industry, coke, on which blast furnace have depended for over 200 years, is becoming more expensive every year, and supplies of suitable coking coals are becoming scarcer.

(2) According to the reducing agents used, direct reduction technology is commonly classed into two types: gas reduction processes and reduction processes with solid reducing agents.

(3) The reducing gases are carbon monoxide, hydrogen or mixtures of two in gas reduction processes.

(4) The shaft-furnace process plays an important part in gas reduction processes.

(5) The main components of the Midrex process are the shaft furnace, the gas reformer, and the cooling-gas system.

(6) The outcome of all direct reduction techniques is sponge iron.

（7）Smelting reduction technology can be subdivided into two main groups：melter gasifier process and iron bath reactors process.

☞ 第五部分　练习题参考答案

Ⅰ. Translate the following expressions into English.

（1）reduction process　（2）rotary kiln furnace　（3）sponge iron
（4）melter gasifier　（5）reaction shaft furnace　（6）go into production

Ⅱ. Fill in the blanks with the words or phrases listed below.

（1）metallurgical　（2）hydrogen　（3）gas　（4）reformer
（5）seal legs　（6）equipment　（7）sponge iron　（8）ore

Ⅲ. Fill in the blanks by choosing the right words form given in the brackets.

（1）contains　（2）needs　（3）including　（4）are　（5）have
（6）which　（7）reducing　（8）enter　（9）molten　（10）combine

Ⅳ. Decide whether the following statements are true or false（T/F）.

（1）F　（2）T　（3）F　（4）T　（5）F

Ⅴ. Translate the following English into Chinese.

在过去的60多年里，高炉工艺取得了巨大的进步并获得显著的成就。然而，今天的高炉也受到一些能够在不污染环境的条件下从矿粉和非焦煤中直接生产金属铁的新技术的强烈挑战。这些新工艺主要是直接还原和熔融还原。

直接还原具有一些优点，包括低投资和使用原始的能源代替焦炭，所以这种技术为许多发展中国家提供了建设国家钢铁工业的可能性。事实上，直接还原与电弧炉相结合，在今后与高炉和转炉技术一样，一定会成为重要的钢铁生产技术。

第 6 单元　炼钢原料

☞ 第一部分　课文参考译文

用于炼钢的主要原料是高炉铁水，此外，还有废钢、海绵铁、造渣剂、合金剂和氧化剂。

1. 高炉铁水（熔融铁水）

高炉铁水是由铁及许多其他化学元素组成的，最常见的元素有碳、锰、磷、硫和硅。生铁可能含有 3%~4.5%的碳（质量分数）、0.05%~2.5%的锰（质量分数）、0.02%~0.06%的硫（质量分数）、0.3%或更多的硅（质量分数）和少量的磷。铁水中大量的杂质元素必须部分或全部除去，事实上，这就是炼钢的主要任务。

2. 废钢和海绵铁

废钢或海绵铁也可以作为炼钢的原料和冷却剂来使用。在世界范围内，废钢占炼钢原料的 40%，所以它是一种重要的原料。废钢的用量根据炼钢生产工艺的不同而变化。在碱性氧气转炉炼钢过程中，废钢量约占 20%；而在电炉生产中，它可以占 100%。

废钢按照来源可以分为以下几类：

（1）循环废钢。循环废钢是在钢铁厂的炼钢和轧钢过程中产生的。循环废钢以各种切头和废品形式存在，通常会很快回到炼钢炉中。

（2）加工废钢。它是用户制造成品时产生的，通常会很快返回到钢厂。

（3）基建废钢。它来源于报废的商品及设备。在某些情况下，这些物质使用 3~4 年后可能又返回到厂里，而大型基建设备的寿命可能有 50 年或更久，但是它们的平均寿命大约为 20 年。

循环废钢及加工废钢通常在没有被杂质元素污染的情况下返回到钢厂，大型基建废钢也具有同样的质量。而许多“短寿命”的基建废钢则由于受到各种涂层的污染，在没有完全与非铁成分分离的情况下以打包的形式返回，这使钢中的杂质含量急剧增加，降低了钢的很多使用性能。

实际上，废钢必须首先分类，而那些不需要的杂质元素，例如非铁金属，必须在特殊的选矿厂去除。目前，这个工作通常用现代的冲压、破碎和切割设备来完成。使用各种商业废钢（如粗钢、废钢片、钢屑）有助于促进其循环利用。

海绵铁在公元 1300 年人类发明高炉之前的几个世纪里，一直是钢铁的主要来源。现在的应用中，海绵铁是指直接还原铁。大部分直接还原铁用来代替电弧炉中的废钢。由天然铁矿石冶炼而成的直接还原铁是一种相对纯净的产品，它的杂质比废钢中的少，从而提高了钢的质量。

3. 造渣剂

就如铁水的（生产）情况一样，使用造渣剂会产生一种反应性好、能吸收杂质元素的低黏度炉渣。在钢铁生产的各个阶段，如精炼阶段、预处理阶段、炉外精炼阶段以及铸钢阶段，都使用造渣剂。造渣剂由石灰、白云石、萤石等组成。石灰（CaO）和白云石（$CaCO_3$，$MgCO_3$）是两种主要的造渣剂，石灰可以通过在回转窑内煅烧碳酸盐获得。造渣剂的一些典型分析见表6-1。

表6-1　石灰、白云石的典型分析

石　灰	CaO	SiO_2	Al_2O_3	MgO
质量分数/%	93.0	1.7	1.2	0.7

萤　石	CaF_2	SiO_2	S
质量分数/%	75~85	10.0（最大）	1.0（最大）

作为造渣剂要特别注意避免使用粉末状材料，因为废气很容易将粉末带走。

4. 合金剂和脱氧剂

只有使用不同种类和数量的合金剂，钢才能拥有某些性能，例如耐腐蚀性、可加工性和高温强度。

用于炼钢的重要合金剂是镍、铬铁、钛铁、钨铁、钒铁、硅铁和钼铁等。

脱氧剂，即与液态钢中溶解的氧进行结合的添加物，常常是在精炼之后立即加入，它们常归于合金剂一类。

5. 氧化剂

氧化剂包括氧气、铁矿石和氧化铁皮等，它们在炼钢过程中起着非常重要的作用。

☞ 第二部分　课程思政

【钢铁航母】“亿吨宝武”

2020年，中国宝武钢产量达到1.15亿吨，实现“亿吨宝武”的历史性跨越，问鼎全球钢企之冠。中国宝武在航空航天、能源电力、交通运输、国家重大工程等众多领域完成了一系列关键材料的研发与制造，解决了一大批“卡脖子”材料难题，有力支撑了我国国民经济的高质量发展。在汽车、家电、建筑等领域，中国宝武不断攀登超越、创新迭代，推出一系列适应新时代发展需要的钢铁精品，为实现人民对美好生活的向往不断贡献着宝武力量，成为当之无愧的“国之重器、镇国之宝”。

☞ 第三部分　延伸阅读

“碳达峰”对炼钢原料供需结构的影响

碳达峰，就是指在某一个时点，二氧化碳的排放不再增长达到峰值，之后逐步回落。

碳达峰与碳中和，简称“双碳”。废钢是炼钢的主要原料之一，在碳达峰碳中和大背景下，钢铁行业绿色转型、节能减排任重道远。废钢作为一种可循环利用的能源，做好废钢的回收利用是钢铁行业发展的一大重点。2021 年 10 月，国务院印发《2030 年前碳达峰行动方案》，其中指出要推动钢铁行业碳达峰，继续压减钢铁产能，提升废钢资源回收利用水平，推行全废钢电炉工艺。2021 年 7 月，国家发展和改革委员会发布了《“十四五” 循环经济发展规划》，其中指出，2020 年我国废钢利用量约 2.6 亿吨，并明确了到 2025 年废钢利用量达到 3.2 亿吨的目标任务。

☞ 第四部分　课文问题参考答案

Answer the following questions.

(1) The raw materials used for steelmaking are hot metal, steel scrap, sponge iron, slag formers, alloying agents and oxidizing agents.

(2) The chemical elements in the hot metal are iron, carbon, manganese, phosphorus, sulphur, and silicon.

(3) According to the text, a most important task of steelmaking is that large quantities of tramp elements within the hot metal still have to be removed partly or completely.

(4) DRI stands for direct-reduced iron.

(5) Slag formers consist of lime, dolomite, fluorspar, etc.

(6) The important alloying agents used in the production of steel are nickel, ferrochromium, ferrotitanium, ferrotungsten, ferrovanadium, ferrosilicon, and ferromolybdenum, etc.

(7) Oxidizing agents consist of oxygen, iron ores and scale, etc.

☞ 第五部分　练习题参考答案

Ⅰ. Translate the following expressions into English.

(1) steel scrap　(2) direct-reduced iron　(3) alloying agent
(4) slag former　(5) capital scrap　(6) dressing plant
(7) production process　(8) circulating scrap　(9) ferrochromium

Ⅱ. Fill in the blanks with the words from the text. The first letter of the word is given.

(1) agent　(2) per cent　(3) sorted　(4) sponge　(5) furnace
(6) stages　(7) waste　(8) oxygen　(9) materials　(10) arises

Ⅲ. Fill in the blanks by choosing the right words form given in the brackets.

(1) of　(2) used　(3) principally　(4) in　(5) least

Ⅳ. Decide whether the following statements are true or false (T/F).

(1) F　　(2) F　　(3) F　　(4) F　　(5) T

Ⅴ. Translate the following English into Chinese.

高炉内的还原剂主要是由燃料中的碳氧化而成的一氧化碳和焦炭中的碳。含有铁氧化物的含铁原料在高炉内通过使用还原剂能被还原为液态生铁。在这个过程中，生铁吸收了3.0%~4.5%的碳（质量分数）。由高炉生产出来的大部分生铁在其仍为液态时，直接被运送到炼钢厂用于炼钢。

第 7 单元　现代炼钢原理

☞ 第一部分　课文参考译文

当前炼钢工艺主要有两种，较为普遍的是以高炉铁水和废钢为原料的氧气炼钢工艺（也称转炉工艺），其次是以高质量的工业废钢或者预还原的球团矿为原料的电弧炉炼钢工艺（平炉工艺，曾占粗钢产量100%，现已缩减到几乎为零）。氧气炼钢工艺主要是通过使用高纯度的氧气，对以铁水（熔融铁水）和废钢组成的金属炉料进行精炼，以便快速生产含碳量和温度合适的钢水。在精炼过程中，还要加入各种造渣剂以去除金属中的硫和磷，使其达到所需要的成分。目前氧气顶吹炼钢工艺（也称 LD 或 LD/AC 顶吹工艺）更被普遍采用，但在新钢厂中逐渐被复合吹炼工艺所代替。电弧炉炼钢的基本原理比较简单，废钢在电弧的直接加热作用下被熔化，电弧产生于石墨电极和废钢之间。精炼是靠金属与碱性炉渣之间的相互作用实现的，其过程与氧气转炉精炼过程一样，但是精炼时间比较长。

炼钢过程中最重要的化学反应是脱碳、杂质元素造渣（脱硅、脱磷、脱锰、脱硫）和脱氧（用硅铁与铝去除多余的氧）。下面这些方程式示出了把铁水和海绵铁转变成钢的化学反应原理：

$$[C] + [O] = CO$$

$$[Si] + 2[O] + 2[CaO] = (2CaO \cdot SiO_2)$$

$$[Mn] + [O] = (MnO)$$

$$2[P] + 5[O] + 3(CaO) = (3CaO \cdot P_2O_5)$$

$$[S] + (CaO) = (CaS) + [O]$$

$$[Si] + 2[O] = (SiO_2)$$

$$2[Al] + 3[O] = (Al_2O_3)$$

脱碳是最重要的化学反应，在这个反应中加入的氧与铁水中的碳反应形成一氧化碳并作为可燃废气逸出，产生的一些热量保留在金属和炉渣中。形成一氧化碳产生的热量大约占碳燃烧潜热的三分之一，当它完全燃烧成二氧化碳时，剩余的热量将全部释放出来。

铁水和废钢中的杂质元素进入炉渣是分两个阶段进行的：第一阶段，把杂质元素氧化，它们在铁液中不能溶解；第二阶段，它们上升到铁水表面并与加入的石灰形成炉渣，钢液和炉渣可以直接得到氧化反应产生的所有热量。

硅能被氧化成二氧化硅并放出热量，二氧化硅加速了渣中石灰的溶解及炉渣的形成进程，释放出的热量可用于熔化废钢。二氧化硅非常稳定，一旦形成，就不会在任何碱性工艺下重新还原。

转炉吹炼期间，锰不可避免地会有很大程度的氧化。锰氧化物可降低碱性渣的熔点，加速熔渣的形成。在某些条件下，一些锰可能从炉渣还原到金属中，但对这一工艺来说并不重要。

尽管一些硫会作为气体逸出，但是它一般与生石灰直接反应形成硫化钙。作为气体逸出的硫量受炉渣表面气氛的强烈影响。脱硫的另一个影响因素是渣量，渣量越多，钢中的含硫量越少。发生这种情况是因为硫化钙在炉渣中有固定的溶解度，所以单位金属的渣量越大，它从金属中吸收的硫量就越多。

磷可以被氧化形成磷的氧化物（P_2O_5）。如果炉渣中含有足够的石灰和铁氧化物，它将保留在渣中。但如果还原条件增强，炉渣温度升高或者炉渣碱度降低，P_2O_5就会被还原，磷将重新返回到金属中去。

用过量的氧进行精炼时，一些氧将留在钢液中。脱氧期间，硅或铝常常被加入到液态钢液中除去多余的氧。脱氧产物进入渣中。

☞ 第二部分　课程思政

【大师足迹】中国冶金物理化学之父——魏寿昆院士

1936 年，魏寿昆院士完成了海外的学业后回到祖国，投身冶金事业。20 世纪 40 年代，他针对高磷铁水，提出了小型贝塞麦炉的去磷程序及贝塞麦炉与马丁炉双联的操作方法，指导小转炉脱磷。为了解决耐火材料问题，他进行了“四川白云石去钙提镁之研究”和“人造镁氧烧制镁砖的研究”，提出了利用二氧化碳选择性溶解去钙提镁的措施，使氧化镁的纯度高达 99.5%，符合制造高质量镁砖的要求，并阐明了提纯氧化镁的机理。他一生治学严谨，教书育人，为中国的高等教育事业和冶金科学发展做出了开创性贡献。

☞ 第三部分　延伸阅读

铁碳相图中的奥秘

铁碳相图是研究铁碳合金最基本的工具。这实际上是 Fe-Fe_3C 相图，铁碳合金的基本成分也应该是纯铁和 Fe_3C。铁具有同素异形转变，即在固态下具有不同的结构。铁和碳的不同结构可以形成不同的固溶体，Fe-Fe_3C 相图上的固溶物都是间隙固溶体。由于 α-Fe 和 γ-Fe 晶格中孔隙的不同特征，两者的碳溶解能力也不同。铁碳合金中有三种相，即铁素体、奥氏体和渗碳体。

铁素体：C 溶于 α-Fe 中形成的间隙固溶体称为铁素体，用符号 F 或者 α 表示。铁素体的强度、硬度不高，但具有良好的塑性和韧性。

奥氏体：C 溶于 γ-Fe 中形成的间隙固溶体称为奥氏体，用符号 A 或者 γ 表示。奥氏体的力学性能与其溶碳量及晶粒大小有关，一般具有良好的塑性和韧性，所以，奥氏体为基体的铁碳合金易于锻压成型。

渗碳体：Fe 与 C 形成的金属化合物，具有复杂晶格的间隙化合物，用 Fe_3C 表示。渗碳体硬度很高，而塑性与韧性几乎为零，脆性很大。

☞ 第四部分 课文问题参考答案

Answer the following questions.

（1） Yes, there are. They are the oxygen processes （also called converter processes） and the electric-arc process.

（2） The most important chemical reactions during refining are decarburization, slagging of tramp elements （desiliconization, demanganization, dephosphorization, desulphurization） and deoxidation （removal of residual oxygen by ferrosilicon and aluminum）.

（3） The added oxygen reacts with the carbon inside the hot metal to form carbon monoxide and escapes as a combustible waste gas.

（4） As silicon is oxidized, it forms silica and produces heat.

（5） Manganese and silicon are unavoidably oxidized to a considerable degree during converter blowing.

（6） Sulphur will normally react directly with the burnt lime to form calcium sulphide.

☞ 第五部分 练习题参考答案

Ⅰ. Translate the following expressions into English.

（1） dephosphorization （2） the carbon electrode （3） interaction
（4） basic slag （5） equation （6） waste gas
（7） desiliconization （8） converter process （9） refining

Ⅱ. Fill in the blanks with the words from the text. The first letter of the word is given.

（1） produce （2） Decarburization （3） slag （4） stable （5） blowing
（6） phosphoric （7） chemical （8） Basic （9） removed （10） melting

Ⅲ. Fill in the blanks by choosing the right words form given in the brackets.

（1） heated （2） to （3） does （4） or （5） oxidizes
（6） into （7） added （8） to bring

Ⅳ. Decide whether the following statements are true or false （T/F）.

（1） F （2） T （3） F （4） F （5） T
（6） T （7） F （8） F （9） T （10） T

Ⅴ. Translate the following English into Chinese.

目前，有两种主要的炼钢流程：（1）高炉—转炉流程；（2）电炉流程。在高炉和转炉生产的第一种流程中，高炉还原铁矿石的目的是为了生产作为炼钢阶段原料的生铁。在第二种流程中，废钢和直接还原铁被熔化成钢。用这些方法产生的钢随后通过二次冶金被进一步精炼。

第 8 单元　转炉生产

☞ 第一部分　课文参考译文

氧气顶吹转炉是一种梨形转炉，形状与贝塞麦和托马斯转炉相似。转炉通常用菱镁矿或者白云石作碱性炉衬。氧气顶吹转炉装置还包括氧枪，副枪和提升转移装置，结构相当复杂（见图 8-1）。氧气转炉炼钢工艺如图 8-2 所示。

经过多年的生产实践，转炉通常采用下列顺序进行装料。上炉出钢完毕，首先装入废钢（废钢预先称量，并被运至转炉，由起重机吊起，倾斜倒入炉中），废钢加完后，立刻将需要量的铁水兑入炉中。成渣剂和其他添加剂（粒度大小为 20~25mm）以两种不同形式加入。

（1）这些材料在吹炼过程中由转炉顶部连续加入；

（2）部分材料在吹炼开始时加入，剩余部分在吹炼过程中连续加入；第二种方法最为普遍。

兑完铁水，转炉摇至垂直位置，氧枪下降，吹炼开始。吹炼过程中，不同炼钢厂氧枪与铁水的距离在 7 英尺与 2 英尺 6 英寸之间变化。

在供热期间，高速氧射流应尽可能穿入金属熔池中，在相对小的区域与铁水作用。熔池循环从这一热点开始，当大量的 CO 开始产生时，被进一步加强。这样渣钢混合物就产生了，这极有助于加快炼钢操作的速度。

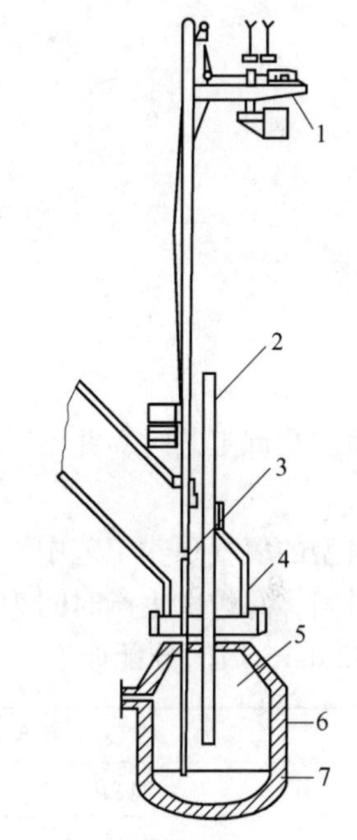

图 8-1　顶吹氧气炼钢装置
1—氧枪；2—副枪；3—烟罩；4—转炉；5—转换器；6—炉壳；7—炉衬

相反在其他期间，氧气射流必须作用于熔池表面而不是深层，以便加速渣中石灰的溶解并形成一定碱度的流动渣。氧气射流对熔池表面的作用应避免金属和渣的喷溅对炉衬造成的危害。

吹炼需要 10~20min，吹炼一完成，就要进行取样，以便把合金成分与预定标准进行对比。与此同时还要测量钢液的温度，出钢温度一般在 1600~1650℃之间。取样分析需要 1 分多钟。

出钢时倾斜转炉，钢水通过出钢口流入浇注钢包。在出钢过程中或出钢之后，炉渣浮在钢水表面。为了除去炉渣，转炉倾斜到另一边，以便钢水从转炉的边缘流出。在特殊情况下，转炉安装特殊的出钢装置以阻挡炉渣流出。

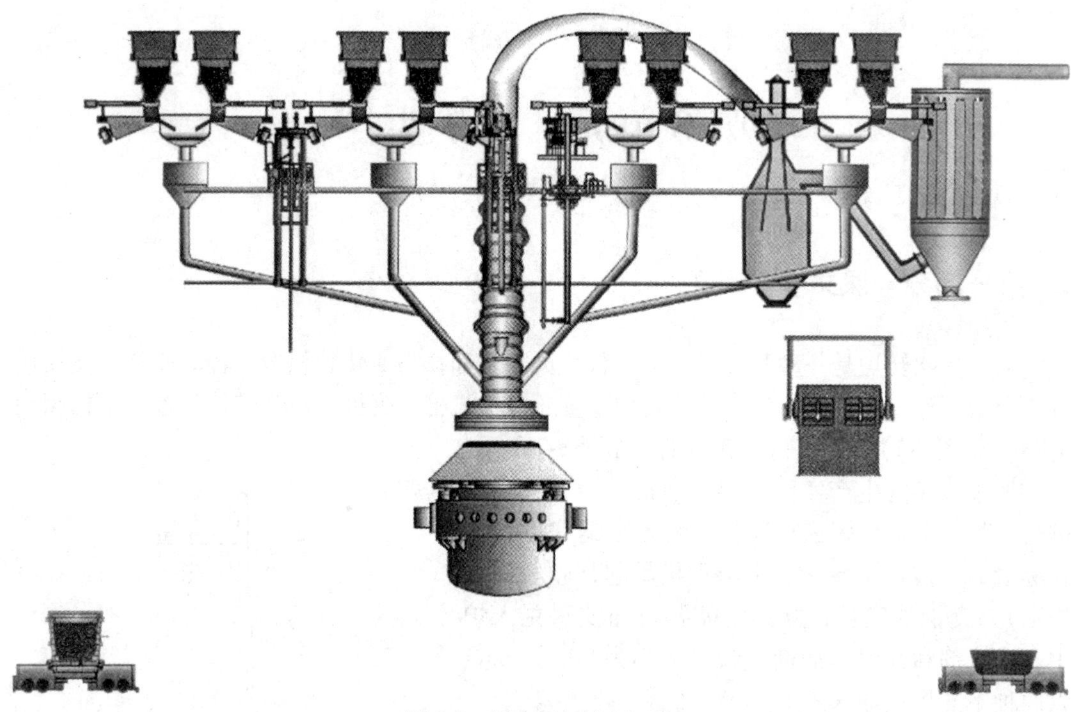

图 8-2　顶吹氧气炼钢工艺

在完成吹炼工艺时，并不是所有的炉渣都要除去，部分渣还会留在炉内进行下次吹炼的循环。

在典型的氧气顶吹工艺中，吹炼时间为 10～20min，加上装料与出钢、出渣，测温以及取样时间，每炉次生产时间为 30～50min。例如，一座 300t 的现代转炉，其冶炼周期平均约为 42min，时间分配如下：

加入废钢和兑铁水	10min
吹氧	17min
倾炉取样	6min
出钢，出渣，修补炉衬	9min

☞ 第二部分　课程思政

【钢魂铁志】双良精神

20 世纪 50 年代，李双良就是闻名全国冶金行业的“工业炉渣爆破能手”。1983 年退休后，李双良主动请缨，不要国家一分钱，带领渣场职工治理渣山。经过 10 年的努力，他搬掉了沉睡半个多世纪的高 23m、占地 2.3km^2、总量达 1000 万立方米的渣山，累计回收废钢铁 130.9 万吨。此外，他自创设备生产各种废渣延伸产品，带领职工在渣山四周建起防尘护坡墙，墙内建有花坛、假山、鱼池，并种植花、树 7 万多株，自此，起风扬沙的渣山蜕变成一座色彩斑斓的大花园。“当代愚公”成为李双良的代名词，传遍大江南北。“双良精神”也被确定为太钢的企业精神，成为太钢人宝贵的精神财富。

☞ 第三部分　延伸阅读

人工智能在钢铁工业智能制造中的应用

钢铁工业是国民经济的基础工业，是决定国家发展水平的最基本要素之一。中国已经成为世界第一钢铁制造大国，但在整体创新能力、素质和产品竞争力方面存在“大而不强”的问题。加快钢铁工业的转型升级，向钢铁强国转变，已成为新时期中国经济社会发展的重大战略任务。智能制造将先进的制造技术与新一代信息技术深度融合，贯穿于制造过程的各个环节，是中国钢铁工业创新发展的一个新机遇，钢铁工业未来重点发展方向将是基于新一代人工智能技术的智能制造。

☞ 第四部分　课文问题参考答案

Answer the following questions.

（1）The converter is basic lined usually with magnesite or dolomite.

（2）The combination of devices for top blowing of oxygen include the lance, a standby lance, and lifting and transfer mechanisms.

（3）It is first charged with scrap (a scrap tray is preliminarily weighed and delivered to the converter; the tray is lifted by the crane and inclined to dump the scrap into the vessel). Immediately after scrap is charged, the required quantity of molten pig iron is poured into the converter. Slag formers and other additions (with the particle size of 20mm to 25mm) are charged by two different methods.

（4）The sample analysis takes a little more than a minute.

（5）During tapping, the steel bath should have a temperature of 1600℃ to 1650℃.

（6）No, not all of the slag has been removed on completion of the blowing process and so part of it is still here for the next blowing cycle.

（7）The duration of a heat in a modern 100t converter is around 42 minutes on the average.

☞ 第五部分　练习题参考答案

Ⅰ. Translate the following expressions into English.

（1）pear-shaped converter　（2）standby lance　（3）oxygen lance

（4）sample analysis　（5）on the contrary　（6）tap-to-tap time

（7）pouring ladle　（8）tapping hole　（9）oxygen jet

Ⅱ. Fill in the blanks with the words from the text. The first letter of the word is given.

(1) dolomite (2) sequence (3) scrap (4) vertical (5) penetrate
(6) harmful (7) steel (8) holding (9) tilting (10) surface

Ⅲ. Fill in the blanks by choosing the right words form given in the brackets.

(1) from (2) are (3) had (4) could (5) took
(6) to (7) measured (8) at

Ⅳ. Decide whether the following statements are true or false (T/F).

(1) F (2) T (3) F (4) F (5) T

Ⅴ. Translate the following English into Chinese.

(1) 顶吹氧气转炉系统是由固定炉底、砖衬、水冷氧枪组成。转炉中要装入一定量的废钢和铁水。氧枪下降到炉内，高速氧射流被吹入液态金属中。吹氧期间，石灰作为熔剂被加入以便熔化氧化的杂质。合金在冶炼末期加入。

(2) 复吹工艺的特殊优点是：通过废钢的快速溶解能达到均匀熔化的目的；钢液循环加快了25%；铁和合金元素的回收率会更高；提高了钢水化学成分的准确性；延长了转炉寿命。从工艺和冶金观点来看，复吹工艺具有进一步发展的潜力。

第 9 单元　电弧炉炼钢工艺

☞ 第一部分　课文参考译文

前面章节中所描述的炼钢工艺对于生产普碳钢和一些低合金钢起到令人满意的作用，但是它们不适合生产含有大量合金元素的钢，因为这些合金元素形成的氧化物比铁氧化物更稳定。就工具钢、模具钢和不锈钢这些高级合金钢而言，必须在严格控制的条件下进行精炼，这样才能把杂质减少到最低程度。由于燃料在炼钢炉中燃烧时不可避免地使钢水受到污染，这使炼钢工作者们意识到从技术上来说，电炉工艺比平炉和贝塞麦工艺更为理想。

电弧炉主要是由带有出钢口和炉门的炉壳、带有电极和移动装置的炉盖组成（见图 9-1）。炉膛既可用酸性炉衬，也可用碱性炉衬（酸性电弧炉主要用于铸钢厂，碱性电弧炉现在用于生产合金钢和特殊钢）。短而粗的电弧是电炉的主要热源，靠它把电流由电极传给金属。

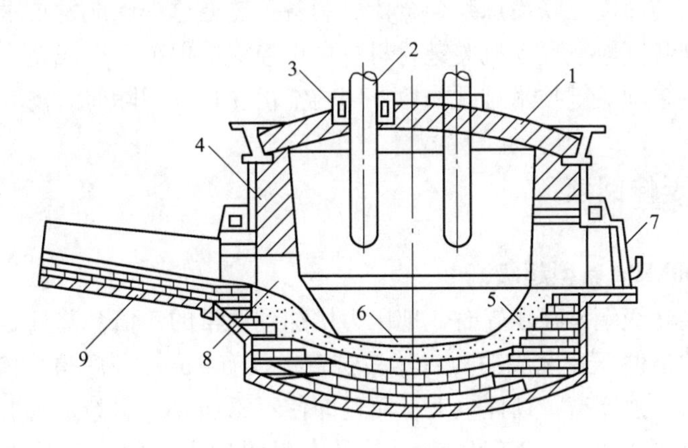

图 9-1　电弧炉

1—炉盖；2—电极；3—水冷圈；4—炉墙；5—炉衬；
6—熔池；7—炉门；8—出钢口；9—出钢槽

碱性电弧炉炼钢过程通常可分为以下几个步骤进行。

（1）装料。先将石灰铺在炉底，再把成分已知的废钢放入炉中。除了废钢和海绵铁外，（装入电弧炉的）炉料还包括矿石、造渣剂、还原剂（碳）以及铁合金。在氧化之前或氧化过程中，这些物料可以通过炉门加入。装完料后，炉门前砌上耐火材料炉门坎以防止熔化后的金属从炉门溅出，关上炉门，电极降到废钢上方开始熔炼。

（2）熔化。电极首先熔化它底部和周围的废钢，然后继续穿入金属料，在炉膛内形成金属熔池。从电极穿井和炉底熔池形成开始，底部废钢受电弧的加热作用而熔化。在碱性电弧炉操作中，熔化期是其成本消耗最大的阶段，因为此阶段的电能和电极消耗最多。

(3) 氧化或者净化。在氧化阶段，通常向熔池吹入氧气，氧气的吹入可以加速熔炼期并直接氧化磷、硅、锰、碳以及金属铁。在氧化期，碱性电弧炉中发生的反应与碱性氧气转炉中的反应相似。当渣中的氧化物与熔池中的碳反应时就产生了一氧化碳，引起热沸腾，氢气、氮气以及非金属化合物将以气体形式逸出。

(4) 脱氧或者精炼（还原期）。还原期的主要任务是脱氧，脱硫，调整钢液成分、温度和进行合金化。炉渣控制是还原期非常重要的因素。

现在，氧化和还原阶段的冶金过程被下道工序或在钢包中进行的炉外精炼所代替，这使得高能量的电弧炉完全用于熔化废钢。

(5) 出钢。出钢时，电极抬高以便在倾斜的位置清空熔池。打开出钢口，炉子用一机械控制机构倾斜，使钢液从炉内流入钢包。出钢过程中用铸钢吊车控制钢包，减少钢流暴露于空气及对钢包耐火材料的腐蚀。根据具体操作，可以选择在出钢之前、之后或出钢时出渣。

☞ 第二部分　课程思政

【钢铁雄心】瞄准关键核心技术，突破“卡脖子”难题

中国宝武的高纯净度三联冶炼某合金棒材制备工艺通过中国航发技术评审，初步实现进口替代，为实现我国航空发动机关键材料自主可控提供保障；完成超大型固体火箭发动机壳体研制，解决多项关键技术难题，填补国内空白，达到国际前沿技术水平。

☞ 第三部分　延伸阅读

中国研发 2200MPa 新型超级钢

2200MPa 新型超级钢是在压轧时，把压力增加到通常的 5 倍，并且是在提高冷却速度和严格控制温度的条件下开发成功的。其晶粒直径仅有 1μm，为一般钢铁的 1/10～1/20。因此，其组织细密，强度高，韧性也大，而且即使不添加镍、铜等元素也能够保持很高的强度。超级钢的研制成功，对于我国来说是具有里程碑意义的。有了超级钢，我国的航母、军舰和核潜艇的综合能力将有飞跃式的提升。

☞ 第四部分　课文问题参考答案

Answer the following questions.

(1) The essential components of the electric-arc furnace are the furnace shell with tapping device and work opening, the removable roof with the electrodes and tilting device.

(2) The main tasks of reduction period are deoxidation, desulphurization, the composition and temperature adjustment and alloy added. Slag control is a very important factor in reduction period.

(3) Electric arcs are the principal source of heat in electric-arc furnace.

(4) Besides scrap or sponge iron, the charge also includes the ores, fluxes, reducing agents (carbon) and alloying elements in the form of ferroalloys.

(5) Yes, they can.

(6) The slag may be tapped before, with, or after the steel, depending on the particular operation.

☞ 第五部分　练习题参考答案

Ⅰ. Translate the following expressions into English.

(1) electric-arc furnace　(2) die steel　(3) stainless steel
(4) steel foundry　(5) oxidizing stage　(6) plain carbon steel
(7) melting phase　(8) high-grade alloy steel　(9) controlled condition

Ⅱ. Fill in the blanks with the words from the text. The first letter of the word is given.

(1) furnace　(2) arcs　(3) alloy　(4) lime　(5) fluxes
(6) bath　(7) reduction　(8) single　(9) clear　(10) crane

Ⅲ. Fill in the blanks by choosing the right words form given in the brackets.

(1) divided　(2) difference　(3) removed　(4) requires　(5) producing

Ⅳ. Decide whether the following statements are true or false (T/F).

(1) F　(2) F　(3) T　(4) F　(5) T

Ⅴ. Translate the following English into Chinese.

如平炉一样，电炉与贝塞麦炉的区别在于它采用外部热源来加热熔池。因此，就温度的控制而言，它不依靠发生在金属中的反应发热来提高温度。电炉是靠穿过钢的电流来产生它的热量。因此金属没有任何被燃料中硫和其他元素污染的可能性。

第 10 单元　炉外精炼

☞ 第一部分　课文参考译文

过去，在转炉或电弧炉精炼过程中生产的钢被称为“成品”，用于铸造和轧制。现在，精炼工艺之后通常进行炉外精炼。炉外精炼的目的是生产“纯净”钢，使它能够满足对表面、内部和显微纯度及力学性能等方面的严格要求。炉外精炼的任务主要是脱气（减少钢中氢气、氧气和氮气的含量）、脱碳、去除不必要的非金属夹杂物（主要是氧化物和硫化物）等。

炉外精炼方法很多，大致可分为常压下炉外精炼（非真空处理）和真空下炉外精炼（真空处理）两类。相应的技术有：使用多孔喷枪进行搅拌处理，或者进行电磁搅拌；向钢液添加精炼剂（喷枪喷射固体、喂线），感应加热和电弧加热；借助于各种技术进行真空脱气等。炉外精炼工艺可以在钢包、钢包炉中进行，有些情况下甚至可以在电弧炉中进行。

真空处理法由于具有多功能性和特殊优点，在炉外精炼中享有优先权。这里将主要介绍真空处理的一些方法。

真空处理是基于以下原理考虑的：当钢液凝固时，溶解在钢中的气体只有一部分能够逸出，因此，所有的钢都含有一定量的气体：氢气、氮气和氧气。这些气体会导致钢发生脆化、缩孔、出现夹杂物和其他凝固后所不希望的现象，这将降低钢材的技术性能。比如，氢在固体钢中的溶解对力学性能有害，它使塑性降低，而强度并没有相应增加，也导致了高应力部位的裂纹；而氮则会降低钢的深冲性能。如果降低外部压力，溶解在钢液中的气体就会逸出。除了脱气外，其他冶金反应也能在真空处理中发生，比如脱碳。真空处理如果与搅拌结合，则能够加快和促进冶金反应。

目前炼钢中有多种真空处理法，比如真空循环脱气法、真空钢包吹氩法、川崎顶吹氧真空脱气法、电弧加热真空精炼法、真空电弧加热脱气法、真空吹氧脱碳精炼法等。

RH 法是真空循环脱气工艺。这种方法如图 10-1 所示。脱气室下部设有与其相通的两根循环流管，钢液通过一根连续吹氩的管子进入真空室进行脱气。重力的作用又导致钢液从另一根管子离开并返回到钢包中。该工艺的平均循环流速通常是每分钟 12t，20min 处理完 100t 的钢包，而对于大约 40t 钢包来说，温度损失是 40~50℃。

就电弧加热真空精炼法而言，其主要设备包括一个钢包炉、一个可移动的感应加热器、一个蒸汽喷射器和一个带加热电极的真空盖。进行精炼时，将带有钢水的钢包放在感应加热器中，并运到真空脱气站。降压到 200μmHg。对 30t 的炉次，15min 后加入脱氧剂。真空处理后，钢包移到加热站，如图 10-2 所示，用电极包盖盖住钢包。加入熔剂造碱性渣，加入合金满足合金化要求。通过电弧加热后出钢，通常将钢水进行连铸。这种方法灵

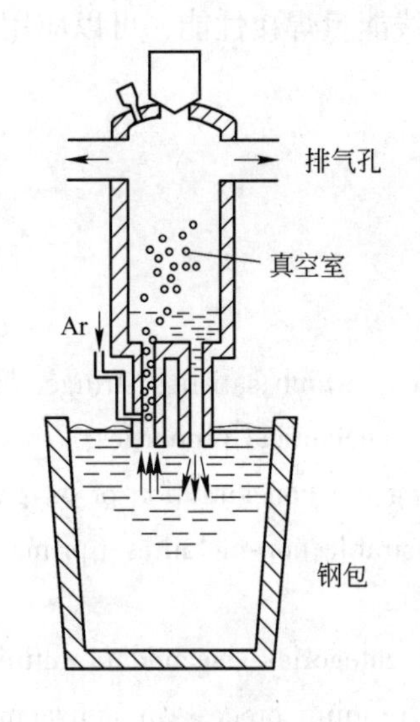

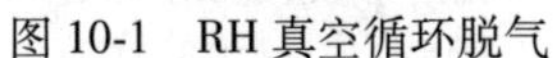
图 10-1　RH 真空循环脱气

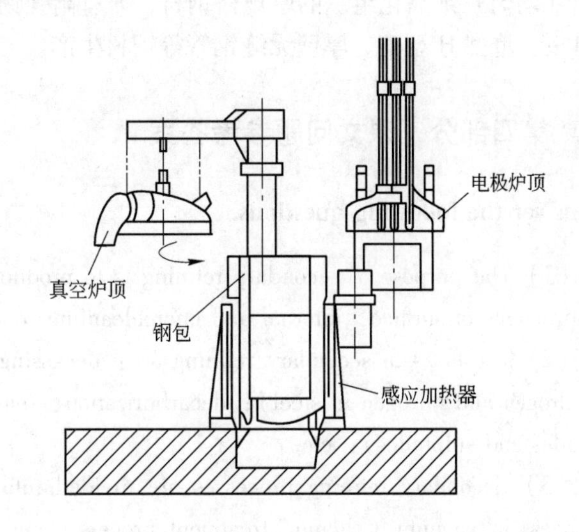

图 10-2　ASEA-SKF 精炼炉

活，适合高质量特殊钢，且允许与连铸机匹配。

许多炉外精炼工艺所达到的冶金功能会相互重叠，为了确定采取何种工艺，以实现对给定混合产品所要求的不同的冶金功能，钢厂必须考虑该工艺本身的实际能力和灵活性。另外，还必须考虑与工序有关的车间内的其他操作、所预期的生产能力、相关投资以及运行成本等。

☞ 第二部分　课程思政

【钢铁意志】保尔 · 柯察金名言

人最宝贵的东西是生命，生命属于我们只有一次，一个人的生命应该这样度过：当他回首往事时，他不因虚度年华而悔恨，也不因过去碌碌无为而羞耻——这样，在他临死时，可以说：我整个生命和精力已献给了世界上最壮丽的事业——为人类的自由和解放而斗争。

☞ 第三部分　延伸阅读

炼钢新工艺

1. 钢包底喷粉高效精炼新工艺

开发炉外精炼过程中钢包底喷粉技术。底喷粉过程无铁损，搅拌动力学条件优于顶喷粉，配套技术成熟，易实现。改造投资低，不改变原工艺。可以建立超低硫洁净钢生产平台，取得良好的除硫效果。

2. 厚规格结构钢的微观组织均匀细化控制

将氧化物冶金技术实施的冶炼过程控制，与后续轧制过程中的轧制与冷却控制相结合，在冶炼过程控制的基础上，实施一定的终轧温度控制、冷却速率控制，可以获得全断面（均匀）细晶化组织的厚规格钢材，兼有高强韧性与可大线能量焊接性能，可以应用于厚板、重型 H 型钢、厚壁无缝钢管等钢材生产。

☞ 第四部分　课文问题参考答案

Answer the following questions.

(1) The purpose of secondary refining is to produce clean steel, which satisfies stringent requirements of surface, internal and microcleanlines quality and of mechanical properties.

(2) The tasks of secondary refining are: degassing (decreasing the concentration of oxygen, hydrogen and nitrogen in steel); decarburization; removing undesirable non-metallics (primarily oxides and sulphides); etc.

(3) Secondary refining processes are divided into two broad categories: secondary refining process in vacuum (vacuum treatment process) and secondary refining process in nonvacuum (nonvacuum treatment process).

(4) Secondary refining processes can be carried out in the ladle, the ladle furnace, in some instances, even in the electric-arc furnace.

(5) Vacuum treatment of steel enjoys high priority because of its versatility and the special advantages in second refining.

(6) Nitrogen lowers the ability of steels to undergo deep drawing operations.

(7) The average circulation rate is usually some 12t per min and 20 min are required to fully treat a 100t ladle of steel in RH process.

(8) In the case of ASEA-SKF, the equipment comprises a ladle-furnace, a mobile induction heater, a steam ejector and a vacuum cover fitted with electrodes for heating.

☞ 第五部分　练习题参考答案

Ⅰ. Translate the following expressions into English.

(1) mechanical property　(2) clean steel　(3) vacuum treatment
(4) electromagnetic stirring　(5) Finkle process　(6) induction furnace
(7) vacuum chamber　(8) external pressure　(9) metallurgical reaction
(10) RH process　(11) circulation leg　(12) mobile induction heater
(13) steam ejector　(14) the vacuum degassing station
(15) metallurgical function

Ⅱ. Fill in the blanks with the words listed below.

(1) applied　(2) moulds　(3) partly　(4) undergo　(5) dissolved
(6) combination　(7) vacuum　(8) outflow　(9) heater　(10) specification

Ⅲ. Fill in the blanks with appropriate prepositions.

（1） of　　（2） in　　（3） by　　（4） in　　（5） on
（6） for　　（7） to　　（8） with

Ⅳ. Decide whether the following statements are true or false（T/F）.

（1） T　　（2） F　　（3） F　　（4） F　　（5） T

Ⅴ. Translate the following English into Chinese.

最初，真空脱气被用来大规模地处理特殊质量的镇静钢，后来则扩展到提高其他镇静钢的质量。最近，通过真空处理后的非镇静钢也具有相当多的优点，并且在这个领域可能将有更大的发展。脱气对镇静钢和非镇静钢的影响将在下面的内容中被讨论。

第11单元　连续铸钢

☞ 第一部分　课文参考译文

自从200多年前首次生产液态钢以来，几乎一直不变的操作工艺是：通过模铸把钢水浇注成矩形的毛坯，再将毛坯经过随后的冷、热加工获得理想的形状。这些毛坯或钢锭起初重量很轻，但是随着炼钢产量的增加，钢锭的重量也增加，现在生产15~20t的钢锭已经是很普通的事情。这些钢锭在初轧厂被轧制成方坯或板坯，随后再进一步加工。

模铸有以下特点：

（1）大量的资金投资在钢锭模、底板和机车上。

（2）可观的资金也投资在铸车的脱模、整模所需的建筑物和起重机上，同时均热炉、预热炉以及初轧机对轧制钢锭也是必要的。

（3）钢锭含有偏析，它是所有钢的缺陷。而且为了一定质量，模铸时需要生成初轧后作为废钢而造成较大损失的“保温帽”。钢锭越大，偏析就越严重。

（4）每个钢锭轧制后需要切头切尾。

这些要求使炼钢工作者产生了把钢水直接铸造成板坯或方坯的设想，这样就减少了上面提到的投资和加工费用，也能把产量的损失和产品中的偏析减小到最低程度。连续铸钢是20世纪50年代在欧洲开发的，此后发展很快。现在世界上连铸的年生产能力超过了10亿吨，相当于钢总产量的89.7%。

最初的连铸机是立式连铸机，它从地面13m以上的高度浇注，铸坯垂直拉下并逐渐被水冷却（见图11-1）。为了降低厂房的高度，首先研制出的连铸系统是将钢水倒入立式结晶器中，并且在弯曲之前让钢水完全凝固。这种系统随后发展为弧形连铸机，这是目前常

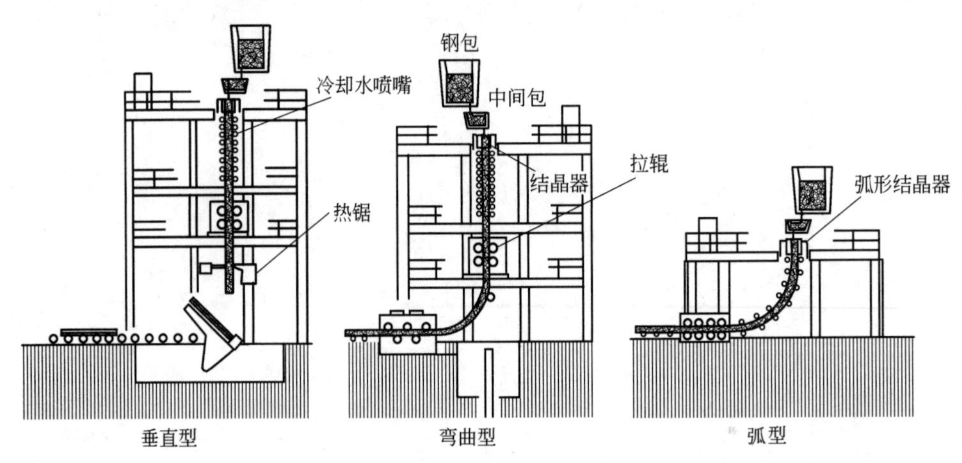

图11-1　连铸机类型

用的设备。

在一个典型的连铸车间，由电炉（或碱性氧气转炉）生产的脱氧钢水被倒入钢包，送到连铸跨，并置于康卡斯连铸机上方的托架上。浇注开始前，用链环做的引锭杆插入结晶器底部，并停留在由机械导辊组成的弧形空腔中。塞棒或滑动水口控制流入中间包的钢水量。当中间包内的钢水达到规定的高度时，打开水口，钢水注入放在托架上有内部水冷装置的结晶器内。结晶器中的钢水遇到引锭杆便同它凝结在一起。当结晶器中的钢液面达到一定的高度时，两种操作立即开始。结晶器开始在垂直方向做往复振动，引锭杆顺连铸机向下拉出。当凝固的铸坯从垂直状态过渡到水平状态时，用一系列辊子托住。为了操作和维修方便，辊子被分组呈扇形布置。在辊子间还要进一步喷水冷却。

在矫直机的出口处装有一个位置较低的辊子，它使引锭杆同铸坯脱离。引锭杆既可以放到机器的一边，也可以提升到机器上方的托架上。铸坯沿出料端辊道连续运行到火焰切割站。在预定尺寸处，火焰切割机将按需要的长度自动切割铸坯。一旦连铸机开始工作，就根据炼钢炉供应的钢水量连续地铸出钢坯。

连铸厂的最终产品是正方形断面的方坯（达到 250mm×250mm），大约 300mm 厚、2500mm 宽的板坯。也生产包括圆棒材，八角断面的棒材在内的其他形状的钢锭。

20 世纪 60 年代以来，连铸就平稳地取代了模铸。下面所列出的连铸机类型同样反映出这一技术的历史发展。连铸机的类型有立式、立弯式、弧形、椭圆形连铸机。从一种形式到另一种形式的转化都伴随着总高度的大大降低。现在，连续铸钢方面的进一步发展包括有固定铸锭模的水平连铸、有传动带或旋转轧辊（双轧辊）的连铸等。

总之，使用连铸系统的理由是：

（1）与初轧相比，降低投资费用。

（2）与传统的铸锭相比，提高了生产能力。

（3）在整个铸坯长度上钢的成分较均匀；中心质量比较好，尤其是板坯；具有较高的内表面质量，这样就节省了其他昂贵的表面处理工序。

（4）高度的自动化。

（5）利于保护环境。

☞ 第二部分　课程思政

【钢铁脊梁】C919 大飞机

C919 是中国第一架按照国际适航标准自主研发的喷气式客机，拥有自主知识产权。该项目于 2007 年获得批准，C919 客机于 2017 年完成首飞。在完成所有适航认证后，C919 于 2022 年 9 月获得了中国民航局颁发的型号证书。第一架飞机将于 2022 年底交付。C919 的研制成功并获得型号证书，标志着中国有能力独立研发世界级的喷气式客机，这是中国大型飞机行业发展的一个重大里程碑。在过去的 15 年中，中国成功地走出了一条以本土设计和系统集成为特色的发展道路，吸引了全球公开招标，并帮助提高了国内技术的比重。我国还培养了一支在大型飞机领域坚持坚定信念、乐于奉献、勇于突破、敢于在艰苦战斗中竞争、具有国际视野的人才队伍，取得了丰硕成果，积累了宝贵经验。

☞ 第三部分　延伸阅读

近终形连铸

近终形连铸是近代钢铁工业发展中一项重大的高新技术，包括薄板坯连铸连轧、带钢连铸、异型坯连铸、管坯连铸、线材铸轧及喷射沉积等。近终形连铸是力求浇铸尽可能接近最终产品尺寸的铸坯，以便进一步减少中间加工工序，节省能源、减少贮存和缩短生产时间。这些技术与传统工艺相比，流程短、设备简单、建设投资省、能耗低、成才率高、生产成本低。

☞ 第四部分　课文问题参考答案

Answer the following questions.

(1) Continuous casting of steel was developed in Europe in the mid 1950s.

(2) The first continuous casting machines for steel were aligned vertically.

(3) The end products of continuous casting plants are square section billets (up to 250mm×250mm), and slabs which are about 300mm thick by 2500mm wide. Other shapes are also produced, including round bars and octagonal sections.

(4) Since the 1960s, continuous casting has steadily displaced ingot casting.

(5) Five types of continuous casting machines are mentioned in this passage. They are vertical, bending, straightening, circular bow, and oval bow.

(6) The primary features of continuous-casting systems are:

1) lower investment outlay compared with that for a blooming train;

2) more productivity than with conventional ingot-casting;

3) high degree of consistency of steel composition along the whole length of the strand; better core quality, especially with flat strands; high inherent surface quality, leading to savings on an otherwise expensive surfacing process;

4) high degree of automation;

5) friendlier to the environment.

☞ 第五部分　练习题参考答案

Ⅰ. Translate the following expressions into English.

(1) ingot casting　(2) primary mill　(3) soaking pit

(4) hot top　(5) the bow-type installation　(6) stopper rod

(7) dummy bar　(8) the discharge roller　(9) square section billet

(10) stationary mould　(11) moving belt　(12) continuous-casting system

(13) blooming train　(14) high inherent surface quality

（15） better working condition

Ⅱ. Fill in the blanks with the words from the text. The first letter of the word is given.

（1） blooms　（2） invested　（3） larger　（4） bottom　（5） vertically
（6） electric　（7） dummy　（8） traveling　（9） overall　（10） ingot

Ⅲ. Fill in the blanks by choosing the right words form given in the brackets.

（1） molten　（2） successful　（3） perform　（4） to　（5） limited

Ⅳ. Decide whether the following statements are true or false（T/F）.

（1） F　（2） F　（3） F　（4） F　（5） T

Ⅴ. Translate the following English into Chinese.

钢水既可以在连铸机上进行连铸，也可以用钢锭模铸锭。钢水中含有溶解的氧，可以添加像铝和硅这样的强脱氧剂将氧脱出。根据钢水的凝固方式（它取决于钢水的成分和使用脱氧剂的程度），钢锭分为三种基本类型，即镇静钢锭、沸腾钢锭和半镇静钢锭。生产镇静钢，要脱净全部溶解的氧，钢水凝固没有一点气体逸出。同镇静钢相反，制造沸腾钢是利用了非脱氧钢凝固时产生的化学反应。溶解的氧和碳结合生成一氧化碳气体，从钢的凝固前沿跑掉。第三种就是半镇静钢，顾名思义，钢水在凝固之前只脱除了一部分氧。

第12单元　钢的分类

☞ 第一部分　课文参考译文

依据GB/T 13304.1—2008，钢按化学成分分为非合金钢（原国标中的普碳钢）和合金钢两类。

1. 非合金钢

从理论上来讲，钢是铁和碳的合金，但在商业生产过程中，不可避免的还存在其他杂质元素，如锰、磷、硫和硅等，含碳量（质量分数）在0.021%～2.11%范围时为非合金钢。非合金钢的强度主要随含碳量的变化而变化，在含碳量（质量分数）低于0.09%范围内，含碳量越高，抗拉强度就越高，并且热处理后所达到的硬度也越高。遗憾的是当含碳量增加时，它的塑性和韧性就降低，并且可淬性也很低。此外，在过高和过低的温度之下，非合金钢的性能也会降低。多数环境下，非合金钢容易被腐蚀。

非合金钢按照碳含量分为低碳钢、中碳钢和高碳钢三类。低碳钢的含碳量低于0.3%（质量分数），拥有良好的可锻性和焊接性，但淬硬性差，不能淬火达到有效的性能。低碳钢一般在热轧和冷轧后使用，常用来制造需要进行多次成型加工的低强度零件。中碳钢的含碳量（质量分数）为0.25%～0.60%，他们常用于锻件以及制造较高强度并需要一定韧性的部件。低碳钢和中碳钢的组织结构都是铁素体和珠光体两部分，只是，随着含碳量的增加，结构中铁素体相对含量减少，珠光体相对含量增加。高碳钢的含碳量（质量分数）超过0.6%，其韧性和塑性很低，但具有高的硬度和耐磨性，淬火后形成高碳马氏体，进一步提高其硬度和耐磨性，主要制造刃具、刀具、量具等小型工具。

非合金钢容易冶炼，价格低廉、工艺性好，具有较好的使用性能，能满足许多场合的需要，因而在机械工程领域得到广泛应用。当需要提高钢材的性能时，可以通过在钢中加入一种或几种合金元素实现。

2. 合金钢

在非合金钢的基础上，有目的地在冶炼的过程中加入一定量的合金元素，含有合金元素的钢叫做合金钢。常用的合金元素有铬、镍、钼、钒、钨、钛、硼、铜、锰［$w(\mathrm{Mn})>1.0\%$］、硅［$w(\mathrm{Si})>0.5\%$］等。合金钢可以分为两种类型：添加元素低于10%的低合金钢和添加元素超过10%（通常为15%～30%）的高合金钢。

合金钢之所以具有各种特殊的性能，取决于所加入金属的种类及其数量。钢中合金元素的作用如下：

（1）铬。铬能提高钢的硬度、强度、耐磨性、耐热性和耐腐蚀性能。1913年，谢非尔德的哈里·布雷尔利试用合金钢制作枪筒，在他认为不适用而丢在一边的试样当中，有一种含铬 14%（质量分数）的钢，数月之后，他发现大多数钢都生锈了，而这种铬钢还保持着它的光泽，这就是不锈钢。铬是不锈钢的主要合金元素，与镍配合提高钢的抗氧化性、热强性和抗腐蚀性。

铬合金钢也可以用来制造轮轴、螺栓、弹簧及机车的双头螺栓，还可以作为工具钢制作模具、凿子、钻头、锉刀、剪刀等。

（2）镍。镍是最古老和最常用的合金元素之一。加镍不但能提高强度、屈服点、硬度和韧性，而且能够增加淬透层深度。镍能够提高耐蚀性，因此它不是“不锈”钢或耐蚀钢的主要成分之一。镍钢可以被用来制造汽车、其他运载工具，包括飞机，以及大量高强度齿轮、轴系部件和其他重型产品。

（3）钼。钼的通常用量（质量分数）少于 0.25%。钼作为一种合金元素，即使是微量也能对钢的物理性质产生很大的影响。钼能提高强度、硬度和耐磨性。钼钢易于热处理、锻造和机加工。

（4）钒。钒的使用含量（质量分数）通常少于 0.25%。作为一种合金元素，钒能提高疲劳强度、极限强度、屈服点、韧性、冲击强度和抗振能力。铬钒钢具有很好的韧性和很高的强度。根据含碳量的不同，他们广泛应用于大型铸件、轮轴、弹簧和工具中。

（5）钨。钨主要和铬一起用作高速工具钢的元素，这种钢的含钨量（质量分数）为 14.00%~18.00%，含铬量（质量分数）为 2.00%~4.00%。高速工具钢具有这样的特性：切削时即使加热到发红的程度，刀刃仍能保持很锋利。

轧辊用钢添加 Mo、W，使钢在回火时基体产生二次硬化，提高耐热耐磨性，如 $9Cr_2Mo$。

（6）钛。钛有助于提高强度、硬度和耐磨性。

（7）铜。加入铜能增强耐蚀性，改善力学性能。

（8）锰。锰用来同硫化合，有助于降低硫的某些不良影响；也可以同碳化合，用于增加硬度和韧性。锰具有增加淬硬层深度的性能。锰还能降低轧制和锻造时的脆性，以改善锻造质量。锰合金钢常用来制造轮轴、锻件、齿轮、传动轴和枪管等。

（9）硅。少量的硅能改善韧性。硅主要是与其他合金元素化合，以提高冲击强度。

在 GB/T 1299—2014《工模具钢》中新增塑料模具用钢类型，在 $3Cr_2Mo$ 和 $3Cr_2MnNiMo$ 基础上，增加了 $4Cr_2MnlMoS$、$8Cr_2MnWMoVS$ 等 19 种新材料，初步形成我国塑料模具用钢体系。如用于精密冷冲模具的 $8Cr_2MnWMoVS$，用于制作大中型镜面塑料模具的 2CrNiMoMnV 等。

☞ 第二部分　课程思政

【国之重器】中国“天眼”——世界最大单口径射电望远镜

“中国天眼”学名为 500m 口径球面射电望远镜，是世界上最大的单口径射电望远镜，具有我国自主知识产权。它主要用于实现巡视宇宙中的中性氢、观测脉冲星等科学目标和空间飞行器测量与通信等应用目标。2016 年“中国天眼”落成。2020 年 1 月通过国家验收正式启用，2021 年 3 月底面向全球开放。截至 2022 年 7 月，500m 口径球面射电望远镜已发现660余颗新脉冲星，进入成果爆发期，极大拓展了人类观察宇宙视野的极限。

☞ 第三部分 延伸阅读

炼钢新技术

1. 高效脱硫铁水预处理技术

开发强力搅拌、喷吹的高效铁水预处理方式，短时间内将硫含量降低到极低水平。

2. 氧化物冶金技术制造大线能量焊接用钢

利用氧化物冶金技术，可以开发大线能量焊接用的碳锰钢、HSLA、高强钢等。这一技术与传统的“纯净化”“洁净化”思路相反，利用炼钢过程中对夹杂物的属性（分布、成分和尺寸等）的有效控制，在后续的凝固、轧制、冷却、使用过程中改善钢材的组织，从而获得需要的组织和新的性能，例如大线能量焊接用钢等。

☞ 第四部分 课文问题参考答案

Answer the following questions.

(1) Plain-carbon steels are generally classed into three subgroups, based on carbon content. They are low-carbon steels, medium-carbon steels and high-carbon steels.

(2) The structures of plain-carbon steels usually are ferrite and pearlite.

(3) Low-carbon steels are usually used for low-strength parts requiring a great deal of forming.

(4) High-carbon steels are used for high-strength parts such as springs, tools, and dies.

(5) Alloy steels are formed by the addition of one or more of the following elements: chromium, nickel, molybdenum, vanadium, tungsten, titanium, and copper, as well as manganese, silicon, and small amounts of other alloying elements.

(6) Nickel is one of the oldest and most common alloy elements.

(7) Medium-carbon steels contain between 0.3 and 0.6 per cent carbon.

(8) Alloy steels are divided into two types. They are low alloy steels and high alloy steels.

☞ 第五部分 练习题参考答案

Ⅰ. Translate the following expressions into English.

(1) tensile strength (2) medium-carbon steel (3) axle

(4) wear resistance (5) mechanical property (6) chromium alloy

(7) yield point (8) corrosion-resisting steel (9) molybdenum steel

(10) fatigue strength (11) high-speed tool steel (12) sharp cutting-edge

(13) forging quality (14) gun barrel (15) plain-carbon steel

Ⅱ. Fill in the blanks with the words from the text. The first letter of the word is given.

(1) composition　(2) carbon　(3) increased　(4) weldability
(5) requiring　(6) formability　(7) upgraded　(8) properties
(9) springs　(10) alloy　(11) strength　(12) amounts

Ⅲ. Fill in the blanks by choosing the right words form given in the brackets.

(1) with　(2) which　(3) conditions　(4) turn　(5) acceptable
(6) used　(7) are　(8) without　(9) tempering　(10) treatment

Ⅳ. Decide whether the following statements are true or false (T/F).

(1) F　(2) T　(3) T　(4) F　(5) T

Ⅴ. Translate the following English into Chinese.

合金元素的基本作用是提高钢的物理性能和力学性能，即增强钢的抗腐蚀性、硬度、耐磨性或强度。自从冶金成为一门科学以来，已经开发了一千多种合金。每一种合金都有其独特的性能。

第 13 单元　硫化矿铜的火法精炼

☞ 第一部分　课文参考译文

铜主要是以硫化矿物的形式存在于地壳中，并常常与硫化亚铁并存，例如

辉铜矿	Cu_2S（黑色）
黄铜矿	CuS · FeS（淡黄色）
斑铜矿	CuS · FeS（浅蓝色）

上述矿物的浓度在矿体内是比较低的，典型矿物中的铜含量（质量分数）约为0.5%~2%。

世界上约90%的铜主要是以硫化物矿石形式存在。其主要杂质是铁，可能还伴有少量的镍、金、银和微量的锌、锡、铅、钴、砷、锑、硒、碲、铋（低品位矿石中也含有氧化物，但可以用湿法冶金来去除这些杂质）。世界上一些主要的铜产地是赞比亚、加丹加（刚果）、智利、加拿大、美国（蒙大拿州）和澳大利亚。

1. 矿石的准备处理

如上所述，硫化铜矿很少，含铜量（质量分数）低至0.5%，而浮选可将经过破碎和研磨后的矿石用来制作含（质量分数）20%~30%的铜矿物。

2. 提取

人们曾经将硫化铜完全熔融至氧化铜后，再用碳将其还原成金属。但是由于生产出的铜杂质含量太多，而且冶炼时有大量金属铜进入渣液造成损耗，因此这种方法已不再使用。现在使用的方法取决于高温下的几个关系：

（1）铜和硫的亲和力比铁大，铁和氧的亲和力比铜大。

（2）铁被氧化成氧化铁后再与二氧化硅结合形成硅酸盐渣液，从而与铜分离。

火法提取铜主要有两个阶段：

（1）锍（冰铜）熔炼形成包含所有炉料中铜的金属硫化物熔体和完全不含铜的熔渣。

（2）锍转化成为粗铜。随着锍熔炼开始进行，精矿粉可能进行局部烧结以便降低含硫量。这个操作可在多炉膛焙烧炉或流化—闪速焙烧炉中进行。无论使用哪种炉子，重要的是都要确保所有的铜和部分铁继续以硫化物形式存在，以便在锍的转化过程中产生热量。

3. 锍熔炼

正如以上所述，在锍的冶炼过程中反射炉正在被闪速炉所取代。干燥的精矿被吹入炉

中后与氧气或富氧结合并燃烧。熔炼期间，一部分铁的硫化物被氧化成氧化铁后与二氧化硅形成硅酸铁炉渣：

$$2FeS + 3O_2 \xlongequal{} 2FeO + 2SO_2$$

$$2FeO + SiO_2 \xlongequal{} 2FeO \cdot SiO_2$$

其余的铁仍以 FeS 形式存在，并与铜的硫化物形成熔融的混合物（锍）。渣液比锍轻并且与锍不相溶，所以可将其分离。含（质量分数）25%～50%铜的锍与含有贵重金属（如金、银）的矿石进入转炉，燃烧反应为加热和熔化锍以及渣液提供了所需的大部分热量。排出的气体中 SO_2浓度比较高（通常为10%以上），可用于生产硫酸、硫或液态二氧化硫。

4. 锍到粗铜的转化

锍到粗铜的转化包括用空气或氧气来氧化熔锍，从而去除铁和硫，同时生产含铜量（质量分数）约为98%的粗铜。

熔锍被注入卧式转炉并通过风口吹入空气。FeS 氧化产生的热量足以使该过程自动进行。这种转化分两阶段进行：

（1）FeS 的消除或矿渣的形成阶段。由于 FeO 比 Cu_2O 更稳定，因此 FeS 优先于 Cu_2S 氧化，形成的氧化铁与二氧化硅熔融形成硅酸铁渣。任何的 Cu_2O 都会与 FeS 发生如下反应：

$$Cu_2O + FeS \xlongequal{} Cu_2S + FeO$$

倾斜转炉将炉渣排出并且加入更多锍。这个过程反复进行直至该转炉充满熔融硫化铜矿。

（2）粗铜形成阶段。当所有铁硫化物被氧化以后，铜开始产生。首先，Cu_2S 被氧化为 Cu_2O：

$$2Cu_2S + 3O_2 \xlongequal{} 2Cu_2O + 2SO_2$$

然后发生反应：

$$Cu_2S + 2Cu_2O \xlongequal{} 6Cu + SO_2$$

粗铜的纯度约为98%，并含有0.02%～0.1%（质量分数）的硫。如果凝固的话，金属是多孔的，并且因为排出二氧化硫，所以引起金属表面有气泡。

锍中的任何贵金属在转化过程中都不会被氧化，而是进入粗铜中。这种粗铜可以通过火法冶炼来精炼。

5. 火法精炼粗铜

这个内容涉及杂质氧化的控制以及铜的脱氧问题。

在反射炉中空气被鼓入熔融粗铜中，更多的杂质（如铁、铅、锌）被氧化并进入渣中。添加适量的碳酸钠、硝酸钠和石灰到炉渣中，有助于去除其他杂质，如砷、锑、锡。

铜也开始氧化，继续鼓风一直持续到铜中的含氧量（体积分数）达到约0.9%为止，以确保脱硫效果。但是，铜有如此高的含氧量将易碎，所以还需要脱氧。

炉渣被清除后用树干将铜液搅拌，这种方法被称为“插树还原”——青木中释放的碳氢化合物会还原氧化铜。青木还原除气是一个关键的操作，其目的是使含氧量（体积分数）下降到0.03%~0.06%。如果含氧量不能充分降低，铜的力学性能就会很差。但是如果含氧量过低，那么铜凝固后将会是多孔的，因为在凝固过程中还原性气体和氧化亚铜会相互作用而形成气泡：

$$Cu_2O + H_2 = 2Cu + H_2O$$

青木还原除气以后，即称铜已经具备了成为“工业纯铜”的条件。火法精炼铜含0.03%~0.06%的氧（质量分数），非常适合加工。如果铜被过度还原，则它必须被重新氧化后再还原。

火法精炼铜的终脱氧可以通过加入磷，获得含14%磷的铜磷合金来实现。不过，少量的磷[$w(P) = 0.05\%$] 留在铜中会影响铜的导电性能。为避免这种情况，可以用锂作为脱氧剂。

火法精炼生产出的铜的纯度约为99.5%，要获得更高纯度的铜，需要采用电解精炼。

☞ 第二部分　课程思政

【冶金回眸】青铜时代

中国的青铜器时代从夏开始，经历商、西周到春秋时期，前后持续了1500多年的时间。大量出土的青铜器物表明，中国创造了灿烂的青铜文明。这些青铜器物不仅有丰富的政治和宗教内涵，而且还具有很高的艺术价值。今藏于中国历史博物馆的大盂鼎是中国青铜器时代的代表性作品之一，它是西周康王时期的作品，距今大约有3000多年。

☞ 第三部分　延伸阅读

《关于完整准确全面贯彻新发展理念做好碳达峰碳中和工作的意见》——推动产业结构优化升级

加快推进农业绿色发展，促进农业固碳增效。制定能源、钢铁、有色金属、石化化工、建材、交通、建筑等行业和领域碳达峰实施方案。以节能降碳为导向，修订产业结构调整指导目录。开展钢铁、煤炭去产能“回头看”，巩固去产能成果。加快推进工业领域低碳工艺革新和数字化转型。开展碳达峰试点园区建设。加快商贸流通、信息服务等绿色转型，提升服务业低碳发展水平。

☞ 第四部分　课文问题参考答案

Answer the following questions.

(1) About 90% of the world’s primary copper comes from sulphide ores.

(2) There are two main stages in the pyrometallurgical extraction of copper:

1) Matte smelting to form a molten sulphide melt, which contains all the copper of the charge, and a molten slag free from copper.

2）Conversion of the matte into blister copper. As a preliminary to matte smelting，partial roasting of the concentrate may be carried out in order to decrease the sulphur content.

（3）Converting consists of oxidising the molten matte with air or oxygen.

（4）When all the iron sulphide has been oxidised，copper production begins.

（5）Some of the important sources of copper are，Zambia，Katanga（Congo），Chile，Canada，U. S. A（Montana）and Australia.

☞ 第五部分　练习题参考答案

Ⅰ. Translate the following expressions into English.

（1）blister copper　（2）green wood　（3）sulphuric acid
（4）chalcopyrite　（5）fluo-solids roaster　（6）copper oxide
（7）horizontal converter　（8）reverberatory furnace　（9）electro-refining

Ⅱ. Fill in the blanks with the words from the text. The first letter of the word is given.

（1）Iron　（2）preliminary　（3）porous　（4）trunks
（5）poling　（6）higher　（7）weak　（8）Copper

Ⅲ. Fill in the blanks by choosing the right words form given in the brackets.

（1）removed　（2）to bring　（3）reducing　（4）suitable　（5）oxidised

Ⅳ. Decide whether the following statements are true or false（T/F）.

（1）F　（2）F　（3）T　（4）F　（5）T

Ⅴ. Translate the following English into Chinese.

如果冶炼像铅和锌这样低沸点的金属，则使用矩形横截面的鼓风炉。这种矩形截面的鼓风炉只在两个长边上设有风口。这样风口的温度就不会很高，而且减少了低沸点金属挥发而造成的损失。当使用鼓风炉来冶炼金属硫化物时，只要有足够的焦炭来提供必要的炉内气氛和热量就可以了，因为这种冶炼不需要还原反应。

第 14 单元　铝的生产

☞ 第一部分　课文参考译文

在大多数国家，铝主要应用于五个领域：建筑和结构，容器和包装，运输，电导体，机器和设备。消耗方式国与国之间差别较大，这主要取决于各个国家工业和经济的发展水平。

铝是从铝土矿中获取的，铝土矿是含有 40%～60%铝水化合物和其他诸如铁氧化物、二氧化硅和钛杂质的矿石。铝土矿最大、最著名的储量是在澳大利亚北部、圭那亚和巴西发现的。从铝土矿中制备铝包括两个截然不同的过程，这两个过程是在不同的地点进行的。首先，纯氧化铝（Al_2O_3）从铝土矿中几乎毫无例外地用拜耳法提取出来。这个过程实际上就是在高达 2400℃的温度下，用强碱（NaOH）溶液加热浸煮被粉碎的铝矾土。绝大部分氧化铝被萃取后只剩下被称为红泥的不可溶残渣，主要成分是铁氧化物和二氧化硅，可以通过过滤去除。冷却后，溶液用 $Al(OH)_3$晶体进行处理，以便发生逆向（水解）化学反应。氢氧化铝发生沉淀，而烧碱（NaOH）被回收利用。整个过程可用化学反应方程式表达：

$$Al(OH)_3 + NaOH = NaAlO_2 + 2H_2O$$

然后，将氢氧化铝在 1200℃下，在回转窑中煅烧去除结晶水，获得氧化铝的细小粉末。氧化铝具有很高的熔点（2400℃），并且是电的不良导体。成功生产铝的关键在于用熔融冰晶石（Na_3AlF_6）溶解氧化铝。典型的电解液含有 80%～90%的冰晶石（质量分数）和 2%～8%的氧化铝（质量分数），以及其他诸如金属铝或氟化钙等添加剂。

电解反应的准确机理还不清楚，导电离子可能是 Na^+、AlF_4^-、AlF_6^{3-}和一种或更多种比较复杂的离子，如 $AlOF_3^{2-}$。在阴极处，氟化铝离子失去电子而产生金属铝和 F^-离子。相反，在阳极处，复杂的离子分解出游离氧而形成 CO_2。总反应如下：

$$2Al_2O_3 + 3C = 4Al + 3CO_2$$

首次铝的大量生产是 1855 年在法国实现的，当时 H. Sainte-Clarie Deville 用钠还原铝的氯化物制出铝。拜耳-霍尔-埃鲁特法是现今世界通用的方法，它由四步组成：

（1）从铝土矿中提取纯氧化铝；

（2）从石油焦炭和煤沥青中制备碳阳极；

（3）在电解槽中还原氧化铝以制备熔融铝；

（4）将回收废料加入熔融金属中，随后进行提纯、合金化和铸造。

添加到电解槽中的氧化铝是在高温高压下，将粉碎的铝矿石放入碱性或酸性溶液中浸煮，生成纯无水氧化铝，然后通过分离得到。用氢氧化钠浸煮，可优先溶解氧化铝。由于氧化铝发生沉淀会与残渣一起丢弃，这样在一定压力下，过饱和偏铝酸钠由于其自身的稳定性，会呈现出在最可能低的温度下溶解尽可能多的氧化铝。一些多余的水分被还原到大气中。这种“溶出”浆液分三步排除未被溶解的残渣：（1）用除沙槽去除粗糙部分；（2）在大直径浅槽中利用沉积作用去除大部分残余物；（3）压力过滤器去除沉积后残留的细小铝矾土残余颗粒。最后得到清澈的褐色液体。冷却后，所要求尺寸的三氧化二铝颗粒由澄清的、过饱和的偏铝酸钠溶液沉淀生成，细粉末被循环使用，从-0.15~+0.03mm的颗粒（有少量过大的和8%~10% -0.045mm的颗粒）在回转式或者喷射式的烧结窑中煅烧，进而形成供入熔炼室的氧化铝。

在更好的反应室设计中，碳阳极首先通过混合和成型，将石油焦和煤沥青制成块状，熔化焙烧，再把浇注好的铸铁电连接放入阳极预制好的插孔中。

第三步是在电解槽中，把氧化铝加入到由氟化铝、氟化钙和其他添加剂调整过的熔融冰晶石中。这些添加剂在给定的温度下可以调整熔炼室上面绝热壁的硬度。同时，这些添加剂可以调整界面张力和存在于熔融状态下的离子的行为，从而减少生成电极产物的逆反应，在950~980℃可电解得到熔融铝和碳氧化物。熔融铝被定期从熔炼室中用真空虹吸法抽出，并运送到敞口保温炉内以便熔化回收废铝并合金化。氧化物、溶解的氢气和不需要的微量元素通过净化系统在流动的熔体中被去除，在净化系统中，金属过滤并与反应性气体接触，而后运输到连铸系统形成锭状成品。碳氧化物中含有少量的氟化物，它在阳极以颗粒状和气体形式释放出来。这些氟化物被1∶100的空气稀释后，进入有罩废气排出系统以达到环保的要求。氟化物通过湿式或干式洗涤系统得以回收并返回电解槽，干净的碳氧化物释放到大气中。

电解槽排列成长队形，称为电解槽系列。它尽可能以阳极相称排列，这样一个室的阳极可以通过电流与另一个室的阴极顺序地联系起来。一个加工厂由400~700个熔池构成。在预先烧结好的电解槽系列中，每个熔池的阳极每天更换一次。用来辅助此项操作的是最简单的起重机，这就要求操作者用手工汽锤打碎附着在阳极上面的硬壳，然后移开夹具，并固定起重机绳索来装入或移走阳极。而采用最精密的起重机时，只要求操作者坐在有空调的工作室中，通过操纵机器人手臂来完成上述操作。

☞ 第二部分　课程思政

【企业风采】走进中国铝业

中国铝业是中国有色金属行业的领先企业。中国铝业一直以维护、开发和利用国家战略资源为己任，在国防工业、航空航天工业、轨道交通及民用高端合金生产中的应用发挥了重要作用。中国铝业为中国的建筑、交通和国防工业提供了关键的铝型材，用于开发中国的第一颗人造卫星、第一座核反应堆和核潜艇、长征火箭、神舟飞船、嫦娥月球探测器、商用飞机、航空母舰和高速列车。今后，中国铝业将加快结构调整和转型升级，成为具有国际竞争力和全球经营能力的世界级企业。

☞ 第三部分　延伸阅读

纳米铝粉

纳米铝粉因具有较高的能量密度、高放热量、成本廉价，来源广泛等优点应用于推进剂和铝-水制氢中。添加一定量的纳米铝粉于推进剂中，能显著提高燃烧速率和比冲。由于纳米铝粉较小的粒径和较大的比表面能，导致纳米铝粉在空气中极易与氧气反应，表面被氧化生成致密氧化层，导致活性铝含量降低。氧化层的厚度限制了推进剂中燃速的进一步提高，同时也增加了铝-水制氢过程中反应的诱导时间。因此，为了维持纳米铝粉的高活性，需要对纳米铝粉进行表面包覆，维持纳米铝粉的高活性。

☞ 第四部分　课文问题参考答案

Answer the following questions.

(1) In most countries aluminium is used in five major areas: building and construction; containers and packaging; transportation; electrical conductors; machinery and equipment.

(2) The Bayer-Hall-Heroult process is the process used worldwide today.

(3) The key to the successful production of aluminium lies in dissolving the alumina in molten cryolite (Na_3AlF_6).

(4) The largest known bauxite reserves are found in Northern Australia, Guyana and Brazil.

(5) An aluminium making plant will consist of between 400~700 cells.

(6) Yes. The Bayer-Hall-Heroult process is the process used worldwide today, it consists of four steps:

1) the extraction of pure alumina bauxite;

2) the manufacture of carbon anodes from petroleum coke and coal tar pitch;

3) the reduction of alumina in electrolytic cells to produce molten aluminium;

4) the melting of recycled scrap with smelted metal followed by purification, alloying and casting.

☞ 第五部分　练习题参考答案

Ⅰ. Translate the following expressions into English.

(1) coal tar pitch　(2) holding furnace　(3) iron oxide

(4) pressure filter　(5) sand trap　(6) sodium hydroxide

(7) interfacial tension　(8) electrolytic cell　(9) electrical conductor

Ⅱ. Fill in the blanks with the words from the text. The first letter of the word is given.

(1) Aluminium　(2) liquor　(3) calcined　(4) uncertain　(5) slurry

(6) cells　(7) vacuum　(8) dry　(9) rows　(10) crane

Ⅲ. Fill in the blanks with appropriate prepositions.

(1) of　(2) with　(3) from　(4) by　(5) from
(6) of　(7) In　(8) in　(9) about　(10) to

Ⅳ. Decide whether the following statements are true or false (T/F).

(1) F　(2) F　(3) T　(4) F　(5) F

Ⅴ. Translate the following English into Chinese.

1913 年，谢菲尔德的哈里·布雷尔利试用合金钢制作枪筒，在他认为不适用而丢在一边的试样当中，有一种含铬 14%的钢。数月之后，他发现大多数钢都生锈了，而这种铬钢还保持着它的光泽。这就导致了不锈钢的发展。不锈钢是指含有 12%或更多铬的合金钢。它具有非常高的抗腐蚀性，在高温下，大多数不锈钢具有较好的力学性能。因为强调它高温使用，它们常常被称为抗热抗腐蚀钢。

第15单元　锌冶金

☞ 第一部分　课文参考译文

大量的贵重金属主要是通过湿法冶金技术来生产的。这包括了大部分的铜，几乎所有的铝、金和铂以及部分铅、锌和镍等贱金属。湿法冶金是指从矿石中提取和回收金属的生产过程，其中水溶液起主要作用。湿法冶金有两种不同的工艺：将矿石中的有价金属溶入水溶液，通过浸出法进行提取；经过适宜的溶液净化或浓缩步骤之后（或两者联合使用），从溶液中回收有价金属。湿法冶金的优点为：适合处理低品位矿石（铜、钛、金、银），可处理成分和浓度完全不同的材料，适于分离高度相似的材料（从锆中分离铪）；与火法冶金相比，其运行规模灵活，物料处理简单，操作及环境控制良好。

锌是软金属之一，密度比铁稍小，熔点低（419.5℃）。锌有较好的抗腐蚀性能和合金性能。其因具有良好的物理化学性能而广泛应用于工业中。锌主要用于防止钢生锈。制造黄铜和锌合金压铸件的锌用量占其总消耗量的80%。锌的其他用途包括（制造）用于蓄电池和建筑物的轧制锌板以及用作防护涂料的锌粉。此外，轧制锌板还可用于照相凸版、风管和某些管衬。金属锌存在于硫化锌矿床中，广泛分布于全世界。最重要的矿山在加拿大、澳大利亚、秘鲁和美国。我国锌冶金发展迅速，已经发展成为全球最大的锌冶炼生产国。

最初锌的生产方法是火法冶炼。此法包括反应罐直接还原法/蒸馏法、电弧炉炼锌法以及密闭鼓风炉熔炼法。

将焙烧硫化矿生成的氧化锌与无烟煤或类似的碳质材料一同在蒸馏罐中加热至1100℃左右，生成锌的蒸气，再通过邻近蒸馏炉的冷凝器将其变为液态金属。但这种方法由于能耗大、污染严重，目前已被淘汰。英国埃文茅斯的帝国熔炼公司从20世纪40年代开始，经过多次工厂试验，于1957年宣布成功开发了鼓风炉炼锌。密闭鼓风炉法是将锌和铅的硫化物焙烧成氧化物，再将此氧化物与一定比例的焦炭一起装入鼓风炉中。经过预热的热风从水冷鼓风口进入炉中。炉渣与含有来自炉料的贵金属和铜的熔融铅从炉底放出并进行分离。锌以蒸气形式混入含有一氧化碳和二氧化碳的炉气中排出炉外，在铅雨冷凝器中急速冷却，锌蒸气被铅吸收。然后含锌的铅液从冷凝器中抽出并冷却，锌就从铅液中析出并浮在铅液的上边。将锌液倒出并进行铸造，冷却的铅返回冷凝器循环使用。

电弧炉炼锌法的特点是能够利用电能直接加热炉料并连续蒸馏生产锌。它使用的原料与蒸馏法相似。该法可以处理多金属锌精矿。

电解锌法是在第一次世界大战期间发展起来的，现在世界锌产量的4/5以上是用此法生产的。湿法炼锌是将锌精矿依次经焙烧—浸出—净化—电解沉积而获得金属锌。

1. 锌精矿的焙烧

闪锌矿（ZnS）与铁闪锌矿（mZnS · nFeS）是炼锌的主要矿物原料。最重要的焙烧反应是硫化锌精矿的反应，且包括与焙烧大气的化合反应。

可能发生的反应包括：

$$2ZnS + 3O_2 = 2ZnO + 2SO_2 \quad \text{（死焙烧）}$$

$$6ZnS + 11O_2 = 2(ZnO \cdot 2ZnSO_4) + 2SO_2 \quad \text{（死焙烧）}$$

$$ZnS + 2O_2 = ZnSO_4 \quad \text{（硫酸化焙烧）}$$

其他需要考虑的平衡反应包括：

$$S + O_2 = SO_2$$

$$2SO_2 + O_2 = 2SO_3$$

通过焙烧可使精矿中的硫化锌转变为可溶于稀硫酸的氧化锌，且希望焙烧产物中含有少量的 $ZnSO_4$，以补偿浸出和电解过程中的硫酸损失。

2. 锌焙砂的浸出

通过在沸腾炉中焙烧锌精矿而获得的锌焙砂由氧化锌、其他金属氧化物以及脉石组成。锌焙砂的浸出是指通过硫酸溶剂萃取固态焙砂中的锌及其他有价金属元素的过程。它使固态焙砂中的有价金属溶解到溶液中，与脉石等不溶物分离。

通过浸出可以达到两个目的：

（1）使物料中的锌尽可能完全溶解到浸出液中。

（2）使有害杂质尽可能不溶解而进入渣中。

3. 浸出液的净化

浸出液常含有砷、锑、铜、镉、钴、镍及其他杂质，电解前必须净化除去。这些杂质的标准电极电势均比锌高，可用锌粉置换净化。

4. 锌的电解沉积

锌的电解沉积是湿法炼锌的最后一道工序，是从含有硫酸的硫酸锌水溶液（电解液）中电积提取纯锌的过程。其反应为：

$$ZnSO_4 + H_2O = Zn + H_2SO_4 + \frac{1}{2}O_2$$

电解槽是一个长方形槽体，Pb-Ag 合金作为阳极（正极），阴极（负极）由压延铝板轧制而成，它们浸入电解液里。氧气在阳极上释放出来，锌电解沉积到铝板上，从铝板上剥下，经熔化，铸成锌板。这样生产的锌的纯度在 99.95% 以上，需要时还可以达到 99.99%。电解沉积过程的能耗巨大，极大影响了湿法冶锌的投资成本。

☞ 第二部分　课程思政

【工程伦理】二氧化硫逸出事故

2019年，某公司锌冶炼厂发生一起二氧化硫逸出事故，但该企业未向生态环境部门报告设备故障及维修情况，也未及时采取降低负荷、抢修烟道泄漏处等措施减少污染物排放，导致部分烟气从焊缝逸散至外环境。

一个处于现代社会中的企业，追求利润当然无可厚非，但是如果企业只注重利润，甚至不惜牺牲环境和社会资源来获取利润，企业就没有存在的价值了，也无法推动社会经济的持续繁荣。在最大限度地保护环境、资源和社会效益的前提下追求企业利润，才是一个现代企业安身立命的首要准则。

☞ 第三部分　延伸阅读

从粉煤灰中提取珍贵的锌

粉煤灰中含有锌等贵重金属，直接填埋会造成贵金属的流失。瑞典查尔姆斯理工大学的研究人员开发了一种提取这些贵金属的方法，减少污染的同时还能对金属进行回收，即利用酸洗处理这种废物，将锌从灰中分离出来。然后，锌可以被提取、洗涤并加工成原料。试验中，他们发现粉煤灰中存在的70%的锌可以回收利用。在提取锌后，再次焚烧残灰，以分解二噁英，将其中90%变成底灰，可以作为建筑材料等使用。

☞ 第四部分　课文问题参考答案

Answer the following questions.

(1) Two distinct processes are involved in hydrometallurgy.

(2) The important mines are in Canada, Australia, Peru and the U. S. A.

(3) The method can be used to deal with multiple metal zinc concentrate.

(4) Leaching of zinc calcine is the process of extracting zinc and other metal values from a solid calcine by means of sulfuric acid solvent.

(5) The electrolytic deposit is the last process step in hydrometallurgy of zinc.

☞ 第五部分　练习题参考答案

Ⅰ. Translate the following expressions into English.

(1) zinc metallurgy　(2) hydrometallurgy of zinc　(3) recover the metal values
(4) melting point　(5) physical and chemical properties　(6) zinc sulfide ore deposit
(7) zinc oxide　(8) mineral raw material　(9) dead roast
(10) sulfating roast　(11) zinc calcine　(12) sulfuric acid solvent

(13) leaching solution (14) electrode potential (15) the electrolytic deposit of zinc

Ⅱ. Fill in the blanks with the words from the text. The first letter of the word is given.

(1) extraction (2) erosion (3) account (4) concentrate (5) dilute
(6) gangue (7) dissolve (8) purified (9) sulfate (10) consumption
(11) powder (12) anode (13) deposited (14) slabs (15) purity

Ⅲ. Fill in the blanks by choosing the right words form given in the brackets.

(11) roasted (2) proportioned (3) Preheated (4) through (5) separated
(6) as (7) containing (8) directly (9) to (10) with

Ⅳ. Decide whether the following statements are true or false (T/F).

(1) F (2) T (3) F (4) T (5) F
(6) F (7) F (8) T (9) T (10) F

Ⅴ. Translate the following English into Chinese.

早在人们认识到锌是一种金属之前，罗马人就把含碳酸锌的菱锌矿与铜矿混合在一起冶炼而获得黄铜。发现黄铜之后的几百年里，锌仍未被分离出来。中国首先炼制出了锌，并在17世纪初出口欧洲，英国是欧洲第一个发展锌生产技术的国家，威廉·钱皮恩于1738年开始以商业规模冶炼金属锌。

第 16 单元　金的提取

☞ 第一部分　课文参考译文

金，其次是银，称为“贵金属”，因为它们能够长期暴露在大气中而不失去光泽，并且能够反复熔化而不损失多少重量。由于这些特点，金、银起初用来制作首饰，后来则用于制造货币。在所有金属中，金的加工性最好，它是延展性最强的金属之一，能够被锻造成半透明状态的金箔。在一些飞机上，把夹有一层不足万分之五毫米厚的金膜的夹层玻璃嵌入经过热处理的挡风玻璃中，以阻挡太阳的有害射线。1 金衡盎司黄金能拉拔成长度超过 80 千米的细丝。黄金可以完全抵抗大气腐蚀和氧化。使人类能够在月球上迈出第一步的极其重要的电子系统，采用金涂层使其免受起飞冲击波的影响。金还在月球旅行的计算机电路上得到应用。纯金对许多用途来说太软了，因此需要加入其他合金元素使其硬化，铜是最常用的合金元素，另外，银、镍、钯和锌是制作珠宝首饰的合金元素。

天然纯金块在自然界中很难找到，其一般与伴生矿物结合在一起。一种矿石的经济价值取决于它所含有金属的价格及其提取的难易程度。铁质量分数小于 20%的铁矿石即为贫矿，而含金 1/100000 的金矿则被看作富矿。重力选矿是最早的金提取方法，它基于一种选别力，这种力的大小随相对密度而变化。另一种在重力法中起作用的力是通过在液体或半流质媒介中发生分离来阻止作用在矿石颗粒上的相对运动。大多数金矿采用氰化法来处理。经济地生产氰化钾这种化学制剂，并利用它来提取金，是一项重大的发现，它对金矿的发展和世界各地的国际贸易产生了巨大的影响。

氰化法浸金工艺可以归纳为三类：堆浸法、渗滤浸出法和搅拌浸出法。

1. 堆浸法

这种方法于 16 世纪首先在德国的哈尔茨山应用。依据堆浸矿石的吨数，将一块大约 300 英尺×400 英尺的地面凿平后，在上面覆盖一层沥青。然后用自卸卡车将低品位矿石堆入其中，达到 3~6m 高度。将氰化液喷淋在矿堆顶部，浸出液则从矿堆底部汇集流出。有时，在矿堆内以一定间距放入中空垂直管，以促进溶液流动，同时也有助于空气流通而加快浸出过程。

堆浸法用于处理低品位的金矿（每吨矿含金 1~2g）。堆浸操作通常需要 30~90 天，金的回收率为 50%~70%。

2. 渗滤浸出法

将待浸出物料放入一个装有假底的槽中，槽底覆盖一层过滤介质。溶剂从槽顶加入，从物料中渗透通过。槽中通常采用一个逆流系统。加入最后一个槽中的新矿体和加入第一个槽中的贫液连续不断地从一个槽泵入另一个槽，直到泵入最后一个接近饱和的槽中。

此法特别适合处理疏松多孔的砂矿，但不适用于抗渗的、易于堆成块状的物质。

通常矿物颗粒的大小而非实际矿物大小，是控制良好渗滤的主要因素。因此，若有大量残渣存在，此法不会得到令人满意的结果。此法的优点是使溶剂消耗量最小化，可生产出高品位的母液，无需使用昂贵的浓缩器或过滤器。浸出完成后，要手工清槽，然后才能进行另一批生产。

3. 搅拌浸出法

用搅拌器浸出矿浆、精矿等物料的方法，通常是在水中将物料磨碎（达到最小的粉尘状），形成具有适宜尺寸的颗粒。金矿被破碎成细粉后，搅拌混入盛有很稀的氰化物溶液的大槽中，金即被氰化物溶解。矿浆浓度在 40%～70%之间变化。加入溶浸剂后不断地搅拌矿浆。搅拌浸出可以通过以下几种方式完成：

（1）机械搅拌浸出：此法通常采用螺旋桨或叶轮，常用于小型浸出槽。

（2）空气搅拌浸出：浸出槽是一个直径约 12 英尺、高 45 英尺、带有 60°圆锥底的圆柱形槽体。两端开口的金属管置于中心部位。压缩空气通过金属管送入，促使槽内矿浆循环流动。

（3）空气—机械联合搅拌浸出：对于大规模的浸出，广泛使用多尔搅拌器。

浸出后的矿浆由含金溶液和尾矿组成。将溶液过滤，并从溶液中沉淀出金。

☞ 第二部分　课程思政

【大国工匠】滴水掘金浇筑“稀贵”人生——潘从明

潘从明说：“中国的贵金属储量仅占世界总储量的 0.39%左右。如果没有先进技术，一些贵金属往往只是普通商品开采的副产品，只能作为废物处理——这对中国来说是一个巨大的损失。”他的发明“镍阳极泥中铂、钯、铑和铱的绿色高效提取技术”大大简化了铂族贵金属的提纯过程，不仅节省了成本，而且更加环保。该技术可从镍矿废料中提取八种贵金属，每种贵金属的纯度可达 99.99%。

他说：“我从不害怕失败。只有通过失败，我才知道我需要采取另一种方法才能成功。”潘从明的努力为他赢得了全国劳动模范和全国技术专家等国家级奖项。

☞ 第三部分　延伸阅读

微生物冶金

微生物冶金技术又称微生物浸矿技术或微生物浸出技术，是指多数微生物可以通过各类途径对不同品位和不同种类的矿物产生作用，把矿物中的有价元素转化为溶液中的离子，利用微生物的这种特性，融合湿法冶金工艺，形成了现代微生物冶金技术。微生物冶金技术尤其适用于处理贫矿、废矿、表外矿，或是用于堆浸和就地浸出难采、难选、难冶矿，同时，微生物冶金技术具有易操作、低成本、低能耗、小污染等优点，在实际生产中得以广泛应用。在国外，生物提取铜和铀，以及含砷金矿的生物预氧化已走向产业化。

☞ 第四部分　课文问题参考答案

Answer the following questions.

(1) Because they could be exposed to the atmosphere for a long time without tarnishing and could be melted repeatedly without much loss in weight.

(2) Most gold ores are treated by cyanide process.

(3) Cyanide gold leaching process can be grouped into three classes: lump leaching, percolation leaching, pulp agitation leaching.

(4) Yes, it is.

(5) The material to be leached is placed in a tank equipped with a false bottom covered with a filtering medium.

☞ 第五部分　练习题参考答案

Ⅰ. Translate the following expressions into English.

(1) associated mineral　(2) cyanide gold leaching　(3) lump leaching
(4) percolation leaching　(5) pulp agitation leaching.　(6) the tank for leaching
(7) recovery of gold　(8) at regular intervals　(9) a false bottom
(10) a countercurrent system　(11) pile into a block　(12) pregnant solution
(13) pulp density

Ⅱ. Fill in the blanks with the suitable words listed below. Change the form where necessary.

(1) exposed　(2) Cyanide　(3) recovery　(4) filtering
(5) stirred　(6) Pulp　(7) agent　(8) mechanical
(9) propeller　(10) cylindrical　(11) circulation

Ⅲ. Fill in the blanks by choosing the right words form given in the brackets.

(1) be leached　(2) medium　(3) through　(4) to　(5) to pile
(6) percolation　(7) present　(8) Its　(9) manually　(10) introduced

Ⅳ. Decide whether the following statements are true or false (T/F).

(1) F　(2) F　(3) T　(4) F　(5) T　(6) T

Ⅴ. Translate the following English into Chinese.

从一种典型富矿中生产1金衡盎司（31.3g）黄金，必须从矿脉上爆破约3t的矿石，把矿石运至竖井，再提升到地面，破碎成均匀的粒度，然后经过搅拌、过滤和处理，用氯化钾把金分离出来。

第 17 单元　金属的成型工艺

☞ 第一部分　课文参考译文

几乎所有的工业产品都具有金属部件或是由含有金属部件的机器制造的。所有的金属部件都是成型件。金属材料的成型方法与工艺是零部件设计的重要内容，也是制造者们极度关心的问题，更是材料加工过程中的关键因素，金属的成型工艺归纳起来有以下几种：铸造、塑性成型、机械加工成型、接合成型、粉末冶金、注射成型、半固态成型以及 3D 打印成型。

1. 铸造

铸造是最早产生的液态金属成型工艺，现在仍是一种重要的工艺。铸造是将熔化的金属导入型腔或铸模中，在那里金属一旦凝固就会变成一个由铸模轮廓确定其形状的物件。其工艺流程为：液态金属→充型→凝固收缩→铸件。

铸造工艺的特点主要表现为以下几个方面：

（1）可生产形状复杂的铸件，特别是内腔形状复杂的铸件。

（2）适应性强，合金种类不受限制，铸件大小几乎不受限制。

（3）材料来源广，废品可重熔，设备投资低。

（4）废品率高、表面质量较低、劳动条件差。

金属铸造的方法主要有砂型铸造、熔模铸造、压力铸造、低压铸造、离心铸造、金属型铸造、真空铸造、挤压铸造、消失模铸造和连续铸造等。

2. 塑性成型

塑性成型就是利用材料的塑性，在工具及模具的外力作用下来加工制件的少切削或无切削的工艺方法。它的种类有很多，主要包括锻造、轧制、拉拔、挤压、冲压等。为了更好地理解各种成型操作的机械学原理，我们将简要讨论每一种方法。

锻造技术又可细分为锤锻和模锻。锤锻是人类所知的最古老的金属加工工艺之一。早在公元前 2000 年，锻造即被用于生产武器、工具和珠宝。这一工艺是手工使用简单的锤子完成的。今天的锻造材料是在两个或多个模具间受到挤压以改变其形状和尺寸。根据情况不同，模具可以是开式的［见图 17-1(a)］，也可以是闭式的［见图 17-1(b)］。

轧制是最广泛使用的金属塑性变形工艺。虽然早在 16 世纪人们已经能够把铅、金、银之类的软金属轧制成片了，但是直到 18 世纪，轧制法才大规模地用来轧制较硬的金属。根据记载，1755 年在伯明翰开始有轧机，当时可以热轧出 7. 5cm 宽条材，其厚度减小到原来的四分之一，相应地延展了长度。轧钢的具体内容我们将在以后的章节中讨论。

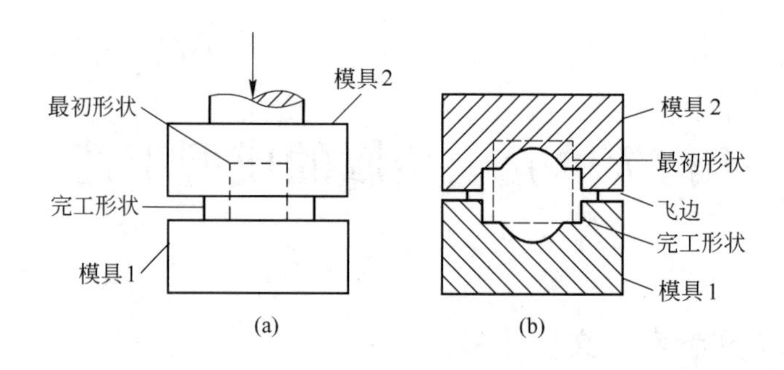

图 17-1　锻造工序

（a）开式模锻；（b）闭式模锻

在拉拔工艺中，金属丝的截面或者是条钢或钢管的截面由于工件被拉过模具的锥形孔而减小。当截面需要减小很多时，也许有必要通过几次拉拔来完成此操作。

挤压法同从牙膏管里挤牙膏和从口袋里往糕点上挤糖衣花饰的原理一样。挤压是最具潜力和最有用的金属加工工艺之一，而且有大量不同的应用方式。挤压在热态和冷态都可以进行。对大多数普通金属进行冷挤压需要有惊人的压力才行，但一般来说设计的设备都是进行热挤压的。挤压法非常经济，是生产线、棒、管和带材的生产方法。

冲压是靠压力机和模具对板材、带材、管材和型材等施加外力，使之产生塑性变形或分离，从而获得所需形状和尺寸的工件（冲压件）的成型加工方法。全世界的钢材中，有60%~70%是板材，其中大部分经过冲压制成成品。汽车的车身、底盘、油箱、散热器片，锅炉的汽包，容器的壳体，电机、电器的铁芯硅钢片等都是冲压加工制成的。仪器仪表、家用电器、自行车、办公机械、生活器皿等产品中，也有大量冲压件。

3. 机械加工成型

机械加工成型是在零件生产过程中，直接用刀具在毛坯上切除多余金属层厚度，使之满足图纸要求的尺寸精度、形状和位置相互精度、表面质量等技术要求的加工过程。常用的方法有：车削、铣削、刨削、钻削、镗削、磨削、齿轮加工和数控加工等。

4. 接合成型

近年来，金属的接合方法如钎焊、焊接等收到了重视，这些方法可以使“装配”结构既快又效率高，而且往往比“整体制造”便宜。金属接合的主要方法有：铆接、钎焊、电焊、熔焊、等离子喷射焊接等。焊接一向被描述为整个冶金业的缩影，因为它要求具有金属熔炼、铸造和锻造方面的综合知识。

5. 粉末冶金

粉末冶金是由以粉末形式的各种金属和合金制造各种各样零件和形状的方法。粉末冶金的三个主要步骤是：（1）金属粉末的准备处理；（2）粘结或压制成型；（3）在惰性气氛下将金属粉末烧结或加热到接近液体的状态。粉末冶金主要被用在制造工具上。粉末冶金除了能够生产不需要再进行机加工的成品部件这一优点外，也对再循环提供了极大的可能性。

6. 注射成型

注射成型是将金属粉末与其粘结剂的增塑混合料注射于模型中的成形方法。它先将所选粉末与粘结剂进行混合，然后将混合料进行制粒再注射成形所需要的形状。注射成型工艺分为四个独特的加工步骤（混合、成型、脱脂和烧结），从而实现零部件的生产，并针对产品特性决定是否需要进行表面处理。

7. 半固态成型

半固态成型工艺是利用金属从液态向固态过渡（固液共存）时的特性，综合了凝固成型和塑性成型的优点，加工温度较低，变形抗力较小，可加工形状复杂，且精度要求高的零部件。目前已成功用于主缸、转向系统零件、摇臂、发动机活塞、轮毂、传动系统零件、燃油系统零件和空调零件等制造方面。

8. 3D 打印成型

3D 打印是一种快速成型技术。该技术以数字模型文件为基础，运用粉末状金属或树脂等材料，通过逐层打印的方式来构造物体。见表 17-1。

表 17-1　几种 3D 打印成型技术比较

技术	FDM 熔融挤出成型	SLA 光固化成型	SLS 选择性激光烧结	3DP 三维印刷
材料	蜡、ABS、PC、PPSF、尼龙、PLA 等热塑性材料，以丝状供料	液态光敏树脂	材料广泛：热塑性材料、树脂裹覆砂（覆膜砂）、聚碳酸酯、金属粉末、陶瓷粉末	陶瓷粉末、金属粉末
方式	熔融挤出成型	紫外光（UV）照射（波长 x = 325mm、强度 ω = 30mW）达成光聚合反应固化成型	高强度 CO_2 激光器烧结成型	微滴喷射粘接剂（如硅胶）成型
层厚精度/mm	0.15~0.4	0.016~0.15	0.08~0.15	0.013~0.1
支撑	必要时需要	需要	无需	无需
特点	使用维护简单、成本较低，桌面级应用最为广泛	目前精度最高的成型方式	多应用于大型结构设计，例如航空航天等	高速多彩成型

☞ 第二部分　课程思政

【创新驱动】党的二十大报告丨加快实施创新驱动发展战略

坚持面向世界科技前沿、面向经济主战场、面向国家重大需求、面向人民生命健康，加快实现高水平科技自立自强。以国家战略需求为导向，集聚力量进行原创性引领性科技攻关，坚决打赢关键核心技术攻坚战。加快实施一批具有战略性、全局性、前瞻性的国家重大科技项目，增强自主创新能力。加强基础研究，突出原创，鼓励自由探索。

☞ 第三部分　延伸阅读

手撕钢

太原钢铁集团（太钢）是一家领先的不锈钢制造商，通过制造一种名为手撕钢的极薄不锈钢板，实现了重大技术突破。

太钢子公司山西太钢不锈钢精密带材有限公司总经理王天祥表示，过去只有德国和日本等少数国家能够生产350~400mm宽度的薄带。然而，太钢在700多次尝试失败后开发的这款产品厚度仅为0.02mm，相当于一张标准A4纸的四分之一，宽度为600mm，是世界上第一款此类产品。王天祥说，这种手撕钢可应用于航空、新能源和柔性折叠屏幕等广泛领域。

太钢总裁高向明表示："高质量发展意味着做别人做不到或做不好的事情，并使用专业产品来满足需求、创造需求并引领市场。要做到这一点，创新是核心，人才是基础，机制是关键。"

☞ 第四部分　课文问题参考答案

Answer the following questions

(1) Metal forming processes can be grouped into four classes. They are casting, plastic molding, mechanical processing molding, joint molding, powder metallurgy, injection molding, semi-solid molding and 3D printing molding.

(2) The characteristics of casting process are mainly shown in the following aspects: 1. It can produce castings with complex shapes, especially those with complex inner cavities; 2. Strong adaptability. Alloy type and casting size are not limited; 3. Wide material sources, remodeling waste products, low equipment investment; 4. High rejection rate, low surface quality and poor working conditions.

(3) Casting processes include sand casting, investment casting, pressure casting, low pressure casting, centrifugal casting, permanent-mold casting, vacuum casting, die casting, lost foam casting and continuous casting etc.

(4) As early as 2000 B. C., forging was used to produce weapons, implements, and jewelry.

(5) As early as the sixteenth century, soft metals such as lead, gold and silver had been rolled intosheets.

(6) Forging techniques include hammer forging and die forging.

(7) The main types of joining processes are riveting, soldering, welding by electrical, fusion-welding, plasma-jet welding, etc.

☞ 第五部分　练习题参考答案

Ⅰ. Translate the following expressions into English.

(1) metal forming process　(2) forming from liquid metal　(3) mold configuration
(4) intricate shape　(5) metal-casting process　(6) centrifugal casting
(7) mass production　(8) hammer forging　(9) powder metallurgy
(10) conical orifice　(11) one-piece manufacture　(12) plasma-jet welding

Ⅱ. Fill in the blanks with the suitable words listed below. Change the form where necessary.

(1) industrial　(2) liquid　(3) produced　(4) forging
(5) techniques　(6) widely　(7) conical　(8) potential
(9) tools　(10) impact　(11) formed　(12) strength

Ⅲ. Fill in the blanks by choosing the right words form given in the brackets.

(1) obtain　(2) shape　(3) shaping　(4) done　(5) used
(6) lower　(7) called　(8) required　(9) pressing　(10) to follow

Ⅳ. Decide whether the following statements are true or false (T/F).

(1) T　(2) F　(3) F　(4) T　(5) F

Ⅴ. Translate the following English into Chinese.

现代粉末冶金的应用始于 18 世纪，到 20 世纪上半叶就已经广泛使用了。粉末冶金是将金属制成粉末并用以制造有用物品的一门工艺。其生产过程包括将粉末压成坯，然后进行加热（称为烧结）。从根本来讲，粉末冶金比一般方法优越的原因有两个方面：第一，能获得采用别的方法所得不到的性能；第二，可以免去机械加工及其后的加工过程。

第18单元 轧钢简介

☞ 第一部分 课文参考译文

1. 轧钢介绍

轧钢是在两个以上旋转轧辊作用下使金属连续成型的工艺（见图18-1）。轧辊主要通过压力作用于金属，所以轧钢属于压力成型加工方法。

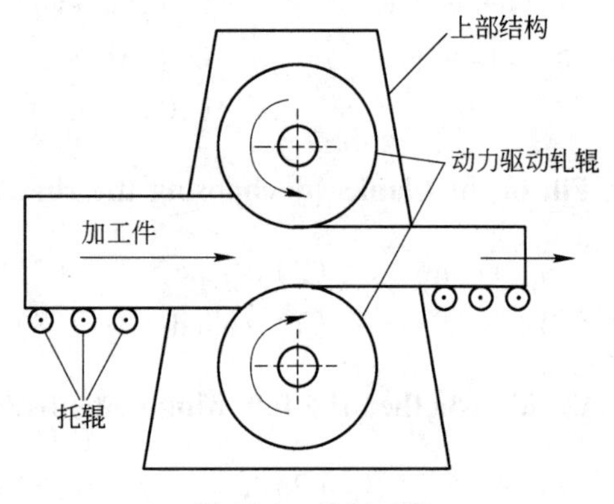

图18-1 轧制工序

从金属学的角度，根据加热温度将轧钢分为热轧和冷轧，所谓热轧是指在再结晶温度以上的加工过程，而在再结晶温度以下的塑性加工过程称为冷轧。

热轧厂一般由加热炉区（在变形前供热）、轧制区和精整区组成。加热炉区把轧件均热到一定温度，在轧制区完成轧制。为了得到需要的形状和性能，需要经过几个道次的轧制，因此，轧制有各种不同的分类，如粗轧道次、中轧道次和精轧道次。成品车间的主要任务是：切割，拉直，表面保护，堆放和储存，观察、检查分类并做标记，收集，捆扎和打包。

冷轧厂一般包括酸洗、轧制（道次）、热处理和精整。冷轧主要用于生产板材，如深冲薄板、镀锡薄板和不锈钢板，使用最广泛的是带钢，带钢可以在两辊、四辊和多辊轧机上冷轧，其次，型钢和管材也可以冷轧。冷轧会产生加工硬化，经常用退火热处理来消除。冷轧和热处理的结合可以使钢材获得规定的技术性能。

与冷轧相比，热轧可以获得更薄的轧件，但是，热轧件表面粗糙度和精度不如冷轧精准。因此，热轧多用来对大钢锭（坯）进行“开坯”，冷轧多用来生产平整且精度高的薄金属板。由于冷轧钢一般比热轧钢硬度高，故冷轧机的轧辊要求硬度较高，且需要较高的动力。

轧钢的产品主要有半成品和成品两种。半成品即板坯、大方坯和热轧宽带钢，成品是轧钢厂已经完成热成形的产品。轧制的成品钢材按照断面形状可分为型材、板材、管材、线材等。

（1）型钢。型钢包括断面是圆形、方形、矩形、八角形或半圆形等的普通型钢以及I、H、U形断面的异型断面钢材和宽缘钢梁板等。

（2）钢筋。钢筋直径为6~8毫米，也是具有圆形断面的钢条，它的表面常常用肋材加强。

(3) 铁路的辅助材料等。这个种类覆盖了建造铁路需要的所有部件。

(4) 板材。板材的断面是矩形的，宽度比厚度大，表面基本是光滑的，也可以用图案装饰。根据厚度，板材分为特厚板（>60毫米）、中厚板（4~60毫米）和薄板（<4毫米）。

(5) 管材。管材是空心断面，虽然断面一般是圆的，但也生产其他形状的断面。

(6) 线材。线材的表面基本是光滑的，断面可以是圆形、椭圆形、方形、矩形，厚度从5毫米到40毫米不等。多数线材需要经过冷拔和冷轧进一步处理。

2. 钢的控制轧制和控制冷却

控制轧制和控制冷却工艺是一项节约合金、简化工序、节约能源消耗的先进轧钢技术。它能通过工艺手段充分挖掘钢材潜力，大幅度提高钢材综合性能。

控制轧制是在热轧过程中通过对金属加热制度、变形制度和温度制度的合理控制，使热塑性变形与固态相变结合，以获得细小晶粒组织，使钢材具有优良的综合力学性能的轧制新工艺。

控制冷却是控制轧后钢材的冷却速度达到改善钢材组织和性能的新工艺。控制轧制与控制冷却相结合能将热轧钢材的两种强化效果（细晶强化和析出强化）相结合，进一步提高钢材的强韧性和获得合理的综合力学性能。

板材的断面是矩形的，宽度比厚度大，表面基本是光滑的，也可以用图案装饰。根据厚度，板材分为特厚板（大于60mm）、中厚板（4~60mm）和薄板（小于4mm）。

管材是空心断面，虽然断面一般是圆的，但也生产其他形状的断面。

线材的表面基本是光滑的，断面可以是圆形、椭圆形、方形、矩形的，厚度从5mm到40mm不等。多数线材需要经过冷拔和冷轧进一步处理。

☞ 第二部分　课程思政

【匠心筑梦】轧钢“牛人”——牛国栋

牛国栋，1975年生，中共党员，太原钢铁（集团）有限公司不锈冷轧厂连轧作业区班长，党的十八大代表和十九大代表，是“全国示范性劳模创新工作室”和“国家级技能大师工作室”带头人。22年，他坚守初心，在12号轧机调试期间，创造了单机架轧机调试时间最短、投产最早、生产产品质量起点最高的纪录。他勇担使命，第一个成功实现大变形量马氏体钢轧制。他依托“牛国栋创新工作室”平台，培养出高级轧钢工92名、高级技师6名，累计创效9000多万元。曾荣获全国五一劳动奖章、中国青年五四奖章、山西省特级劳模、三晋英才等荣誉。

☞ 第三部分　延伸阅读

高品质特殊钢高效率、低成本特种冶金新流程

1. 三次精炼技术

在常规的电炉或转炉流程后面，增加三次精炼，例如真空自耗炉或电渣重熔冶炼，可以获得高洁净度的特种钢材，用于高效和低成本制备航空航天等应用的特殊钢材料和其他高性能金属材料。

2. 新一代特钢洁净化、均质化精炼技术

研发对钢水无污染的、加热和脱氧为特征的新一代特钢钢包洁净精炼技术、高端不锈钢加压增氮冶金新技术、基于导电结晶器的电渣重熔技术等，用于生产高端合金钢材。

☞ 第四部分 课文问题参考答案

Answer the following questions.

(1) Hot-rolling mills are generally divided into the following zones: furnace area (for heat supply prior to deformation), rolling area, and finishing area.

(2) The most important tasks of the finishing shop are: cutting, straightening, surface protection stacking and retrieving, inspection, checking, sorting, marking, collecting, bundling, packing.

(3) Cold-rolling mills generally include the following areas: pickling area, rolling train, heat treatment, and finishing area.

(4) The advantage of hot rolling is that it can be reduced in thickness much more easily.

(5) Finished rolled steel products are classified according to the shape of their cross-sections as: section product, flat product, tube, and wire rod.

(6) Flat products are classified by thickness into: heavy flat steel (>60mm), medium steel plate (4~60mm) and steel sheet (<4mm).

(7) Most wire rod undergoes further treatment by cold drawing or cold rolling.

(8) Two main groups of rolled steel products are semi-finished products and finished products.

(9) Steel is cold rolled mainly for producing flat products such as deep-drawing sheet, tin sheet and stainless sheet.

☞ 第五部分 练习题参考答案

Ⅰ. Translate the following expressions into English.

(1) seamless tube (2) finishing area (3) cold drawing
(4) drawing force (5) thin sheets of metal (6) continuous casting plant
(7) hot forming (8) hollow section (9) reinforcing steel

Ⅱ. Fill in the blanks with the words from the text. The first letter of the word is given.

(1) rotating (2) drive (3) stands (4) preheated
(5) intermediate (6) flat (7) strip (8) thickness
(9) completed (10) rectangular (11) drawing (12) hollow

Ⅲ. Fill in the blanks by choosing the right words form given in the brackets.

(1) widely　(2) thinner　(3) width　(4) along
(5) steeper　(6) in　(7) through　(8) larger

Ⅳ. Decide whether the following statements are true or false (T/F).

(1) T　(2) F　(3) F　(4) F　(5) T

Ⅴ. Translate the following English into Chinese.

用于加热轧件的加热炉用水范围极广，如冷却炉门、滑道管、拱座等。轧钢机的轧辊也需用大量的冷却水以保持轧辊的外形，把热裂的可能性降到最低，从而延长其使用寿命。如果用的是热轧机，则在轧制前需用高压水冲去热轧件上的铁鳞，并使轧件在一次次的轧制过程中都保持表面洁净。轧制带钢的热轧机的输出辊道上方需要进行喷水冷却，从而把带钢冷却到合适的温度以便进行盘卷。

第 19 单元　管材和线材的生产

☞ 第一部分　课文参考译文

轧制产品种类繁多，很难予以概括说明，这里仅就大型轧钢厂管材以及线材的生产加以说明。

1. 管材的生产

100 多年前，钢管开始用于自行车的制造方面。由于它恰当地结合了强度高、机动性强和重量轻等优点，目前管材广泛应用于日常生活、交通以及工业等各个领域。

世界最大的一家管材制造商很通俗地说明了制管的两种基本方法：用带钢卷成一个精确的型；在钢棒上穿一个粗大的孔。

第一种方法首先从带钢卷开始，带钢自动喂入一系列成形辊，这些辊子逐渐把钢带卷成管状，然后对管缝进行感应加热并压焊在一起，于是管缝就焊合了。管子再通过一系列辊子进行定径和矫直。虽然许多这样的钢管是以直管方式定尺交货的，但也有许多管子是按用户的要求加工成弯形、锥形、带外螺纹和内螺纹等各种形状。

第二种方法，即“在钢棒上穿孔”，有很多变更的方法，产品一般称为无缝管。比如，生产无缝管的曼内斯曼法，如图 19-1 所示。首先，用火焰切割机把圆钢定尺切割成坯；其次，用传送带送进加热炉加热；接下来，热管坯经过一台水压机，水压机将管坯一端的中心压出一个锥形凹坑，以便下一步进行穿孔。然后，一根红热的圆钢被两个互相倾斜并朝同一方向旋转的轧辊旋压，并在两辊之间拉向前方，并从如图 19-1 所示的顶杆的外围通过。通过这种方法生产出厚壁管，管子的尺寸根据辊缝的大小和芯棒顶头的尺寸变化而不同。这样生产出的厚管可以在后面的工序中继续加工成薄壁管。如果要求生产最佳表面和强度高的管子，如飞机构件或注射针头，则需要进一步进行冷拔。

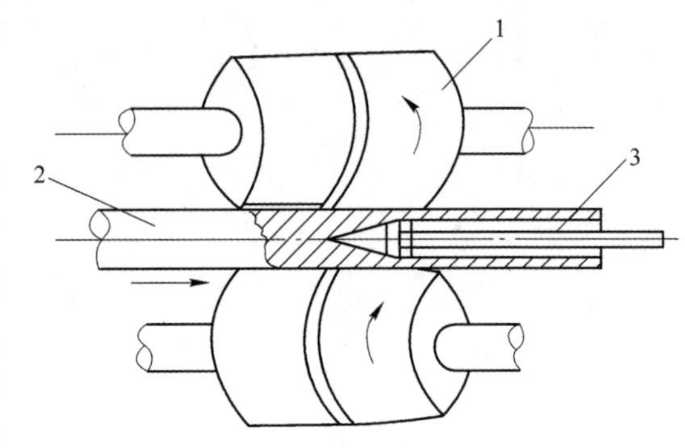

图 19-1　曼内斯曼制管工艺
1—轧辊；2—钢棒；3—圆钢

2. 线材的拉拔

人们使用线材已有千百年的历史了。最初（及以后的很长时间）制造线材的方法是先将金属锤打成板材，然后将板材制成带材，最后再进一步将带材打成圆丝。早在 14 世纪人们就掌握了拉拔线材的方法，但是直到 19 世纪，拉拔线材所用的机械设备却才完善。现代轧钢厂所用的原料都已制成小棒料或小坯料。坯料经过加热后送到一组轧机上，压缩其断面。如需要的是普通尺寸的线材，则将坯料轧成比铅笔还细的线材。加热后的线材通过管道送到卷取机去盘绕成卷。成卷的线材冷却后即送往拉拔厂，拉成各种尺寸的线材。首先，应在酸洗池里除去覆在材料上的氧化皮，再在碱池内脱酸。接着，将端部已压缩得很小的线材送入拉模板的喇叭口或进入用硬钢制成的拉模。大孔进，小孔出，断面便缩小了。这个拉拔过程一直继续下去，线材通过的孔一个比一个更小，直到粗细达到要求为止。金属在被拉得越来越细的同时，也会变得越来越硬、越来越脆，因此必须进行中间退火以使其变软并增加其韧性。在材料通过数个拉模或板孔时，需不断上油加以润滑。图 19-2 所示为线材的拉拔情况。

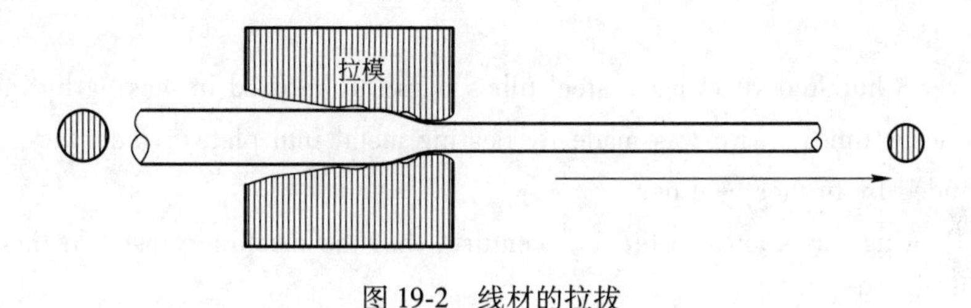

图 19-2　线材的拉拔

☞ 第二部分　课程思政

【钢筋铁骨】国家体育场——鸟巢

国家体育场鸟巢是 2008 年北京奥运会主场馆，也是北京 2022 年冬奥会和冬残奥会开闭幕式的场馆。在国家体育场鸟巢的建设过程中，仅外部钢结构总重就达 4. 2 万吨，这些都要靠 12 对 V 形立柱支撑。专家论证后认为，立柱必须采用强度、低温韧性和防震 3 项指标达到国标最高极限的 Q460E-Z35 钢。而 110mm 厚的 Q460E-Z35，全世界没有任何一个厂家生产过。河钢舞钢为了国家的荣誉，勇担使命，承担了研制 Q460E-Z35 钢板的重任。经半年多轮试制，2005 年 10 月，3600t 鸟巢用钢全部交货，其中共计 680t Q460E-Z35 钢板成为鸟巢最坚固的“钢筋铁骨”。

☞ 第三部分　延伸阅读

高速线材生产的质量控制

线材作为钢铁生产中一个重要的产品，具有重复性、连续性、时效性的特点，其质量的优劣直接关系到企业的经济效益，先进技术的应用可以提升产品的质量。

目前，高速线材生产的先进技术有：

（1）蓄热式燃烧技术。该技术具有预热温度高，能够使用低热值燃料，缩短加热时间，节约能耗的优点。

（2）无头轧制技术。无头轧制技术主要是指在钢坯出炉时，将两根钢坯首尾相连焊接在一起进行轧制。

（3）精密轧机。它在轧制精度、表面质量等方面都有了极大的提高，主要分为减定径轧机和双模块轧机。

（4）控制冷却系统。控制冷却系统主要是通过水冷线和风冷线来实现高速线材的生产。

☞ 第四部分　课文问题参考答案

Answer the following questions.

（1）The two basic ways of making tubes that one of the world's largest manufacturers has modestly described are wrapping an accurate hole in strip steel and pushing a very strong hole through a steel bar.

（2）Just over a hundred years ago，steel tubes began to be used in making bicycles.

（3）In the early times，wire was made by beating metal into plates which were then cut into strips and rounded by further beating.

（4）Wire drawing was known in the 14th century，but the machinery used in this process was perfected until the 19th.

（5）In modern rolling steel mill the raw material for making wire is small bars or billets.

☞ 第五部分　练习题参考答案

Ⅰ. Translate the following expressions into English.

（1）steel tube　（2）steel wire　（3）steel bar
（4）coils of steel strip　（5）forming roll　（6）electric induction
（7）torch cutter　（8）heat furnace　（9）hydraulic press
（10）hypodermic needle　（11）high-strength tube　（12）racing car
（13）steel rod　（14）the pointed end　（15）working temperature

Ⅱ. Fill in the blanks with the words from the text. The first letter of the word is given.

（1）bicycles　（2）induction　（3）size　（4）traffic　（5）curl
（6）billet　（7）conveyor　（8）drawing　（9）diameters　（10）wire

Ⅲ. Fill in the blanks by choosing the right words form given in the brackets.

（1）sleep　（2）turn　（3）takes　（4）charged　（5）an
（6）cross　（7）with　（8）steps　（9）being　（10）use

Ⅳ. Decide whether the following statements are true or false（T/F）.

（1）F　（2）T　（3）F　（4）T　（5）F

V. Translate the following English into Chinese.

生产钢丝的原料是直径约为 6mm 的热轧盘条。盘条经退火、冷却和除鳞后，将一端锤尖，然后穿入比盘条直径稍小一点的拉模孔，从模子的另一侧钳住尖端，将盘条拉过模孔，从而减小直径。现代的连续拉拔，可以使钢丝在通过最后一道拉模时其直径已经很小，而运行速度则很高。

第 20 单元　钢的热处理

☞ 第一部分　课文参考译文

钢的热处理工艺就是通过加热、保温和冷却的方法改变钢的组织结构以获得工件所要求性能的一种热加工技术。通过适当的热处理，能够使金属获得某些性能，例如硬度、抗拉强度（抗拉伸能力）和韧性。钢的热处理主要包括淬火、回火、退火和表面热处理。

1. 淬火

淬火是把钢加热后冷却，以提高其硬度、抗拉强度，降低韧性并获得细晶粒组织的方法。下面的简单实验可能有所启发。取两根老式的钢制毛衣针、一个煤气炉、一碗水、一块砂纸和一把钳子。

实验 1：轻轻弯曲根毛衣针，试一试它的韧性和弹性，然后用钳子夹住一端放在火上烧，手旁放一碗水。当毛衣针达到红热状态时，将其端迅速浸入水中（毛衣针在淬火时应当还是红热的），然后即可发现弯曲淬过火的一端硬而脆，而且会折断。

2. 回火

回火是把金属再加热到较低或适中的温度，然后急冷或在空气中冷却。回火是在淬火之后采用的使金属尽可能变硬、变韧的方法，它能消除淬火工件的脆性。

关于钢的淬火和回火有两个要点需要注意：

（1）为了达到完全的硬化组织，必须把钢加热到合适的温度以上，然后进行急冷。淬火之前的合适温度取决于钢的含碳量。

（2）回火不能恢复珠光体组织。要恢复珠光体，必须把钢加热到合适的温度，然后缓冷。

3. 退火

退火是将钢软化，使之消除内应力的方法。这可以使钢更易于机加工。把钢加热到临界温度之上，然后慢慢冷却。最常用的方法是把钢放到炉子里热透，然后把炉子熄灭，让金属慢慢冷却。另一种方法是把金属用黏土包起来，把它加热到临界温度，然后从炉中取出，让它慢慢冷却。

实验 2：取第二根毛衣针烧至红热状态，维持这一温度约 15s，然后从火中慢慢取出，让它逐渐冷却。如果试验这刚刚退过火的一端，它则像一根软金属丝一样很容易弯曲，而且会保持弯曲状态。

4. 表面热处理

表面热处理是使钢的外表面或壳体硬化的方法，主要用于齿轮、凸轮、曲轴及各种轴类等零件在扭转、弯曲等交变载荷下工作，并承受摩擦和冲击，其表面要比心部承受更高的应力。主要有以下几种工艺：

（1）表面淬火。表面淬火是将工件快速加热到淬火温度，然后迅速冷却，仅使表面层获得淬火组织的热处理方法。根据工件表面加热热源的不同，钢的表面淬火有很多种，例如感应加热、火焰加热、电接触加热、电子束加热、电解液加热以及激光加热等表面淬火工艺。

（2）化学热处理。将金属工件放入含有某种活性原子的化学介质中，通过加热使介质中的原子扩散渗入工件一定深度的表层，改变其化学成分和组织并获得与心部不同性能的热处理工艺叫做化学热处理。化学热处理种类很多，根据渗入元素的不同，可分为渗碳、渗氮（氮化）、碳氮共渗（氰化）、多元共渗、渗硼、渗金属等。

（3）形变热处理。形变热处理是一种新的热处理工艺，将塑性变形和热处理有机结合在一起的一种复合工艺。该工艺既能提高钢的强度，又能改善钢的塑性和韧性，同时还能简化工艺，节省能源。

目前，形变热处理是提高钢的强韧性的重要手段之一。

最先进的表面处理技术之一是激光硬化，它的最大优点是能进行选择区域硬化，而不必硬化整个部件。

和许多其他冶金技术一样，热处理工艺发展趋势是提高产量、降低能源消耗、减少环境污染。

☞ 第二部分 课程思政

【智慧钢铁】数智重塑未来钢铁

数字技术正在改变世界，极大地改善组织的运作方式。今天，钢铁和金属制造商面临着巨大的机遇，通过实施数字技术来转变他们的运营模式，使他们能够提高运营效率、客户服务、库存水平和利润率。

钢铁和金属行业的数字化时代已经到来。过去几年，数据采集、存储和分析的成本大幅下降。作为一种数字技术，预测分析已经在速度、成本和实现的便捷性方面展现了其变革运营模型的潜力。它使用先进的自我学习算法来筛选大量数据，生成见解并识别模式。传统的预测方法依赖于人的经验，收集和分析的数据有限，没有建立起深思熟虑的机制。借助传感器和机器学习算法，数字解决方案可以大大提高预测精度，并在计划外停机前留出额外时间来解决潜在问题。

☞ 第三部分 延伸阅读

钢的正火工艺

正火是将钢加热到 AC_3（或 AC_{cm}）以上适当温度，保温以后在空气中冷却得到珠光体类组织（一般为索氏体）的热处理工艺。

正火可以作为预备热处理，为机械加工提供适宜的硬度，又能细化晶粒、消除应力、消除魏氏组织和带状组织，为最终热处理提供合适的组织状态。正火还可作为最终热处理，为某些受力较小、性能要求不高的碳素钢结构零件提供合适的力学性能。

☞ 第四部分 课文问题参考答案

Answer the following questions.

(1) The procedures of heat treatment of steel include hardening, tempering, annealing, and case hardening.

(2) Yes, it can.

(3) The purpose of annealing is softening steel to relieve internal strain.

(4) Case hardening is a process of hardening the outer surface or case of ferrous metal.

(5) One of the very latest surface treatment techniques is laser hardening.

(6) Tempering does not restore the pearlitic structure.

☞ 第五部分 练习题参考答案

Ⅰ. Translate the following expressions into English.

(1) heat treatment (2) tensile strength (3) case hardening
(4) fine grain structure (5) tempering (6) annealing
(7) hardened structure (8) the pearlite structure (9) internal strain
(10) critical temperature (11) soft wire (12) the outer surface
(13) hardening (14) piston pins (15) laser hardening

Ⅱ. Fill in the blanks with the suitable words listed below. Change the form where necessary.

(1) treatment (2) described (3) dip (4) moderate
(5) depends (6) critical (7) hardening (8) cyaniding
(9) nitriding (10) nitrocarburizing (11) laser (12) springy

Ⅲ. Fill in the blanks by choosing the right words form given in the brackets.

(1) tempering (2) consists (3) cooling (4) hardened (5) reheating
(6) called (7) gives (8) carried (9) too (10) having

Ⅳ. Decide whether the following statements are true or false (T/F).

(1) F (2) F (3) F (4) F

V. Translate the following English into Chinese.

金属产品可以分为热轧产品和冷轧产品两种，而这两种产品的表面粗糙度差别很大。在热轧过程中，金属处于炽热状态，呈红色或橘红色。处于这样的高温条件下，加上周围的空气和水分，金属的氧化（即腐蚀）便进行得很快。甚至就在金属轧制成型时，便已在重新生成氧化皮了（有时成为铁鳞）。当轧辊在金属和鳞皮上压过时，非常坚硬的鳞皮被压入金属的表面，因而使表面产生了一些麻点，显得粗糙。如果金属产品的尺寸精度要求很高，表面必须光洁无瑕，那么就需要进行冷轧。由于没有加热，在轧制过程中不会产生氧化铁皮。有时可以很细地喷上一层油，以保护冷轧件的光洁表面。

第 21 单元　钢铁冶金发展趋势

☞ 第一部分　课文参考译文

钢铁材料作为用量最大的传统基础材料，对国民经济建设和国防军工发展起着重要支撑作用。新中国成立以来，我国钢铁工业迅速崛起，取得辉煌成就，钢产量跃居世界第一。钢铁工业快速发展导致供需矛盾、环境污染等问题日益突出，我国钢铁工业发展将面临资源-能源、环境-生态、市场-品牌等方面的制约。因此，未来钢铁工业的发展是以提升钢铁产业科技创新能力和核心竞争力为出发点，开发绿色化、智能化钢铁流程技术，满足国民经济建设、重大工程及高端装备制造等需求，支撑钢铁工业转型升级和可持续发展。

1. 绿色冶金发展

坚持绿色发展是我国发展的必经之路，破解冶金行业绿色低碳转型发展的难点和痛点，共同推广先进适用的工业节能低碳技术，是冶金工业今后发展的必经之路。

（1）资源充分利用。冶金工艺中燃料燃烧、原材料消耗等环节产生大量的固体废弃物，这些固体废弃物除了铁元素之外还有不同类型的其他金属元素，具有较高的二次利用价值，可以采用绿色技术将其适当加工变废为宝。如，高炉炉渣经极冷水淬粒化，作为水泥添加剂。

（2）节能减排。冶金行业能源消耗总量和碳排放量巨大，黑色金属冶炼及压延加工业能源消耗占全国能源消耗比重为 14%，其中钢铁工业能源消耗占全国能源消耗比重为 11%左右、碳排放量占中国总碳排放量的 15%左右。积极开发新型绿色技术，建立资源节约型与环境友好型的企业模式，是冶金工业的必然趋势。

2. 智能冶金发展

钢铁行业实施智能制造（智能化）和信息化发展具备一定优势条件：一是钢铁行业是亟待转型升级的传统行业，具有紧迫的内驱性；二是钢铁行业是大数据产生及应用的典型行业，具有实现的可行性；三是钢铁行业的生产过程属于长流程、规模型的制造产业，具有广泛实施的可复制性。

钢铁冶金生产的智能化主要是以 PLC 和 DCS 以及工业计算机技术为核心，在冶金生产模拟控制的过程中，对生产总线实现系统化管理。全面性的过程控制是冶金智能化发展的主要发展趋势，即在冶金自动化生产中将新型传感技术和数据融合与处理技术进行综合运用，提高生产过程控制的全面性。过程控制全面化又能有效提高生产效率和质量，提升冶金企业的生产力。

☞ 第二部分　课程思政

【人才战略】二十大报告 | 培养造就大批德才兼备的高素质人才

培养造就大批德才兼备的高素质人才，是国家和民族长远发展大计。加快建设世界重要人才中心和创新高地，促进人才区域合理布局和协调发展，着力形成人才国际竞争的比较优势。加快建设国家战略人才力量，努力培养造就更多大师、战略科学家、一流科技领军人才和创新团队、青年科技人才、卓越工程师、大国工匠、高技能人才。

广大青年要坚定不移听党话、跟党走，怀抱梦想又脚踏实地，敢想敢为又善作善成，立志做有理想、敢担当、能吃苦、肯奋斗的新时代好青年！

☞ 第三部分　延伸阅读

广泛践行社会主义核心价值观——二十大报告

社会主义核心价值观是凝聚人心、汇聚民力的强大力量。弘扬以伟大建党精神为源头的中国共产党人精神谱系，用好红色资源，深入开展社会主义核心价值观宣传教育，深化爱国主义、集体主义、社会主义教育，着力培养担当民族复兴大任的时代新人。

推动理想信念教育常态化制度化，持续抓好党史、新中国史、改革开放史、社会主义发展史宣传教育，引导人民知史爱党、知史爱国，不断坚定中国特色社会主义共同理想。坚持依法治国和以德治国相结合，把社会主义核心价值观融入法治建设、融入社会发展、融入日常生活。

☞ 第四部分　课文问题参考答案

Answer the following questions.

(1) Improve the technological innovation ability and core competitiveness of the steel industry.

(2) To persevere in green development and promote industrial energy-saving and low-carbon technologies.

(3) By using green technology.

(4) Firstly, the iron and steel industry is a traditional industry that needs to be transformed and upgraded urgently and has urgent internal drive. Secondly, the iron and steel industry is a typical industry for the generation and application of big data, which is feasible to realize. Thirdly, the production process of iron and steel industry belongs to the manufacturing industry with long process and model, which is widely implemented and reproducible.

(5) To effectively improve production efficiency and quality and increase the productivity of metallurgical enterprises.

☞ 第五部分 练习题参考答案

Ⅰ. Translate the following expressions into English.

(1) iron and steel material (2) low-carbon (3) Ecology
(4) ferrous metal (5) blast furnace slag (6) cement additive.
(7) sustainable development (8) fuel combustion (9) resource-saving
(10) solid wastes (11) data fusion
(12) comprehensive process control

Ⅱ. Fill in the blanks with the words from the text. The first letter of the word is given.

(1) prominent (2) constrained (3) combustion (4) process
(5) Blast, granulated (6) consumption (7) upgraded (8) Intelligent

Ⅲ. Fill in the blanks with appropriate prepositions.

(1) of (2) on (3) in (4) of (5) of
(6) of (7) in (8) to (9) of (10) of

Ⅳ. Decide whether the following statements are true or false (T/F).

(1) F (2) T (3) F (4) T (5) T

Ⅴ. Translate the following English into Chinese.

冶金工艺中燃料燃烧、原材料消耗等环节产生大量的固体废弃物，这些固体废弃物除了铁元素之外还有不同类型的其他金属元素，具有较高的二次利用价值，可以采用绿色技术将其适当加工变废为宝。如，高炉炉渣经极冷水淬粒化，作为水泥添加剂。

Glossary

words

accelerate [æk'seləreit] *v.* 加速，促进 U7
accentuate [æk'sentjueit] *v.* 增强，使……更明显，加重 U8
accessories [æk'sesəriz] *n.* 辅助设备，附件 U18
accompany [ə'kʌmpəni] *v.* 伴随，陪伴 U11
accumulate [ə'kjuːmjuleit] *v.* 积聚，堆积 U19
acid ['æsɪd] *a.* 酸性的 U1
adaptability [əˌdæptə'biləti] *n.* 适应性，可变性，适合性 U15
adapted [ə'dæptid] *a.* 适合的 U9
addition [ə'diʃən] *n.* 增加物，添加剂 U20
additive ['æditiv] *n.* 添加剂 U14
adjoin [ə'dʒɔin] *v.* 邻近，毗连，邻接 U15
adjust [ə'dʒʌst] *v.* 调整，调节 U9
adjustment [ə'dʒʌstmənt] *n.* 调整，调节 U9
adoption [ə'dɔpʃən] *n.* 采用 U11
advance [əd'vɑːns] *n.* 进步，进展 U2
adversely ['ædvəːsli] *ad.* 逆地，反对地 U13
affinity [ə'finiti] *n.* 亲和力，吸引力 U13
agglomerate [ə'glɔməreɪt] *v.* 结块，烧结 U2
aim [eim] *n.* 目标，目的 U4
v. 以……为目标 U4
align [ə'lain] *v.* 排列 U4
alkali ['ælkəlai] *a.* 碱性的 U4
alloy ['ælɔɪ，ə'lɔi] *n.* 合金 U15
alter ['ɔːltə] *v.* 改变，变更 U17
alternative [ɔːl'təːnətiv] *n.* 替换物，抉择 U5
alumina [ə'ljuːminə] *n.* 氧化铝，矾土 U4
aluminum [ə'ljuːminəm] *n.* 铝（元素符号 Al）U7
amenability [əmiːnə'biləti] *n.* 可处理性，可控制性 U15
ammonia ['æməunjə] *n.* 氨，氨水 U20
analyse ['ænəlaiz] *n.* 分析 U6
anhydrous [æn'haidrəs] *a.* 无水的 U14

annealing [æ'ni:liŋ] *n.* 退火 U18
annum ['ænəm] *n.* 年，岁 U11
anode ['ænəud] *n.* [电] 阳极，正极 U14
anthracite ['ænθrəsait] *n.* 无烟煤 U15
antimony ['æntiməni] *n.* 锑（元素符号 Sb）U15
appearance [ə'piərəns] *n.* 外观，外表 U5
applicability [ˌæplikə'biləti] *n.* 适用性，适应性 U15
application [ˌæpli'keiʃən] *n.* 应用，运用 U5
approach [ə'prəutʃ] *n.* 方法，途径 U2
aqueous ['eikwiəs] *a.* 水的，水成的，水状的 U15
arc [ɑ:k] *n.* 弧，弓形 U7
arise [ə'raiz] *v.* 来源于，出现 U6
arrangement [ə'reindʒmənt] *n.* 装置，配置 U3
arsenic ['ɑ:s(ə)nik] *n.* 砷（元素符号 As）U15
article ['ɑ:tikl] *n.* 物品，商品，项目，条款 U1
ash [æʃ] *n.* 灰分，灰烬 U2
asphalt ['æsfælt] *n.* 沥青 U16
assist [ə'sist] *n.* 帮助，辅助 U14
assure [ə'ʃuə] *v.* 保证，担保 U4
atmosphere ['ætməsfiə] *n.* 空气，气氛 U7
autogenous [ɔ:'tɔdʒinəs] *a.* 自生的，自体的 U13
automation [ɔ:tə'meiʃən] *n.* 自动控制，自动操作 U4
automobile ['ɔ:təməbi:l] *n.* [美] 汽车，车辆 U12
axle ['æksl] *n.* 轮轴，车轴 U12
bank [bæŋk] *n.* 炉坡，堤 U9
bar [bɑ:(r)] *n.* 棒材 U17
barrel ['bærəl] *n.* 管状物，圆桶状物 U12
base [beis] *a.* 低劣的 U15
basic ['beɪsɪk] *a.* 碱性的，基本的 U1
basicity [bə'sisiti] *n.* 碱度，碱性 U8
batch [bætʃ] *n.* 一批，一炉 U16
bath [bɑ:θ, bæθ] *n.* 熔池 U1
bauxite ['bɔ:ksait] *n.* 铝矾土，铝土矿，铁铝氧石 U14
bell-less [belles] *a.* 无料钟的 U3
bellow ['beləu] *n.* 风箱 U1
belly ['beli] *n.* 炉腰 U3
belt [belt] *n.* 皮带 U3
bend [bend] *v.*【过去式/过去分词】bent 弯曲 U19
billet ['bilit] *n.* 方坯 U17

bind [baind] *v.* 联合 U6
binder ['baində] *n.* 粘结剂 U2
blast ['blæst] *n.* 冲击波 U16
bleeder ['bliːdə] *n.* 分压器，放散阀 U3
blend [blend] *v.* 混合 U3
blister ['blistə] *n.* 粗铜，气泡 U13
v. 产生气泡 U13
block [blɔk] *n.* 毛坯，粗坯 U11
bloom [bluːm] *n.* 大（初轧）方坯 U11
bluish ['bluːiʃ] *a.* 浅蓝色的，带蓝色的 U13
bolt [bəult] *n.* 螺栓 U12
bore [bɔː] *v.* 穿井，钻孔 U9
bornite ['bɔːnait] *n.* [矿] 斑铜矿 U13
bosh [bɔʃ] *n.* 炉腹 U3
brick [brik] *n.* 砖，砖块 U1
briquet [bri'ket] *v.* 压块 U5
brittleness ['britlnis] *n.* 脆性，脆度 U12
broad [brɔːd] *a.* 主要的，粗略的 U17
bundle ['bʌndl] *n.* 捆，束，包 U6
bundling ['bʌndliŋ] *n.* 捆扎 U18
bunker ['bʌŋkə] *n.* 料仓 U3
burden ['bəːdn] *n.* 炉料，配料 U5
burner ['bəːnə] *n.* 燃烧器 U4
bypass ['baɪpɑːs, 'baɪpæs] *v.* 绕过，使……通过旁道 U4
cable ['keibl] *n.* 钢丝绳，缆，索 U14
cadmium ['kædmiəm] *n.* 镉（元素符号 Cd）U15
calcine ['kælsain] *v.* 焙烧 U6
calcium ['kælsiəm] *n.* 钙（元素符号 Ca）U1
capacity [kə'pæsiti] *n.* 能力，容量，生产量 U9
capital ['kæpɪt(ə)l] *n.* 资金，资产 U6
capitalize ['kæpitəlaiz] *v.* 使……资本化，积累资本 U15
carbonaceous [ˌkɑːbə'neiʃəs] *a.* 碳质的，碳的，含碳的 U15
carbonate ['kɑːbəneit] *n.* 碳酸盐 U2
carburize ['kɑːbjuraiz] *v.* 使……渗碳 U2
case [keis] *n.* 壳，套，表层 U20
caster ['kɑːstə] *n.* 连铸机 U10
casting ['kɑːstiŋ] *n.* 铸造，铸件 U15
catalyst ['kætəlist] *n.* 催化剂，触媒 U5
category ['kætigəri] *n.* 种类，范畴 U10

cathode [ˈkæθəud] *n.* 阴极 U15
caustic [ˈkɔːstik] *a.* 腐蚀性的，苛性的 U14
cavity [ˈkæviti] *n.* （铸造）型腔 U17
cell [sel] *n.* 电解槽，小房间 U14
centrifugal [senˈtrifjugəl] *a.* 离心的 U17
ceramic [siˈræmik] *a.* 陶瓷的 U3
chalcocite [ˈkælkəsait] *n.* ［矿］辉铜矿 U13
chalcopyrite [ˌkælkəˈpaiərait] *n.* ［矿］黄铜矿 U13
chamber [ˈtʃeimbə] *n.* 室，房间 U4
chapter [ˈtʃæptə] *n.* （书的，文章的）章，回 U9
characteristic [ˌkæriktəˈristik] *n.* 特性，特征 U20
charcoal [ˈtʃɑːkəul] *n.* 木炭 U1
charge [tʃɑːdʒ] *n.* 炉料 U15
v. 装料，装炉 U2
checker work [ˈtʃekəwəːk] *n.* 格式装置，砌砖格 U4
chimney [ˈtʃimni] *n.* 烟囱 U4
chisel [ˈtʃizl] *n.* 凿子 U12
chloride [ˈklɔːraid] *n.* ［化］氯化物 U14
chromium [ˈkrəumjəm] *n.* 铬（元素符号 Cr）U12
circular [ˈsəːkjulə] *a.* 圆形的，环形的 U18
circulate [ˈsəːkjuleit] *v.* （使……）运行，（使……）循环 U6
circulation [ˌsəːkjuˈleiʃən] *n.* 循环，流通 U8
civilization [ˌsɪvɪlaɪˈzeɪʃən] *n.* 文明，文化，文明社会 U1
clarify [ˈklærifai] *v.* 澄清，使（液体）纯净，阐明 U14
class [klɑːs] *v.* 把……分类 U5
classical [ˈklæsikəl] *a.* 传统的，古典的 U5
clay [klei] *n.* 黏土，泥土 U20
coarse [kɔːs] *a.* 粗的，粗糙的 U14
coating [ˈkəutiŋ] *n.* 涂层 U16
cobalt [ˈkəubɔːlt] *n.* 钴（元素符号 Co）U15
coils [kɔilz] *n.* 镀锡卷板 U19
coke [kəuk] *n.* 焦炭 U1
combustible [kəmˈbʌstəbl] *a.* 易燃的，可燃的 U7
commence [kəˈmens] *v.* 开始，着手 U8
commensurate [kəˈmenʃərit] *a.* 相称的，相当的 U14
commercial [kəˈməːʃəl] *a.* 商业的，商务的 U6
commercially [kəˈməːʃəli] *ad.* 商业上，通商上 U12
comparatively [kəmˈpærətɪvlɪ] *ad.* 比较地，相当地 U19
compensate [ˈkɔmpenseit] *v.* 补偿，赔偿 U15

completion [kəm'pliːʃ(ə)n] *n.* 完成 U8
composite ['kɔmpəzɪt] *a.* 复合的，合成的 U3
composition [kɔmpə'ziʃ(ə)n] *n.* 构成，合成物 U15
compound ['kɔmpaund] *n.* 混合物，[化] 化合物 U9
compress [kəm'pres] *v.* 压缩 U4
compressed [kəm'prest] *a.* 压缩的 U16
comprise [kəm'praɪz] *v.* 包括，由……组成 U3
concentrate ['kɔnsentreit] *n.* 精矿，精煤 U2
concentration [ˌkɔnsen'treiʃən] *n.* 浓缩，浓度，含量 U13
condenser [kən'densə] *n.* 冷凝器 U15
conductor [kən'dʌktə] *n.* 导体 U14
cone [kəun] *n.* 锥体 U3
cone-shaped ['kəun'ʃeipt] *a.* 锥形的 U19
configuration [kənˌfigju'reiʃən] *n.* 轮廓，外形 U17
conical ['kɔnikəl] *a.* 圆锥的，圆锥形的 U17
conical ['kɔnikəl] *a.* 圆锥形的 U16
considerable [kən'sidərəbl] *a.* 相当大的，相当多的 U7
consistency [kən'sɪstənsɪ] *n.* 一致性 U11
constant ['kɔnstənt] *a.* 恒定的，不变的 U4
constituent [kən'stitjuənt] *n.* 要素，组成部分，成分 U12
container [kən'teinə] *n.* 容器（箱，盆，罐，壶，桶）U14
contaminant [kən'tæminənt] *n.* 致污物，污染物 U6
contaminate [kən'tæmɪneɪt] *v.* 弄脏，污染 U6
contamination [kənˌtæmi'neiʃən] *n.* 污染，污染物 U4
content [kən'tent] *n.* 里面的东西，内容，容量，目录 U3
conventional [kən'venʃənl] *a.* 传统的，常规的 U5
conversion [kən'vəːʃən] *n.* 变换，转化 U13
converter [kən'vəːtə(r)] *n.* 转炉，炼钢炉 U1
conveyor [kən'veiə] *n.* 传送带，输送机 U19
copper ['kɔpə] *n.* 铜（元素符号 Cu）U3
corporation [ˌkɔːpə'reiʃən] *n.* 公司 U15
corresponding [ˌkɔris'pɔndiŋ] *a.* 相应的，对应的 U5
corrosion [kə'rəuʒən] *n.* 腐蚀，侵蚀 U6
countercurrent ['kauntəˌkʌrənt] *n.* 逆流 U5
crack [kræk] *v.* (使……）破裂，裂纹 U10
cracking ['krækiŋ] *n.* 破裂，裂化 U20
crane [krein] *n.* 起重机 U8
crankshaft ['kræŋkʃɑːft] *n.* 曲轴，机轴 U20
creep [kriːp] *n.* 蠕变，蠕动 U4

creep-resistant [kri:pri'zistənt] *a.*　抗蠕变的 U4
critical ['kritikəl] *a.*　关键性的，紧要的 U3
crude [kru:d] *a.*　天然的，未加工的 U7
crush [krʌʃ] *v.*　破碎，压碎，碾碎 U6
crust [krʌst] *n.*　[地质] 地壳，表层 U13
cryolite ['kraiəulait] *n.*　冰晶石 U14
crystal ['kristl] *n.*　晶体，结晶，水晶 U14
crystallization ['kristəlai'zeiʃən] *n.*　结晶化 U2
culture ['kʌltʃə] *n.*　文化，文明 U1
curl [kə:l] *v.*　(使……) 卷曲 U19
current ['kʌrənt] *n.*　电流，水流，气流 U9
curve [kə:v] *v.*　弯，使……弯曲 U11
cyanide ['saɪəˌnaɪd] *n.*　氰化物 U16
cyaniding ['saiənaidiŋ] *n.*　氰化 U20
cycle ['saikl] *v.*　循环 U4
cylinder ['silində] *n.*　圆筒，圆柱体 U3
cylindrical [sɪ'lɪndrɪkəl] *a.*　圆柱（形，体）的 U3
dam [dæm] *n.*　堤，坝 U3
decompose [ˌdi:kəm'pəuz] *v.*　分解，(使……) 腐烂 U20
decoration [ˌdekə'reiʃən] *n.*　装饰，装饰品 U17
define [di'fain] *v.*　限定，规定 U6
deformation [ˌdi:fɔ:'meiʃən] *n.*　变形 U18
degas [di:'gæs] *v.*　脱气，去氧 U10
deleterious [ˌdeli'tiəriəs] *a.*　有害的，有毒的 U10
deliver [di'livə] *v.*　输送 U3
dense [dens] *a.*　稠密的，浓厚的 U15
density ['densiti] *n.*　密度 U3
deoxidation [di:ˌɔksɪ'deɪʃən] *n.*　脱氧，还原 U6
deoxidize [di:'ɔksidaiz] *v.*　脱氧，还原 U6
deoxidizer [di:'ɔksidaizə] *n.*　脱氧剂，还原剂 U10
dependency [di'pendənsi] *v.*　依靠，信赖 U2
dephosphorization [di:fɔsfərai'zeiʃən] *n.*　脱磷 U7
derive [di'raiv] *v.*　来自，出自，源自 U5
descend [di'send] *v.*　下来，下降 U5
description [dis'kripʃən] *n.*　描写，描述 U19
desiliconization [di:silikɔnai'zeiʃən] *n.*　脱硅 U7
desulphurization [di:sʌlfərai'zeiʃən] *n.*　脱硫 U7
detract [di'trækt] *v.*　降低，减损 U6
dictate [dik'teit] *v.*　要求，规定 U4

die [dai] *n.* 硬模，冲模 U12
冲模，钢模 U15
diffuse [di'fju:z] *v.* 扩散，传播 U20
digest [di'dʒest，dai'dʒest] *v.* 浸煮，蒸煮，煮解 U14
digestion [dɪ'dʒestʃ(ə)n] *n.* 溶解，溶出，煮解 U14
dilute [dai'lju:t，di'lju:t] *v.* 冲淡，稀释 U6
a. 淡的，稀释的 U15
dimension [di'menʃən] *n.* 尺寸，尺度 U11
dioxide [dai'ɔksaid] *n.* 二氧化物 U7
dip [dip] *v.* 浸，泡 U15
discard [dis'kɑ:d] *v.* 丢弃，抛弃 U11
discharge ['dɪstʃɑ:dʒ] *v.* 排出，流出 U5
discriminating [dis'krimineitiŋ] *a.* 识别的，有识别力的 U16
displace [dis'pleis] *v.* 取代，置换 U11
dissociate [di'səuʃieit] *v.* 分离，游离，分裂 U14
dissolution [disə'lju:ʃən] *n.* 分解，解散 U7
distinct [dis'tiŋkt] *a.* 清楚的，明显的 U14
独特的，有区别的 U15
distribute [di'stribju:t，'distribju:t] *v.* 分配，分布，散布 U3
ditch [ditʃ] *n.* 沟渠 U16
diverse [dai'və:s] *a.* 不同的，相异的 U19
dolomite ['dɔləmait] *n.* 白云石 U1
dome [dəum] *n.* 圆顶，圆屋顶 U4
downcomer ['daunˌkʌmə(r)] *n.* 下降管 U3
drain [drein] *v.* 排出，流掉 U2
dramatically [drə'mætɪkəlɪ] *ad.* 鲜明地，显著地 U2
drill [drɪl] *n.* 钻头，锥子 U12
ductile ['dʌktail，dʌktil] *a.* 柔软的，易延展的 U16
ductility [dʌk'tiliti] *n.* 延展性，塑性 U10
dump [dʌmp] *v.* 倾倒，倾卸 U3
duration [djuə'reiʃən] *n.* 持续时间，为期 U8
dust [dʌst] *n.* 粉尘，粉末 U6
灰尘，尘埃 U15
dwindle ['dwindl] *v.* 减少，缩小 U7
economical [ˌi:kə'nɔmikəl] *a.* 经济的，节俭的 U17
efficiency [i'fiʃənsi] *n.* 效率，效能 U4
effluent ['efluənt] *a.* 发出的，流出的 U13
effort ['efət] *n.* 努力，成就 U5
electrode [ɪ'lektrəud] *n.* 电极 U7

electrolysis [ɪlek'trɔlɪsɪs] *n.* 电解（作用）U14
【复数】-ses [-siːz]
electrolyte [ɪ'lektrəlaɪt] *n.* 电解，电解液，电解质 U14
electrolytic [iˌlektrəu'litik] *a.* 电解的，电解质的 U15
electromagnetic [ɪlektrəu'mægnɪtɪk] *a.* 电磁的 U10
electrowinning [iˌlektrəu'winiŋ] *n.* 电解冶金法，电积金属法 U15
elevated ['eliveitid] *a.* 提高的，严肃的 U6
eliminate [i'limineit] *v.* 除去 U1
embed [im'bed] *v.* 使……嵌入，使……插入 U16
embrittlement [em'britlmənt] *n.* 变脆，脆化 U10
emerge [i'məːdʒ] *v.* 排（涌，射，冒）出 U19
emergency [i'məːdʒnsi] *n.* 出现，紧急情况 U1
encircle [in'səːkl] *v.* 环绕，围绕，包围 U3
enlightening [in'laitniŋ] *a.* 有启发作用的 U20
enrich [in'ritʃ] *v.* 使……富足，使……富化 U4
equal ['iːkwəl] *a.* 相等的，均等的 U4
equation [i'kweiʃən] *n.* 等式，方程式 U7
equilibria [iːkwi'libriə] *n.* 平衡，均势 U15
【equilibrium [iːkwi'libriəm] 的复数形式】
equip [i'kwip] *n.* 装备，配备 U8
erosion [i'rəuʒən] *n.* 腐蚀，侵蚀 U9
escape [is'keip] *v.* 逸出 U7
evacuate [i'vækjueit] *v.* 排空，抽空 U10
evenly ['iːvənli] *ad.* 均匀地 U3
evolution [ˌiːvə'luːʃən, ˌevə'luːʃən] *n.* 放出（气体），形成 U8
evolve [i'vɔlv] *v.* 放出，发出 U7
excess [ik'ses, 'ekses] *a.* 多余的，过度的，额外的 U14
exclusively [ɪk'skluːsɪvlɪ] *ad.* 排外地，专有地 U14
exert [ig'zəːt] *v.* (on or upon) 对……施加影响 U16
exhaust [ig'zɔːst] *v.* 排气 U4
exit ['eksɪt] *v.* 排出，离去 U3
expansion [iks'pænʃən] *n.* 发展，扩大，扩展 U5
exploit [iks'plɔit] *v.* 利用，使用，开发，开采 U14
expose [iks'pəuz] *v.* 使……面临，暴露 U5
extension [iks'tenʃən] *n.* 延长，扩充 U17
external [eks'təːnl] *a.* 外部的，外面的 U3
extract [iks'trækt] *v.* 萃取，提取 U15
extraction [ɪk'strækʃən] *n.* 萃取，提取 U1
extrude [eks'truːd] *v.* 挤压出 U17

extrusion [eks'tru:ʒən] *n.* 挤压，推出 U17
fabricate ['fæbrikeit] *v.* 制作 U19
facilitate [fə'siliteit] *v.* 促进，帮助 U16
facility [fə'siliti] *n.* 设施，设备 U3
false [fɔ:ls] *a.* 假的 U16
fatigue [fə'ti:g] *n.* 疲劳 U12
ferric ['ferik] *a.* 铁的，三价铁的 U2
ferrite ['ferait] *n.* 铁素体 U12
ferroalloy [ˌferəu'ælɔi] *n.* 铁合金 U6
ferrochromium [ˌferəu'krəumɪəm] *n.* 铬铁（合金）U6
ferromolybdenum [ˌferəumɔ'lɪbdɪnəm] *n.* 钼铁（合金）U6
ferrosilicon [ˌferəu'silikən] *n.* 硅铁（合金）U6
ferrotitanium [ˌferəutai'teiniəm] *n.* 钛铁（合金）U6
ferrotungsten [ˌferəu'tʌŋstən] *n.* 钨铁（合金）U6
ferrous ['ferəs] *a.* 含铁的，亚铁的 U2
ferrovanadium [ˌferəvə'neidiəm] *n.* 钒铁（合金）U6
file [faɪl] *n.* 锉刀 U12
filling ['filiŋ] *n.* 装料 U8
filter ['filtə] *n.* 过滤器，滤光器 U14
filtration [fil'treiʃən] *n.* 过滤，筛选 U14
flat [flæt] *n.* 板片，（窄）带材 U18
flexibility [ˌfleksə'biliti] *n.* 灵活性 U10
flexible ['fleksəbl] *a.* 灵活的 U10
float [fləut] *v.* 飘浮，漂浮 U3
flotation [fləu'teiʃən] *n.* 浮选 U13
flue [flu:] *n.* 烟道，通气管 U4
fluidity [flu(:)'iditi] *n.* 流动性 U2
fluorspar ['flu(:)əspɑ:] *n.* 萤石，氟石 U6
flux [flʌks] *n.* 熔剂，造渣剂 U2
v. 熔化，使……熔解 U13
foil [fɔil] *n.* 箔，金属薄片 U16
footing ['futiŋ] *n.* 立足处，（社会）地位 U1
forbid [fə'bid] *v.* 禁止，阻止 U7
forefather ['fɔ:ˌfɑ:ðə] *n.* 祖先，先人，前辈 U1
forging ['fɔ:dʒiŋ] *n.* 锻件 U12
formability [fɔ:mə'biliti] *n.* 可模锻性，可成型性 U12
formulate ['fɔ:mjuleit] *v.* 用公式表示 U7
foundry ['faundri] *n.* 铸造厂 U9
fraction ['frækʃən] *n.* 小部分，微量 U14

freeze [friːz] *v.* 凝固，凝结 U11
frit [frit] *v.* 熔融（化）U2
froth [frɔθ, frɔːθ] *n.* 渣滓，废物 U13
fuel [fjuːəl] *n.* 燃料 U2
function [ˈfʌŋkʃən] *vi.* 发挥作用，运行 U9
gangue [gæŋ] *n.* 脉石 U2
gaseous [ˈgæsɪəs] *a.* 气体的，气态的 U7
gasifier [ˈgæsifaiə] *n.* 气化器，燃气发生炉 U5
gear [giə] *n.* 齿轮，传动装置 U12
generate [ˈdʒenəˌreit] *v.* 产生，发生 U5
gibbsite [ˈgibzait] *n.* 水铝矿，铝土矿 U14
grate [greɪt] *n.* 箅条，固定筛 U2
gravel [ˈgrævəl] *n.* 碎石，砂砾 U16
gravity [ˈgræviti] *n.* 重力 U5
green [griːn] *a.* 未加工（处理）的，湿的 U14
grind [graind] *v.* 磨快，磨尖锐 U20
grinding [ˈgraindiŋ] *n.* 研磨，制粉 U13
grip [grip] *v.* 紧夹，紧握 U19
hafnium [ˈhæfniəm] *n.* 铪（元素符号 Hf）U15
harden [ˈhɑːdn] *v.* 使……变硬，使……坚强 U2
hardenability [ˌhɑːdənəˈbiliti] *n.* 淬透性，可淬性 U12
hardened [ˈhɑːdənd] *a.* 变硬的，淬火的 U20
hazard [ˈhæzəd] *n.* 危险，冒险，公害 U5
hearth [hɑːθ] *n.* 炉膛，炉缸 U1
heat [hiːt] *n.* 炉次 U8
heavy [ˈhevɪ] *a.* 粗的，迟钝的，沉闷的 U9
hematite [ˈhemətait] *n.* 赤铁矿 U2
hoist [hɔist] *v.* 升起，吊起 U11
hollow [ˈhɔləu] *a.* 空的 U18
hooded [ˈhudɪd] *a.* 有罩盖的，戴头巾的 U14
hopper [ˈhɔpə] *n.* 漏斗，料斗 U3
horizontal [ˌhɔriˈzɔntl] *a.* 水平的 U11
hydrate [ˈhaidreit] *n.* 氢氧化物，水合物 U14
hydraulic [haiˈdrɔːlik] *a.* 水力的，水压的 U19
hydrogen [ˈhaidrəudʒən] *n.* 氢（元素符号 H）U5
hydrometallurgical [ˈhaidrəuˌmetəˈləːdʒikəl] *a.* 湿法冶金的 U15
hydrometallurgy [ˌhaidrəuˈmetlədʒi] *n.* 湿法冶金术 U15
hydrous [ˈhaidrəs] *a.* 含水的 U2
hypodermic [ˌhaipəuˈdəːmik] *a.* 皮下的，皮下注射用的 U19

icing [ˈaisiŋ] *n.* （糕点上的）糖衣 U17
ignite [igˈnait] *v.* 着火，点火 U2
illustrate [ˈiləstreit] *v.* 加插图于，举例说明 U19
ilmenite [ˈilmiˌnait] *n.* 钛铁矿 U2
immerse [iˈməːs] *v.* 浸入（常与 in 连用）U20
immiscible [iˈmisəbl] *a.* 不能混合的 U13
impair [imˈpɛə] *v.* 削弱，损害 U12
impart [imˈpɑːt] *v.* 给予（尤指抽象事物）U20
impeller [imˈpelə] *n.* 叶轮，推进者 U16
imperial [imˈpiəriəl] *a.* 帝国的 U15
impetus [ˈimpitəs] *n.* 推动力，促进 U11
implement [ˈimplimənt] *n.* 工具，器具 U17
impoverish [imˈpɔvəriʃ] *v.* 使……枯竭，使……贫穷 U10
impure [imˈpjuə] *a.* 不纯的，杂质 U13
impurity [imˈpjuəriti] *n.* 杂质 U1
incline [inˈklain] *v.* 使……倾斜，倾向，倾斜 U8
inclusion [inˈkluːʒən] *n.* 夹杂物 U10
indentation [ˌindenˈteiʃən] *n.* 凹痕，压痕 U19
inefficient [ˌiniˈfiʃənt] *a.* 效率低的，效率差的 U1
influence [ˈinfluəns] *v.* 影响，改变 U7
inherent [inˈhiərənt] *a.* 固有的，内在的 U11
initial [iˈniʃəl] *a.* 最初的，开始的 U4
initiate [ɪˈnɪʃɪeɪt] *v.* 开始，发起 U11
injection [inˈdʒekʃən] *n.* 喷吹，喷射 U2
innovation [ɪnəˈveɪʃ(ə)n] *n.* 革新，创新 U1
input [ˈɪnput] *n.* 输入，进料量 U6
inspection [inˈspekʃən] *n.* 检查，修补 U8
install [inˈstɔːl] *v.* 安装 U10
installation [ˌinstəˈleiʃən] *n.* （整套）装置（备），设备（施）U11
intense [inˈtens] *a.* 强烈的，剧烈的 U4
interaction [ˌintərˈækʃən] *n.* 互相作用，互相影响 U7
interfacial [ˌintə(ː)ˈfeiʃəl] *a.* 界面的，分界面的 U14
intermediate [ˌintəˈmiːdjət] *a.* 中间的 U18
internal [inˈtəːnl] *a.* 内在的，内部的 U10
interval [ˈintəvəl] *n.* 时段，（时间的）间隔 U9
间隔，间距 U16
intimate [ˈintimit] *a.* 密集的，致密的 U8
intricate [ˈintrikit] *a.* 复杂的，难懂的 U17
invariably [ɪnˈveərɪəb(ə)lɪ] *ad.* 不变地，总是 U11

involve [in'vɔlv] *v.* 包括，涉及 U8
ion ['aiən] *n.* 离子 U14
ionic [aɪ'ɔnɪk] *a.* 离子的 U14
ironmaking [aiən'meikiŋ] *n.* 炼铁 U1
jet [dʒet] *n.* 射流，气流 U8
jewelry ['dʒu:əlri] *n.* 珠宝，珠宝饰物 U17
justify ['dʒʌstifai] *v.* 证明……是正当的 U1
kiln [kiln] *n.* 窑 U2
ladle ['leidl] *n.* 钢包 U9
laminated ['læmineitid] *a.* 层压的，由薄片叠成的 U16
latter ['lætə] *n.* （两者中）后者 U8
layer ['leiə] *n.* 层，阶层 U8
leach [li:tʃ] *n.* 浸出，过滤 U15
v. 浸出，过滤 U15
lead [li:d] *n.* 铅（元素符号 Pd）U17
lean [li:n] *a.* 贫乏的，歉收的 U13
lightness ['laitnis] *n.* 轻 U19
likewise ['laik,waiz] *ad.* 同样地，照样地 U11
lime [laɪm] *n.* 石灰 U1
limestone ['laimstəun] *n.* 石灰石 U2
limitation [,limi'teiʃən] *n.* 限制，局限性 U6
limonite ['laimə,nait] *n.* 褐铁矿 U2
line [laɪn] *n.* 炉衬，管线 U1
v. 造衬 U4
lining ['lainiŋ] *n.* 衬里，内衬 U15
liquid ['likwid] *n.* 液体 U3
liquor ['likə] *n.* 液，液体，母液 U14
lithium ['liθiəm] *n.* [化] 锂（元素符号 Li） U13
locomotive [,ləukə'məutiv] *n.* 机车，火车头 U11
lower ['ləuə] *v.* 降低，跌落 U8
lump [lʌmp] *n.* 堆，块 U16
machinability [məʃi:nə'biliti] *n.* 机械加工性，切削性 U6
magnesia [mæg'ni:ʃə] *n.* 氧化镁 U1
magnesite ['mægnəsait] *n.* 菱镁矿 U8
magnetite ['mægnitait] *n.* 磁铁矿 U2
magnitude ['mægnitju:d] *n.* 大小，量级 U16
mainstay ['meɪnsteɪ] *n.* 支柱，中流砥柱 U1
malleability [,mæliə'biliti] *n.* 加工性，展延性 U16
mandrel ['mændrɪl] *n.* 顶杆，芯棒 U19

manganese [ˈmæŋɡəniːz] *n.* 锰（元素符号 Mn）U6
manipulate [məˈnipjuleit] *v.* （熟练地）操作，使用 U14
manually [ˌmænjuəli] *ad.* 手动地，用手地 U16
manufacture [ˌmænjuˈfæktʃə] *n.* 制造，加工 U6
markedly [mɑːkidli] *ad.* 显著地，明显地 U8
marking [ˈmɑːkiŋ] *n.* 做记号，做标记 U18
marmatite [ˈmɑːmətait] *n.* 铁闪锌矿 U15
martensite [ˈmɑːtənzait] *n.* 马氏体 U12
mass [mæs] *a.* 大规模的 U17
matte [mæt] *n.* 锍，冰铜 U13
mature [məˈtjuə] *a.* 成熟的 U5
mechanically [miˈkænikəli] *ad.* 机械地 U13
mechanics [miˈkæniks] *n.* 机械学，力学 U17
mechanism [ˈmekənizəm] *n.* 机械装置，机件 U8
melt [melt] *v.* 熔化，溶解 U15
metallic [mi ˈtælik] *a.* 金属的，含金属的 U2
metallurgical [ˌmetəˈləːdʒikəl] *a.* 冶金学的 U1
metalworking [ˈmətəlˌwəːkɪŋ] *n.* 金属加工 U17
methane [ˈmeθein] *n.* 甲烷，沼气 U5
microcleanliness [maikrəuklenˈlinis] *n.* 显微清洁（度）U10
mineralogical [ˌminərəˈlɔdʒikəl] *a.* 矿物学的 U2
miniature [ˈminjətʃə] *n.* 缩小模型，缩小的模型 U17
minimize [ˈmɪnɪmaɪz] *v.* 将……减到最少 U4
minimum [ˈminiməm] *n.* 最小值，最低限度 U16
mode [məud] *n.* 方式，模式 U17
modify [ˈmɔdifai] *v.* 调节，限制 U14
molten [ˈməultən] *a.* 熔化的，铸造的 U15
molybdenum [məˈlibdinəm] *n.* 钼（元素符号 Mo）U12
momentous [məuˈmentəs] *a.* 重要的，重大的 U16
momentum [məuˈmentəm] *n.* 动力，要素 U1
multihearth [mʌltiˌhaːθ] *n.* 多膛焙烧炉 U13
multiple [ˈmʌltipl] *a.* 多样的，许多的 U15
multitude [ˈmʌltitjuːd] *n.* 大量，众多 U12
mutually [ˈmjuːtʃuəli, ˈmjuːtjuəli] *ad.* 互相地 U19
necessitate [niˈsesiteit] *v.* 成为必要 U4
negligible [ˈneglidʒəbl] *a.* 可忽略的，不重要的 U7
nickel [ˈnikl] *n.* 镍（元素符号 Ni）U6
nitrate [ˈnaitreit] *n.* [化] 硝酸盐，硝酸根 U13
nitride [ˈnaitraid] *n.* 氮化物 U20

nitriding ['naitraidiŋ] *n.* 渗氮 U20
nitrocarburizing [naitrəu'kɑːbjuraiziŋ] *n.* 碳氮共渗 U20
nitrogen ['naitrədʒən] *n.* 氮（元素符号 N）U10
nonmetallic ['nɔnmi'tælik] *n.* 非金属物质 U10
a. 非金属的 U10
nosepiece ['nəuzpiːs] *n.* 顶，端 U19
notably ['nəutəbəlɪ] *ad.* 显著地，特别地 U12
notch [nɔtʃ] *n.* 出口，槽口 U3
nugget ['nʌgit] *n.* 天然金块，矿块 U16
obtain [əb'tein] *v.* 获得，得到 U5
octagonal [ɔk'tægənl] *a.* 八边形的，八角形的 U11
offtake ['ɔːfteik] *n.* 煤气导出管，排气管 U3
oil [ɔil] *v.* 给……加油，涂油 U19
optimum ['ɔptiməm] *a.* 最好的，最佳的，最适宜的 U19
ore [ɔː(r)] *n.* 矿石 U1
organ ['ɔːgən] *n.* 风琴，管风琴 U15
orifice ['ɔrifis] *n.* 孔，口 U17
otherwise ['ʌðəwaiz] *a.* 其他方面的，另外的 U11
ounce [auns] *n.* 盎司，英两 U16
outcome ['autkʌm] *n.* 结果，成果 U5
outflow ['autfləu] *n.* 流出，流出物 U10
outlay ['autlei] *n.* 花费，支出 U11
outlet ['autlet] *n.* 出口，出路 U5
outline ['əutlain] *v.* 略述 U11
output ['autput] *n.* 产量，产品 U1
oval ['əuvəl] *a.* 椭圆的 U11
overlap ['əuvə'læp] *n.* 重叠 U10
over-pole [əuvəpəul] *v.* 过度还原 U13
oxide ['ɔksaid] *n.* 氧化物 U2
oxidize ['ɔksiˌdaiz] *v.* （使……）氧化 U7
pack [pæk] *v.* 填塞，塞满 U4
packaging ['pækidʒiŋ] *n.* 包装，包装物 U14
palladium [pə'leidiəm] *n.* 钯（元素符号 Pd）U16
packing ['pækiŋ] *n.* 包装 U18
parameter [pə'ræmitə] *n.* 参数 U3
partial ['pɑːʃəl] *a.* 部分的，局部的 U13
participate [pɑː'tisipeit] *v.* 分担，含有，带有 U6
particle ['pɑːtikl] *n.* 粒子，极小量 U2
particulate [pə'tikjulit, pə'tikjuleit] *a.* 微粒的 U3

pass [pɑːs] *n.* 通道，轨道 U17
passage ['pæsidʒ] *n.* 通道，通路 U10
passivation [ˌpæsi'veiʃən] *n.* 钝化 U5
patent ['peitənt, 'pætənt] *n.* 专利，执照 U1
pearlite ['pəːlait] *n.* 珠光体 U12
pellet ['pelit] *n.* 球团矿 U2
小球 U16
pelletize ['pelitaiz] *v.* 使……成颗粒状 U2
penetrate ['penitreit] *v.* 穿透，渗透 U8
penetration [peni'treiʃən] *n.* 穿过，渗透 U12
percentage [pə'sentidʒ] *n.* 百分率，百分比 U12
percolate ['pɜːkəleit] *v.* 浸透，渗透 U16
perforate ['pəːfəreit] *v.* 穿孔于，打孔穿透 U16
perforated ['pəːfəreitid] *a.* 多孔的，穿孔的 U19
perform [pə'fɔːm] *v.* 完成 U7
performance [pə'fɔːməns] *n.* 生产情况 U1
性能 U15
permeability [ˌpəːmiə'biliti] *n.* 渗透，渗透性 U2
perpendicularly [ˌpəːpən'dikjuləli] *ad.* 垂直地；直立地 U16
petroleum [pi'trəuliəm] *n.* 石油 U14
phenomena [fi'nɔminə] *n.* 现象 U10
phosphate ['fɔsfeit] *n.* 磷酸盐 U1
phosphor ['fɔsfə] *n.* 磷（元素符号 P）U1
phosphoric [fɔs'fɔrik] *a.* 磷的（尤指含五价磷的），含磷的 U1
phosphorous ['fɔsfərəs] *a.* 磷的 U1
photoengraving [ˌfəutəuin'greiviŋ] *n.* 照相凸版，照相凸版印刷 U15
pickling ['pikliŋ] *n.* 酸洗 U18
piece [piːs] *n.* 部件，构件 U20
piercing ['piəsiŋ] *a.* 刺穿的 U19
piling ['pailiŋ] *n.* 打桩，打桩工程 U3
pipe [paip] *v.* 以管输送 U5
n. 管 U15
piston ['pistən] *n.* [机] 活塞 U20
pit [pit] *n.* 池，深坑 U3
pitch [pitʃ] *n.* 树脂，沥青 U13
plain [plein] *a.* 普通的，平常的 U9
plant [plɑːnt] *n.* 机器，设备 U17
platinum ['plætinəm] *n.* 铂，白金；U15
pliers ['plaiəz] *n.* 钳子 U20

plug [plʌg] *n.* 炉底砌块，塞子 U3
plus [plʌs] *v.* 加上 U8
pocket ['pɔkit] *n.* 凹处，小块地区 U13
pole [pəul] *v.* （青木）还原（除气）U13
poling ['pəuliŋ] *n.* 插树还原，还原，除气 U13
pool [puːl] *n.* 坑，池 U1
porosity [pɔː'rɔsiti] *n.* 多孔性，有孔性 U5
porous ['pɔːrəs，'pəurəs] *a.* 多孔的，有气孔的，能穿透的 U13
portion ['pɔːʃən] *n.* 部分，一份 U2
possess [pə'zes] *v.* 拥有，占有 U6
post-treatment [pəust'triːtmənt] *n.* 炉外精炼，后处理 U1
potassium [pə'tæsiəm] *n.* 钾（元素符号 K）U16
potential [pə'tenʃ(ə)l] *a.* 潜在的 U7
n. 势 U7
potline ['pɔtlaɪn] *n.* （制铝用的）电解槽系列 U14
practice ['præktɪs] *n.* 实际操作 U3
preceding [pri(ː)'siːdiŋ] *a.* 在前的，前述的 U9
precious ['preʃəs] *a.* 贵重的，宝贵的 U13
precipitate [pri'sipiteit] *v.* 使……沉淀，沉淀 U14
n. 沉淀物 U16
predetermine ['priːdi'təːmin] *n.* 预定，预先确定 U11
predominant [pri'dɔminənt] *a.* 主要的，支配的 U15
preferentially [ˌprefə'renʃəli] *ad.* 先取地，优先地 U14
preheat [priː'hiːt] *v.* 预热 U5
preliminarily [pri'liminərili] *ad.* 预先地 U8
preliminary [pri'liminəri] *a.* 预备的，初步的，准备工作 U13
prescribe [pris'kraib] *v.* 指示，规定 U8
present [pri'zent] *v.* 介绍，引见 U5
preset ['priː'set] *v.* 事先调整，预调 U4
press [pres] *v.* 压，按 U6
pressure ['preʃə(r)] *n.* 压，压力 U18
principal ['prɪnsɪp(ə)l] *a.* 主要的，首要的 U6
priority [prai'ɔriti] *n.* 优先，优先权 U10
probable ['prɔbəbl] *a.* 很可能的，大概的 U14
procedure [prə'siːdʒə] *n.* 工序，过程 U17
prodigious [prə'didʒəs] *a.* 巨大的，庞大的 U17
productivity [ˌprɔdʌk'tiviti] *n.* 生产力，生产率 U11
profile ['prəufail] *n.* 剖面，外形 U18
progressively [prə'gresivli] *ad.* 逐渐的，渐进的 U4

propeller [prə'pelə] *n.* 螺旋桨，推进器 U16
property ['prɔpəti] *n.* 性质，性能 U15
proportion [prə'pɔːʃən] *n.* 比例，部分 U7
proportioned [prə'pɔːʃənd] *a.* 相称的，成比例的 U15
provided [prəː'vaidid] *conj.* 假若，倘若，倘使 U7
pulp [pʌlp] *n.* 矿浆 U16
pulverize ['pʌlvəraiz] *v.* 使……成粉末 U2
pulverized ['pʌlvəraizd] *a.* 粉状的 U2
purification [ˌpjuərifi'keiʃən] *n.* 净化，提纯，精制（炼）U14
pyrite ['paɪəraɪt] *n.* 黄铁矿 U2
pyrometallurgical [pairɔˌmitæ'ləːdʒikəl] *a.* 火法冶金的 U13
pyrometallurgy [ˌpairəume'tæləːdʒi] *n.* 火法冶金学，高温冶金学 U15
pyrrhotite ['piərəutait] *n.* 磁黄铁矿 U2
quench [kwentʃ] *v.* 淬火 U12
ramp [ræmp] *n.* 斜面，斜台 U11
range [reindʒ] *v.* 在……范围内变化 U18
ratio ['reiʃiəu] *n.* 比，比率，比值 U4
reciprocate [ri'siprəkeit] *v.* （使……）往复（运动），来回 U11
reckon ['rekən] *v.* 估计，认为 U16
recommend [rekə'mend] *v.* 使……成为可取，推荐 U4
recompress [rekəmp'res] *v.* 再压缩 U5
recover [ri'kʌvə] *v.* 回收，恢复 U15
recovery [ri'kʌvəri] *n.* 回收，恢复 U15
rectangular [rek'tæŋgjulə] *a.* 矩形的，成直角的 U11
recycle ['riː'saikl] *v.* 使……再循环，反复应用 U5
reducible [ri'djuːsəbl] *a.* 可还原的 U2
refining [ri'fainiŋ] *n.* 精炼 U7
refractory [rɪ'fræktərɪ] *n.* 耐火材料 U1
regenerative [ri'dʒenərətiv] *a.* 蓄热的 U1
regeneratively [ri'dʒenərətivli] *ad.* 蓄热地，再生地 U1
regulate ['regjuleit] *v.* 调节，校准 U4
reject [ri'dʒekt] *v.* 拒绝，排斥 U15
reinforce [ˌriːin'fɔːs] *v.* 增强，加强 U18
relieve [ri'liːv] *v.* 减轻，解除 U20
remainder [ri'meində] *n.* 残余，剩余物 U7
remote [ri'məut] *a.* 遥远的，边远的 U3
remove [ri'muːv] *v.* 消除，除去 U20
reoxidation [riɔksi'deiʃən] *n.* 重新氧化 U5
represent [reprɪ'zent] *v.* 代表 U5

requirement [ri'kwaiəmənt] *n.* 需求，要求 U4
reserve [ri'zəːv] *n.* 储备（物），储藏量 U14
reset ['riːset] *v.* 重放，重新安排 U11
residual [ri'zidjuəl] *n.* 剩余，残留，残渣 U6
residue ['rezidjuː] *n.* 剩余物，残余，残渣 U14
resist [ri'zist] *v.* 抵抗，抗拒，承受 U4
resistance [rɪ'zɪstəns] *n.* 抵抗力，阻力 U12
restore [ris'tɔː] *v.* 恢复，使……回复 U20
restrictive [ris'triktiv] *a.* 限制性的 U12
retain [ri'tein] *v.* 保持，保留 U12
retort [ri'tɔːt] *n.* 曲颈瓶，蒸馏罐 U15
retrieve [ri'triːv] *v.* 保持，保存，收回 U18
reusable [riː'juːzəbl] *a.* 可以再度使用的 U17
reverberatory [ri'vəːbərətəri] *a.* 反射的 U13
n. 反射炉 U13
reverse [ri'vəːs] *v.* 颠倒，倒转，倒退 U14
rib [rib] *v.* 加肋于，加肋材于 U18
rigidly ['ridʒidli] *ad.* 严格地 U9
rim [rim] *n.* 边，轮缘 U8
riveting ['rivitiŋ] *n.* 铆接（法）U17
roast [rəust] *v.* 煅烧，焙烧 U13
robot ['rəubɔt, 'rɔbət] *n.* 机器人，遥控设备 U14
rock [rɔk] *n.* 岩石 U2
roll [rəul] *v.* 轧，轧制 U10
rolled [rəuld] *a.* 轧制的，滚制的 U15
roller ['rəulə] *n.* 辊子 U11
rotate [rəu'teit] *v.* 旋转，自转 U19
routine [ruː'tiːn] *a.* 日常的，常规的 U1
runner ['rʌnə(r)] *n.* 流槽 U3
sample ['sæmpl] *n.* 样品，例子 U8
sandpaper ['sændpeɪpə(r)] *n.* 砂纸 U20
satisfactorily [sætɪs'fæktərɪlɪ] *ad.* 满意地 U9
satisfy ['sætisfai] *v.* 满足，使……满意 U10
saturated ['sætʃəreitid] *a.* 饱和的 U16
scale [skeil] *n.* 氧化铁皮，锈皮 U6
scarce [skɛəs] *a.* 缺乏的，不足的 U5
scrap [skræp] *n.* 废钢，废料 U6
scrub [skrʌb] *v.* 使（气体）净化，擦洗 U3
scrubber ['skrʌbə] *n.* 气体洗涤器 U5

seal [siːl] *v.* 密封 U3
seamless ['siːmlɪs] *a.* 无缝的 U19
sector ['sektə] *n.* 部分，部门 U1
sedimentation [ˌsedimen'teiʃən] *n.* 沉淀，沉降 U14
segment ['segmənt] *n.* 弧形，扇形体 U11
segregation [ˌsegri'geiʃən] *n.* 偏析作用 U11
semis ['semis] *a.* 半成品的 U18
separate ['sepəreit] *v.* 分开，隔离 U3
separation [sepə'reiʃən] *n.* 分离，分开 U6
sequence ['siːkwəns] *n.* 顺序，序列 U8
连续，一连串 U15
severe [si'viə] *a.* 剧烈（猛烈）的 U12
shaft [ʃɑːft] *n.* 传动轴，杆状物 U12
竖炉 U15
shafting ['ʃɑːftiŋ] *n.* 轴系，制轴材料 U12
shallow ['ʃæləu] *a.* 浅的 U1
shear [ʃiə] *v.* 剪切，修剪 U6
n. 大剪刀，剪床 U12
sheet [ʃiːt] *n.* （一）片，（一）张，薄片 U17
shred ['ʃred] *v.* 切割，切碎 U6
siderite ['sidəˌrait] *n.* 菱铁矿 U2
significant [sig'nifikənt] *a.* 重大的，重要的 U12
silhouette [ˌsilu(ː)'et] *n.* 轮廓 U3
silica ['silikə] *n.* 硅石，二氧化硅 U1
silicate ['silikit] *n.* 硅酸盐 U1
silicon ['silikən] *n.* 硅（元素符号 Si）U1
sinter ['sintə] *v.* 烧结 U1
n. 烧结矿 U1
siphon ['saifən] *v.* 用虹吸管吸 U14
size [saiz] *v.* （管材，轧管）定径 U19
skimmer ['skimə] *n.* 撇渣器 U3
skinwall [skɪnwɔːl] *n.* 燃烧室的衬墙，隔墙 U4
skip [skip] *n.* 料车 U3
slab [slæb] *n.* （大，初轧）板坯 U11
平板，厚片 U15
slag [slæg] *n.* 炉渣，矿渣 U1
slagging ['slægiŋ] *n.* 造渣 U7
slim [slim] *a.* 细长的 U3
slop [slɔp] *v.* 溢出，溅溢 U9

slotted [slɔtid] *a.* 有槽的 U19
slurry ['sləːri] *n.* 不溶解物的悬浮液 U14
smelt [smelt] *v.* 熔炼，冶炼 U15
smelter ['smeltə] *n.* 熔炉，熔炼工 U13
socket ['sɔkit] *n.* 穴，孔，插座 U14
soda ['səudə] *n.* 碳酸水，纯碱，氢氧化钠 U14
sodium ['səudɪəm] *a.* 可浸渍的 U13
n. [化] 钠（元素符号 Na）U14
soldering ['sɔldəriŋ] *n.* 软钎焊，低温焊接 U17
solidify [sə'lidifai] *v.* （使……）凝固 U10
solubility [ˌsɔlju'biliti] *n.* 溶解性 U7
soluble ['sɔljubl] *a.* 可溶的，可溶解的 U7
solution [sə'luːʃən] *n.* 溶液，溶解 U15
solvent ['sɔlvənt] *n.* 溶剂 U16
sort [sɔːt] *v.* 分类，拣选 U6
spar [spɑː] *n.* 晶石 U15
specification [ˌspesifi'keiʃən] *n.* 规格，技术要求 U4
spin [spin] *v.* 自旋（转），旋转 U19
【过去式/过去分词】spun
splash [splæʃ] *n.* 溅上的斑点，溅泼的量 U15
v. 飞溅，泼 U8
sponge [spʌndʒ] *n.* 海绵，海绵状物 U1
spring [spriŋ] *n.* 弹簧，弹性 U1
springiness ['spriŋinis] *n.* 富于弹性，弹性 U12
springy ['spriːŋi] *a.* 有弹性的 U20
squeeze [skwiːz] *v.* 挤出，挤压 U17
squirt [skwəːt] *v.* 喷出 U17
stable ['steibl] *a.* 稳定的 U7
stack [stæk] *n.* 炉身，堆，一堆 U3
stacking ['stækiŋ] *n.* 堆垛 U18
stage [steɪdʒ] *n.* 阶段 U6
standby ['stændbaɪ] *a.* 备用的 U8
stationary ['steiʃ(ə)nəri] *a.* 固定的 U11
stave [steɪv] *n.* 板，冷却壁 U3
steelmaking [s'tiːlmeikiŋ] *n.* 炼钢 U1
steep [stiːp] *a.* 急剧上下的 U1
stepwise [stepwaiz] *a.* 逐步的，逐渐的 U18
stick [stik] *vi.* 粘住，粘贴 U2
stir [stəː] *v.* 搅和，搅拌 U10

stockline [ˈstɔklain] *n.*	料线 U3
storage [ˈstɔridʒ] *n.*	储藏，储存 U2
straighten [ˈstreitn] *v.*	（使……）弄直，伸直 U11
strain [strein] *n.*	张力，应变 U17
strand [strænd] *n.*	铸坯 U11
strength [streŋθ] *n.*	强度 U2
stretch [stretʃ] *v.*	伸展 U20
strike [straɪk] *v.*	打动，穿透 U7
stringent [ˈstrindʒənt] *a.*	苛刻的，必须严格遵守的 U10
strip [strip] *v.*	剥夺，剥去 U15
n.	带钢，棒 U17
structural [ˈstrʌktʃərəl] *a.*	结构（上）的 U2
stud [stʌd] *n.*	双头螺栓，柱头螺栓 U12
subdivide [ˈsʌbdiˈvaid] *v.*	再分，细分 U5
subgroup [ˈsʌbgruːp] *n.*	小群，子群 U12
subsequent [ˈsʌbsikwənt] *a.*	随后的，后来的 U11
substitute [ˈsʌbstitjuːt] *n.*	替代品，代用品 U6
sufficient [səˈfiʃənt] *a.*	充足的，足够的 U7
sulfate [ˈsʌlfeit] *n.*	硫酸盐 U15
v.	使……成硫酸盐，用硫酸处理，硫酸盐化 U15
sulfur [ˈsʌlfə] *n.*	硫（元素符号 S）U2
sulfuric [sʌlˈfjuːrik] *a.*	（正）硫的，含（六价）硫的 U15
sulphide [ˈsʌlfaid] *n.*	硫化物 U2
supersaturate [sjuːpəˈsætʃəreit] *v.*	[化] 使……过度饱和 U14
supersede [ˌsjuːpəˈsiːd] *n.*	代替，取代 U13
surfacing [ˈsəːfisiŋ] *n.*	表面加工，表面处理 U11
surplus [ˈsəːpləs] *a.*	过剩的，剩余的 U7
swell [swel] *n.*	膨胀 U2
sword [sɔːd] *n.*	剑，刀 U1
tank [tæŋk] *n.*	盛液体、气体的大容器（槽，桶，箱）U14
tap [tæp] *v.*	使……流出 U6
tapered [ˈteipəd] *a.*	锥形的 U19
taphole [ˈtæphəul] *n.*	出（铁，钢，渣）口 U3
tapped [ˈtæpid] *a.*	内螺纹的 U19
tar [tɑː] *n.*	焦油，柏油 U14
tarnish [ˈtɑːniʃ] *v.*	失去光泽，变灰暗 U16
tempering [ˈtempəriŋ] *n.*	回火 U20

tensile ['tensail] *a.*	可拉长的 U20
tension ['tenʃən] *n.*	压力，张力，牵力 U14
term [təːm] *v.*	称为，把……叫作 U17
theoretically [θiə'retikəli] *ad.*	理论上，理论地 U12
thermal ['θəːməl] *a.*	热的，热量的 U4
thickwalled [θɪkwɔːld] *a.*	厚壁的 U19
thoroughly ['θʌrəlɪ] *ad.*	十分地，彻底地 U20
throat [θrəut] *n.*	（炉）喉 U3
throughput ['θruːput] *n.*	生产量，生产能力 U10
tilt [tilt] *v.*	（使……）倾斜 U8
titania [tai'teiniə] *n.*	二氧化钛，氧化钛 U14
titanium [tai'teinjəm, ti'teinjəm] *n.*	钛（元素符号 Ti）U2
tonnage ['tʌnidʒ] *n.*	吨，产量 U10
toothpaste ['tuːθpeɪst] *n.*	牙膏 U17
tough [tʌf] *a.*	有韧性的 U20
toughness ['tʌfnis] *n.*	韧性，坚韧 U12
trace [treis] *n.*	少许，有点 U13
track [træk] *n.*	轨道，铁轨 U18
train [trein] *n.*	滚道，轧机组，机列 U18
transfer ['trænsfəː(r)] *n.*	转移，迁移 U7
transformation [ˌtrænsfə'meiʃən] *n.*	变化，转化 U7
translucent [trænz'ljuːsənt, træns'ljuːsənt] *a.*	半透明的 U16
transport [træns'pɔːt] *v.*	传送，运输 U3
tray [trei] *n.*	底板，托（支）架 U8
trench [trentʃ] *n.*	沟槽，沟道 U3
trihydrate [trai'haidreit] *n.*	［化］三水合物 U14
truncated ['trʌŋkeitid] *a.*	截短了的 U3
trunk [trʌŋk] *n.*	树干，躯干 U13
	汇集管 U15
tulip ['tjuːlip] *n.*	喇叭形 U3
tundish ['tʌndɪʃ] *n.*	中间包，中间罐 U11
tungsten ['tʌŋstən] *n.*	钨（元素符号 W）U12
turbulent ['təːbjulənt] *a.*	涡旋的，狂暴的 U4
tuyere [twiː'jɛə] *n.*	风口，鼓风口，风嘴 U3
typically ['tipikəli] *ad.*	代表性地，作为特色地 U8
ultimate ['ʌltimit] *a.*	最后的，根本的 U12
ultrafine [ˌʌltrə'fain] *a.*	极其细小的 U2
undergo [ˌʌndə'gəu] *v.*	经受，经历，遭受 U10
undertake [ˌʌndə'teik] *v.*	进行，从事 U3

uniform [ˈjuːnifɔːm] *a.* 均匀的，均质的 U5
upgrade [ˈʌpgreid] *v.* 使……升级，提升 U12
uptake [ˈʌpteik] *n.* 上升管 U3
utilize [juːˈtɪlaɪz] *v.* 利用 U7
vacuum [ˈvækjuəm] *n.* 真空，空间 U10
valve [vælv] *n.* 阀门 U4
vanadium [vəˈneɪdɪəm] *n.* 钒（元素符号 V）U12
variation [ˌvɛəriˈeiʃən] *n.* 变更，变化 U17
vary [ˈvɛəri] *v.* 变化，不同 U8
vehicle [ˈviːikl] *n.* 交通工具，车辆 U12
versatility [ˌvəːsəˈtɪlətɪ] *n.* 多功能性 U10
vertical [ˈvəːtikəl] *a.* 垂直的，直立的 U8
vertically [ˈvəːtikəli] *ad.* 垂直地 U4
vibration [vaiˈbreiʃən] *n.* 振动，颤动 U12
virgin [ˈvəːdʒin] *a.* 原始的，未被玷污的 U6
viscosity [visˈkɔsiti] *n.* 黏度，黏性 U6
void [vɔid] *n.* 空隙，缩孔 U10
volume [ˈvɔljuːm, ˈvɔljəm] *n.* 体积，大量 U7
weapon [ˈwepən] *n.* 武器，兵器 U17
weld [weld] *v.* 焊接，熔接 U4
weldability [weldəˈbiliti] *n.* 可焊性，焊接性 U12
whereas [(h)weərˈæz] *conj.* 然而，反之 U4
windshield [ˈwindʃiːld] *n.* 挡风玻璃 U16
withdraw [wiðˈdrɔː] *v.* 取回，提取 U11
workpiece [ˈwəːkpiːs] *n.* 工件 U17
wrap [ræp] *v.* 卷，缠绕 U19
yellowish [ˈjeləuɪʃ] *a.* 微黄色的 U13
yield [jiːld] *n.* 产量，收益 U11
屈服（点）U12
zirconium [zəːˈkəuniəm] *n.* 锆（元素符号 Zr）U15

Phrases & Expressions

a function of 随……而变 U12
a series of 一系列的 U3
a wide range of 各种各样的，种种的 U17
account for （指数量等）占 U6
air stream 风流，空气流 U4
alloying addition 合金剂 U6
alloying agent 合金剂 U6

alumina trihydrate 氢氧化铝 U14
aluminium chloride 氯化铝 U14
be subject to 受……影响的，易于……U12
as a whole 总的说来 U7
banks of retort 蒸馏罐 U15
be classified as 被分成……类 U5
be commensurate with 与……相称，与……相当 U14
be referred to as 叫作，称为 U3
be seeded with 做孕育处理，使孕育 U14
benefit from 获益，得益于 U1
Bessemer process 酸性转炉法 U9
blast furnace 高炉 U1
blast volume 风量 U1
blooming train 初轧机组 U11
blower house 鼓风机室 U13
blow-in 开炉 U3
blow-off 排出，喷出，停炉 U14
bottom plate 底板 U11
breaking down 粗轧，开坯 U18
brought back 恢复 U5
built-up construction 组合结构，装配结构 U17
bustle pipe 环形风管 U5
calcium sulphide 硫化钙 U7
capital cost 投资费用 U1
carbon anode 碳阳极 U14
carbon electrode 石墨电极 U7
carbon monoxide 一氧化碳 U5
carry out 实现，完成，实行 U18
case hardening 表面硬化 U20
cast house 出铁场 U3
cast iron 铸铁 U1
casting bay 连铸跨 U11
charge solid 固体炉料 U5
checker brick 格子砖 U4
checker chamber 蓄热室 U4
checker mass 蓄热室 U4
circulation leg 循环流管 U10
clean steel 纯净钢 U10
coal tar pitch 煤焦油沥青 U14

coke oven	焦炉 U4
coke rate	焦比 U2
coke ratio	焦比 U1
cold working	冷加工 U11
combustion chamber	燃烧室 U4
conductor of electricity	导电体 U14
conical orifice	锥形孔 U17
consist in	在于 U1
continuous casting	连铸 U11
cooling plate	冷却板 U3
cored wire	芯钢丝，焊条芯 U10
cut-off station	切割站 U11
date back	追溯 U1
deep-drawing	深冲，深压 U18
dependent on	依靠，由（随）……决定的 U4
die casting	加压铸造，压力铸造 U17
die forging	模锻 U17
die steel	模具钢 U9
direct reduced iron（DRI）	直接还原铁 U5
discharge rollers	卸料辊道辊子 U11
double bell	双料钟 U3
dressing plant	选矿厂 U6
dummy bar	引锭杆 U11
dump truck	自动倾卸（自卸）卡车 U16
duration of a heat	冶炼（一炉）时间 U8
electrical conductor	导电体 U14
electric-arc furnace	电弧炉 U1
electrode potential	电极电势 U15
electrolytic deposit	电解沉积 U15
finished article	成品 U6
finishing area	精整区 U18
finishing shop	成品车间 U18
fire refining	火法精炼 U13
flash furnace	闪速熔炼炉，闪速炉 U13
flat product	板材 U18
free running slag	易流动渣 U8
fusion welding	熔焊 U17
gas reformer	气体转化炉 U5
go into production	投产，开始生产 U5

go out	熄灭 U1
gravity feed	重力给料 U5
green pellet	生球 U2
green wood	生（湿）木材 U13
gun barrel	枪管 U12
hammer forging	锤锻 U17
hammer into shape	锤打成型 U1
hard coal	无烟煤 U2
hardened structure	硬化组织 U20
heavy duty product	重型产品，大型产品 U12
holding furnace	保温炉 U14
horizontal converter	卧式转炉 U13
hot blast main	热风管 U3
hot blast stove	热风炉 U1
hydraulic press	水压机，液压机 U19
impact resistance	冲击阻力 U12
in accordance with	根据，按照 U11
in conjunction with	与……一道，结合 U13
in detail	详细地 U17
in miniature	小型，小比例 U17
in terms of	以……的观点，就……而说 U1
in the case of	在……的情况 U6
ingot casting	铸锭，模铸锭 U11
interfacial tension	界面张力 U14
iron bath reactor	铁浴反应炉 U5
iron notch	铁口 U3
iron runner	铁沟 U3
iron-bearing material	含铁原料 U2
be termed as	被……称为，被……叫作 U17
knitting needle	编织针 U20
lead splash condenser	铅雨冷凝器 U15
liberated heat	释放的热量 U7
lime shell	石灰涂层 U5
loose scrap	粗钢 U6
mass production	大量生产 U17
melter gasifier	熔融气化炉 U5
melting phase	熔炼期 U9
mixer line	混风管 U3
mixer valve	混风阀 U4

mobile induction heater	移动感应加热器 U10
mould set	全套锭模，整套锭模 U11
natural gas	天然气 U4
on blast	送风，鼓风 U4
on the contrary	（与此）相反，反之 U8
one-piece manufacture	整体制造 U17
open hearth	平炉 U1
ounce troy [trɔi] *n.*	金衡盎司 U16
outlet pipe	流出管，排出管 U5
pearlitic structure	珠光体组织 U20
permanent-mold casting	金属模铸造，硬模铸造 U17
physical properties	物理性能 U12
pig iron	生铁 U1
plain carbon steel	普碳钢 U12
plasma-jet welding	等离子喷射焊接 U17
play an important part in	在……中起重要作用 U5
porous lance	多孔氧枪 U10
pour off	倒出 U13
pouring ladle	钢包 U8
powder metallurgy	粉末冶金 U17
preheating furnace	预热炉 U11
pressure filter	压力过滤器 U14
pressure relief valve	减压阀，溢流阀 U3
primary mill	初轧厂 U11
prior to	居先，在前 U18
process scrap	加工废钢，边角废料 U6
profiled section	异型断面 U18
pulverized coal injection	喷煤 U2
raw material	原料 U2
reformer tube	重整管 U5
rejected material	废品，废料 U6
result in	导致，终于造成……结果 U1
retort process	反应罐直接还原法，蒸馏过程，干馏过程 U15
ring main bustle	热风围管 U3
rolling mill	轧机 U17

rolling stock	轧件 U18
rotary kiln furnace	回转窑 U5
rotating chute	旋转溜槽 U3
rotating roll	旋转轧辊 U18
roughing train	粗轧机组 U18
sand trap	除沙槽 U14
seal leg	料封管 U5
seamless tube	无缝管 U19
secondary refining	炉外精炼，二次精炼 U10
sectional steel	型钢 U18
serve as	用作，充当 U1
settle out	沉积 U7
shock-cool	急速冷却 U15
shaft-type furnace	竖炉 U2
sheared end	切头 U6
slag former	造渣剂 U6
slag notch	渣口 U3
slag volume	渣量 U7
slide gate	滑动水口 U11
snap off	突然折断 U20
soaking pit	均热炉 U11
soft coal	烟煤 U2
soft metal	软金属 U17
specific gravity	相对密度 U16
sponge iron	海绵铁 U5
stainless steel	不锈钢 U9
standby lance	副枪 U8
stopped rod	棒塞 U11
straightening unit	矫直装置 U11
sulfating roast	硫酸化焙烧 U15
sulphur holding power	脱硫能力 U2
surface finish	表面光洁度 U18
take the place of	代替 U5
tap off	分出，抽出 U13
taphole drill	开（铁）口机 U3
tapping hole	出钢口 U8
tap-to-tap time	出钢时间 U8

teeming crane	铸锭吊车 U9
tensile strength	抗拉强度，拉伸强度 U12
torch cutter	火焰切割机 U19
tough pitch copper	火法精炼铜，工业纯铜 U13
tramp element	杂质元素 U6
traveling grate	移动床 U2
ultimate strength	极限强度 U12
vacuum pump	真空泵 U10
vacuum roof	真空炉顶 U10
volume ratio	体积比，容积比 U4
water colled conducting collar	水冷圈 U9
water glass	水玻璃（硅酸钠）U5
water of crystallization	结晶水 U2
wide-flanged beam	宽缘钢梁 U18
wire feeding	喂丝，喂线 U10
wrought iron	熟铁 U1
yield point	屈服点 U12
zinc dust	锌粉 U15
zinc sulfide concentrate	硫化锌精矿 U15

Proper Names

Abraham Darby	亚伯拉罕·达比 U1
Avonmouth [eivɔn'mauθ] *n.*	埃文茅斯（在英国布里斯托尔附近）U15
Birmingham ['bəːmiŋhəm]	伯明翰（城市名）U17
Brazil [brə'zil] *n.*	巴西 U14
Congo ['kɔŋgəu] *n.*	刚果 U13
Dorr agitator	多尔搅拌器 U16
Germany ['dʒəːmənɪ] *n.*	德国 U16
Guyana [gai'ɑːnə, gai'ænə] *n.*	圭亚那（拉丁美洲）U14
Harz [hɑːrts] Mountains *n.*	哈尔茨山（德国中部山）U16
Henry Bessemer ['henri 'besimə]	亨利·贝塞麦（1813~1898 年，首创酸性转炉炼钢的英国工程师）U1
Katanga [kə'tæŋgə] *n.*	加丹加（扎伊尔沙巴地区 Shaba 的旧名）U13
Montana [mɔn:tænə] *n.*	蒙大拿州（美国州名）U13
Nilson [nilsn]	尼尔森 U1
Percy Gilchrist ['pəːsi gilkrist]	珀西 ·吉尔克里斯特 U1
Peru [pə'ruː] *n.*	秘鲁（拉丁美洲国家名）U15
Sidney Thomas ['sɪdnɪ tɔməs]	西德尼·托马斯 U1

LD	Linz Düsenverfahren 首字母的缩写 U1
the LD process	顶吹氧气炼钢工艺 U1
Victorian [vik'tɔːriən] *a.*	维多利亚女王时代的 U1
Zambia ['zæmbiə] *n.*	赞比亚 U13

参 考 文 献

[1] 卜玉坤，金敬红，战东平，等．材料英语 1 [M]．北京：外语教学与研究出版社，2000.
[2] 吴非晓，卜玉坤，张坚毅，等．机械英语 1 [M]．北京：外语教学与研究出版社，2000.
[3] 阿瑟·斯特利特，廉·亚历山大．金属与人类文明（上、下）[M]．北京：冶金工业出版社，1981.
[4] 炼钢教研组．炼钢英语读本 [M]．上海：上海冶金工业学校，1986.
[5] 王福明．An Introduction to Iron and Steel（讲义）．北京：北京科技大学，2004.
[6] 严俊仁．汉英科技翻译 [M]．北京：国防工业出版社，2004.
[7] 黄忠廉，李亚舒．科学翻译学 [M]．北京：中国对外翻译出版公司，2004.
[8] 张培基，喻云根，李宗杰，等．英汉翻译教程 [M]．上海：上海外语教育出版社，1980.
[9] 孔庆炎，刘鸿章．新编实用英语综合教程 [M]．北京：高等教育出版社，2002.
[10] 百度文库．有色冶金专业英语 [DB/OL]．[2014-2-8]．http：//www.doc88.com/p-900537304499.html.